HUMAN BIOLOGICAL VARIATION

HUMAN BIOLOGICAL VARIATION

James H. Mielke
University of Kansas

Lyle W. Konigsberg
University of Tennessee

John H. Relethford
State University of New York College at Oneonta

New York   Oxford
OXFORD UNIVERSITY PRESS
2006

Oxford University Press, Inc., publishes works that further Oxford University's
objective of excellence in research, scholarship, and education.

Oxford New York
Auckland Cape Town Dar es Salaam Hong Kong Karachi
Kuala Lumpur Madrid Melbourne Mexico City Nairobi
New Delhi Shanghai Taipei Toronto

With offices in
Argentina Austria Brazil Chile Czech Republic France Greece
Guatemala Hungary Italy Japan Poland Portugal Singapore
South Korea Switzerland Thailand Turkey Ukraine Vietnam

Published by Oxford University Press, Inc.
198 Madison Avenue, New York, New York 10016
http://www.oup.com

Oxford is a registered trademark of Oxford University Press

Library of Congress Cataloging-in-Publication Data

Mielke, James H.
  Human biological variation / James H. Mielke, Lyle W. Konigsberg, John H. Relethford.
    p. cm.
  Includes bibliographical references and index.
  ISBN 13: 978-0-19-518871-4 (pbk. : alk. paper)
  ISBN 0-19-518871-3 (pbk. : alk. paper)
    1. Human population genetics.   2. Human evolution.   3. Variation (Biology)
  4. Human genetics--Variation.   I. Konigsberg, Lyle W.   II. Relethford, John.   III. Title.

QH431.M525 2006
599.9'4--dc22

                                                                          2005040639

Printing number: 9 8 7 6 5 4 3 2 1

Printed in the United States of America
on acid-free paper

Contents

SECTION 2: Variation in Genes, Simple Genetic Traits, and DNA Markers

SECTION 3: Variation in Complex Traits

Preface

People are biologically and culturally very diverse. We often think about these differences in terms of race or ethnic group. Seldom do we pause and think about the variations in terms of their evolutionary origin or their adaptive significance. Human variation can be visible (e.g., differences in skin color, hair form, or nose shape) or invisible (biochemical differences, e.g., blood group antigens and serum proteins or molecular traits). Anthropologists have studied these variations for years and have attempted to understand why populations have different traits or have the same traits but in different frequencies. Research on human variation initially focused on racial classification and the documenting of physical and genetic differences between populations. Today, most of the research focuses on examining variation using evolutionary models and perspectives. The goal of these studies is to understand *why* the differences exist and how they help humans adapt to varying environments rather than to simply document the differences and create racial categories.

The study of human biological diversity is challenging and has historically been fraught with controversy. Chapter 1 opens this book with a brief history of how scientists have studied human diversity over the ages. Examining this history provides insight and an understanding of the different ways scholars have approached this often volatile topic.

Chapters 2 and 3 provide a basic overview of genetics and evolutionary concepts that are vital to understanding how human variation is studied by anthropologists and other scientists. Chapter 2 explains basic terms and concepts such as transmission genetics, the structure and function of DNA, recombination, segregation, and mitochondrial DNA. Chapter 3 provides the population base for our understanding of human diversity by detailing Hardy-Weinberg equilibrium, allele frequencies, and the evolutionary forces of mutation, natural selection, gene flow (migration), and genetic drift that shape the human genome. Following chapters focus on the specifics of human diversity using this evolutionary perspective.

Chapters 4 and 5 explore human diversity by focusing on what are known as the "classical markers" (in contrast to molecular markers) of human variation. These are the blood groups, serum proteins, and red cell enzymes.

No text on human variation would be complete without a discussion of hemoglobin variation (Chapter 6). Many introductory texts detail the relationship between malaria and the sickle cell trait, and this book is no exception. To end the examination of the "classical markers," Chapter 7 explores the human leukocyte antigen (HLA) system and some other polymorphisms of anthropological interest such as phenylthiocarbamide tasting, cerumen diversity, and lactase variation.

Chapter 8 carries our exploration of human diversity to the molecular level by examining DNA markers (e.g., restriction fragment length polymorphisms and variable number of tandem repeats). The diversity and colonization of Oceania and a discussion of the Lemba serve as examples of the uses of these molecular markers in exploring diversity.

Chapter 9 focuses on how we study physical variations that we can see and measure. The quantitative genetic or polygenic model is explained and detailed. This chapter provides the background for understanding how anthropologists use anthropometry or body measurements (Chapter 10) to detail the morphological variation found within and between populations and how it relates to environmental influences such as climate and high altitude. Chapter 11 takes a look at variation in pigmentation, focusing primarily on those features (skin color, hair color, and eye color) that have been used in many racial classifications over the centuries.

A major goal in studying human diversity is to determine the genetic similarity or dissimilarity between populations. Chapter 12 details how anthropologists have studied population structure and population history. Topics such as genetic distance and the cultural and historical influences shaping the genetic structure of a region are discussed in this chapter. We close the discussion of human diversity with an exploration of genetics and behavioral characteristics (Chapter 13).

We thank the many people at Oxford University Press who have made this book a reality. In particular, we are grateful to our sponsoring editor, Jan Beatty, for continued encouragement and support. Thanks also to Talia Krohn, assistant editor; Christine D'Antonio, production editor; and Andrew Pachuta, copyeditor.

We also thank our colleagues who have served as reviewers: C. Loring Brace, University of Michigan; Tom Brutsaert, University at Albany; Robert Corruccini, Southern Illinois University; Herbert Covert, University of Colorado; Trenton Holliday, Tulane University; Richard Jantz, University of Tennessee; Patricia Lambert, Utah State University; Deborah Overdorff, University of Texas; Susan Saul, California State University, Los Angeles; Lynette Leidy Seivert, University of Massachusetts; William Stini, University of Arizona; and Alan Swedlund, University of Massachusetts. Having reviewed manuscripts ourselves, we are aware of the time and effort taken and extend our appreciation.

We thank Henry Harpending, University of Utah, for providing computer code that was invaluable in running simulations of the coalescent model described in Chapter 8. Thanks also go to John Mitchell, La Trobe University, for his comments and corrections on Chapters 4 and 5.

Last but not least, we acknowledge the continued support of our families. John thanks his wife, Hollie, and his sons, David, Benjamin, and Zane, for their love and support. Lyle thanks his wife, Susan, and son, Iain, for their encouragement and tolerance of occasional neglect. Jim thanks his wife, Diane, and children, Evan and Jessica, for putting up with him and keeping him "on track" during this project. He knows he was a "bear" at times, and the nudging did help (even though he sometimes did not appreciate it). He thanks them for their constant support.

Section 1
Background

1

Classifying Human Biological Diversity
A Brief History

The study of human biological diversity is challenging and has historically been fraught with controversy. This chapter briefly examines the history of how human diversity has been explained and studied. Examining this history provides insight and an understanding of the different ways scholars have approached this often volatile topic. Human beings have probably always classified and judged different peoples in some manner. Anthropologists have found that many of the names a group has for itself are translated as "we the people," while others are viewed as "them." This dichotomy of "we" and "them" (civilized/savage, moral/amoral) is often based on cultural characteristics or, in other cases, a combination of cultural and biological traits. The history of racial classifications and how anthropologists and other scientists have viewed human biological variation is linked with the social and intellectual climate of the time. As new data, methods, and theories appear and as paradigms shift, we see humans grappling with the "concept of others" in various ways. Some of these are objective, others judgmental and cruel. In this chapter, we provide a glimpse at how human biological variation has been examined, cataloged, analyzed, and interpreted over the last 500 years. The dominant perspective in the past was to view human variation in terms of race and racial classifications. Even today many people approach human variation in terms of race, sometimes using race in the biological sense and at other times in the sociocultural sense. In some cases, both cultural and physical (often visual) traits are combined to create classifications. Oftentimes it is confusing because the term *race* is equated with a variety of factors, such as skin color, intelligence quotient (IQ), national origin, and even religion. Race is a descriptive concept that provides little, if any, understanding of the dynamics, breadth, and causes of human variation. We ask readers to compare and contrast the racial perspective to studying human diversity with the evolutionary perspective, which is more concerned with explaining and studying the variation by applying evolutionary theory. Today, anthropologists try to understand how much diversity exists, why the differences exist, and how the differences help humans adapt to varying environments rather than to simply document the differences in order to create racial categories.

THE BEGINNINGS OF WESTERN CLASSIFICATION SYSTEMS

As Europeans began exploring the world, naturalists and other writers gradually began to publish information and descriptions of the flora and fauna they had collected from the far reaches of the earth. As these Western explorers traversed the world, they came in contact with peoples who looked and acted differently. Descriptions of these different peoples accumulated rapidly. Many travelers also acquired material culture from the "strange people" they encountered. These acquisitions would later become the ethnographic collections in many European museums.

Jean Bodin (1530–1596) noted differences in human groups:

> . . . the people of the South are of a contrarie humour and disposition to them of the North: these are great and strong, they are little and weak: they of the north hot and moyst, the others cold and dry; the one hath a big voice and greene eyes, the other hath a weake voice and black eyes; the one hath a flaxen and a faire skin. . . . (quoted in Slotkin 1965, 43)

Bodin's characterizations of human groups were purely descriptive, relying on outward appearance and employing the Hippocratic "concept of humors." These early classifications used terms such as *varieties, types, species,* and *nations.* None of these terms especially relied on biological features for classification into groups, but most attempted to provide some explanation for the observed differences. For example, Bodin could not adequately explain the differences and noted that climate alone could not be responsible for the variation.

In 1684, François Bernier (1620–1688) wrote ". . . that there are four or five species or races of men in particular whose difference is so remarkable that it may be properly made use of as the foundation for a new division of the earth" (cited in Slotkin 1965, 94). Bernier then listed the following species or races: Europeans, Africans (Negroes or blacks), Asians (Far Easterners), and Lapps. He did not find the Americans sufficiently different to constitute a separate species/race but suggested that the ". . . blacks of the Cape of Good Hope seem to be a different species to those from the rest of Africa." According to Slotkin (1965), this description may be one of the first racial classifications found in European writings. It was, however, penned anonymously and ignored at the time, only to be resurrected years afterward.

CLASSIFYING THE DIVERSITY

By the eighteenth century, humans were viewed by some Western scholars as "natural" beings who could be described in a similar fashion as one would a dog, cat, or ape. This viewpoint was controversial given the religious influences of the time. Many writers were also grappling with the basic question of whether human races were separate species or just varieties of a single species. This question had broad implications for the spiritual unity of humans, their descent from Adam and Eve, and colonial rights and obligations. Why had some tribes become civilized, while others remained in a "savage" state? For some scholars, human diversity was ancient, permanent, divinely ordained, and part of the "Great Chain of Being" (Lovejoy 1933, Greene 1959). Those who adhered to this view saw all living creatures as occupying a position on a continuous scale from the lowest to the highest and most perfect—that is, from the least perfect atom to the most perfect human. Other writers saw human varieties as products of natural cause. These natural forces (primarily the climate) had acted on humans over the last 6,000 years, producing the diversity. Some scholars were interested in the origins of the diversity, while others were seemingly content with simply classifying.

Carolus (Linné) Linnaeus (1707–1778) is known as the great classifier (Fig. 1.1). Linnaeus placed human beings at the top of the chain of nature but also noted that more than one link separated humans from those immediately below. According to Broberg

Figure 1.1. Carolus Linnaeus (1707–1778).

(1983), Linnaeus was the first to put humans in a classification along with the primates. Linnaeus not only classified all living things but also attempted to classify the varieties or subspecies of humans. Linnaeus separated humans into four basic "varieties" on the basis of geography, color, humor, posture, and customs. Also included in his classification were "wild men" (*Homo sapiens ferus*) and six varieties of *Homo sapiens monstrous*. Linnaeus accepted stories of "troglodytes," who were nocturnal, humanlike animals that lived underground (Smedley 1999). In fact, according to Greene (1959), Linnaeus had trouble distinguishing such entities as the troglodytes and satyrs from real human beings.

Linnaeus had a specific understanding of the taxonomic category *species* and differentiated it from the category *varieties*. Species were essentially unchanged creations, while varieties were groups or clusters within a species that had become altered in appearance. The varieties in his classification (Fig. 1.2) reflected the changes that had occurred over time by external factors such as temperature, climate, and geography (Smedley 1999).

American   "rufus, choleric, rectus"
           (red, choleric, upright)
           regitur consuetudine (ruled by habit)
European   "albus, sanguineus, torosus"
           (white, sanguine, muscular)
           regitur ritibus (ruled by custom)
Asian   "luridus, melancholicus, rigidus"
           (pale yellow, melancholy, stiff)
           regitur opinionibus (ruled by opinion)

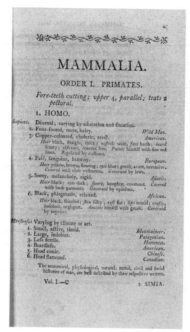

Figure 1.2. Page from Linnaeus's *Systema naturae* (Linné 1806, 8).

African   "niger, phlegmaticus, laxus"
                   (black, phlegmatic, relaxed)
                   regitur arbitrio (ruled by caprice)

Even though there is no ranking implied in Linnaeus's classification, Marks (1995, 50) came to the following conclusion: "It [classification of humans] is rather based on socio-cultural criteria that correlated only loosely with those geographical criteria . . . he was using broad generalizations and value judgments about personality, dress, and custom, to classify the human species." Gould (1994, 1996) argued that Linnaeus simply classified human beings on the basis of the major geographical regions in the world and added *Homo sapiens monstrous* to account for strange, elusive, deformed, and imaginary beings. The descriptions and last characterization do imply a rank order of superiority from European to Asian to American and finally to African. However, this implied ranking is offset by the fact that the American variety, not Europeans, appeared first in the classification itself. These descriptions reflected the conventional beliefs and worldview of the time—that is, the belief of European superiority.

So, why classify at all? These early scientists were primarily concerned with ordering, naming, and classifying the diversity of life found on the earth. Human beings seem to understand their world more easily by creating classifications for lots of things, and creating order among the varieties of humans was no exception. Classifications simplify and bring order to the complexity in the natural world, making it easier to understand and study. Also, European scholars were still separating myth from reality. The unicorn was a real animal for some, tales of monsters and half-human/half-animal beings abounded, and chimpanzees and orangutans were thought to be subhumans.

Figure 1.3. Johann Friedrich Blumenbach (1752–1840).

A SHIFT IN RACIAL CLASSIFICATIONS

Johann Friedrich Blumenbach (1752–1840) has been called the "father of physical anthropology" and the "founder of racial classifications." Blumenbach was a German naturalist and anatomist of the Enlightenment (Fig. 1.3). He was a professor at the University of Göttingen. He did not believe in the Great Chain of Being and classified humans in a separate order (Bimanus) from the other primates (Quadrumana). In 1775, in *De Generis Humani Varietate Nativa* (*On the Natural Variety of Mankind*), Blumenbach listed four races of humans: Europeans, Asians, Africans, and Americans. In a later edition (1795) of his treatise, Blumenbach listed five races: Caucasian, Mongolian, Ethiopian, American, and Malay (Polynesians, Melanesians, and aborigines of Australia). He coined the term "Caucasian" (Slotkin 1965). As the name implies, it is derived from the mountain range between Russia and Georgia. Blumenbach states

> I have taken the name of this variety from Mount Caucasus, both because its neighborhood, and especially its southern slope, produces the most beautiful race of men, I mean the Georgian; and because . . . in that region, if anywhere, it seems we ought with the greatest probability to place the autochthnes [original forms] of mankind. (Gould 1994, 65)

Blumenbach's 1781 and 1795 classifications were similar to that of his teacher Linnaeus (Table 1.1). However, as Gould (1994, 66) argues, ". . . Blumenbach radically changed the geometry of human order from a geographically based model without explicit ranking to a hierarchy of worth, oddly based upon perceived beauty, and fanning out in two directions from a Caucasian ideal." Thus, the shift from a cartographic model (Linnaeus) to one of ranking (Blumenbach) was a major theoretical shift in human

Table 1.1 Comparison of early classifications

Linnaeus 1735	Blumenbach 1770	Blumenbach 1782 and 1795
American	European	Caucasian
European	Mongolian (Asian)	Mongolian
Asian	Ethiopian (African)	Ethiopian
African	American (New World)	American
		Malay

classification (Gould 1996). For Blumenbach, *Homo sapiens* had been created in one place and then spread across the world. Climate, environment, different modes of life, and the transmission of acquired characteristics shaped these peoples into the different races (Greene 1959, 224). Blumenbach thought there were also some unknown factors that mediated the impact of climate and mode of life. He did, however, emphasize that racial variation was superficial and could be changed by moving to a new environment and adopting new patterns of behavior. Thus, racial classifications were also arbitrary and incomplete. Other typological classifications differed in the number of human races. John Hunter, a London surgeon, thought there were seven varieties; Immanuel Kant identified four; and Göttingen professor Johan Christian Polycarp Erxleben and British playwright and naturalist Oliver Goldsmith believed there were six races (Augstein 1996).

Blumenbach had an extensive collection of human skulls. These enabled him to empirically investigate differences rather than merely speculate about varieties based on secondhand observations and traveler's accounts. Blumenbach divided humans into five varieties based on skull shape, preferably as seen from above (On J. F. Blumenbach's *On the Native* 1796). The ideal type was the Caucasian skull, with degeneration in two directions (Fig. 1.4).

Blumenbach saw much diversity in skull shape but also uniformity within a nation. Again, the climate was the major cause of these differences in skull shape. These were not permanent shapes but could be molded if one migrated. Blumenbach was not sure how the climate accomplished these changes and suggested that mode of life and customs also had an influence on the features of the skull (On J. F. Blumenbach's *On the Native* 1796).

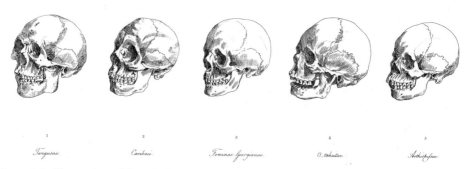

Figure 1.4. Illustration of five skulls from Blumenbach (1865, Plates IV) showing "degeneration" in both directions from the ideal type (*Feminae Georgianae* or Caucasian) in the middle.

Scientific classifications during the seventeenth and eighteenth centuries were not only typological in nature but also ethnocentric and often relied upon subjective descriptions of non-Europeans in contrast with a European ideal. These scientists incorporated the cultural values, ethics, and preconceptions of their times into their explanations and classifications of humans (Greene 1981). The classifications, especially those of the eighteenth century, had common features that had far-reaching social consequences. Classifications were rigid, linking behavioral traits (morals, values, temperament) to physical characteristics. This strengthened popular notions of other nations and races. The hierarchical structure of many of the classifications also implied inequality among the races, thus providing scientific legitimacy to a racial worldview that had social, economic, and political consequences. The ideas of progression and of "civilized" and "savage" peoples gained further scientific support. However, most classifiers believed in the unity of the human species and that "savage" peoples could improve their lot. Finally, these classifications placed humans into the same natural order as everything else that was created. Thus, the inferior physical features and behaviors of these other peoples were "God-given" characteristics (Smedley 1999). As we will see, the criteria, explanations, methods, and worldview for creating classifications of humans were slow to change, even when there were dissenting voices, such as that of Samuel Stanhope Smith.

The Reverend Samuel Stanhope Smith (1750–1819), professor of moral philosophy at the College of New Jersey (later it became Princeton), was opposed to the theory of cultural evolution. He believed that all humans belonged to one species and were descended from the original pair as depicted in the Scriptures. Climate, state of society, customs (culture), and manner of living had caused these peoples to physically change and become the diverse groups inhabiting the world today. He was the first to provide a detailed account of the differentiation of humans following the dispersal from the Garden of Eden.

Smith (1810) refused to accept the validity of racial classifications and suggested that it was probably impossible to draw a line precisely between the various races of humans. He also considered these attempts at classification a useless exercise. Here we see a very early rejection of the race concept. Smith (1810) also argued that, along with climate, cultural behaviors played significant roles in the biological constitution of human populations by modifying, blocking, and changing the effects of the natural environment.

POLYGENISM

Racial theories during the nineteenth century combined several features: (1) humans could be divided into a fixed number of races, (2) moral and intellectual capabilities were not evenly distributed among the races, and (3) mental capabilities were associated with specific racial features. According to Augstein (1996), ". . . 'race' was the be-all and end-all of history." Writers accepting the biblical version of the creation of humans had difficulty in explaining the outward differences between peoples in different regions of the world. Was there a single origin for humans or multiple origins? Had God created the different races of people and placed them in different parts of the world? Or had God created humans only once, with the racial differences appearing later due to a variety of causes such as climate?

Many scholars of the late eighteenth and early nineteenth centuries believed in the doctrine of biological unity of humans and a single origin (**monogenesis**). Human races were the result of changes from this single origin, as described in the Scriptures. The racial distinctions seen across the world were brought about by different climates. Europeans had degenerated the least, while Africans had degenerated the most. One problem with monogenesis was the time factor. If the world had been created in 4004 B.C., as James Ussher and John Lightfoot had calculated from the begats in the Bible, how could all the human races emerge within a mere 6,000 years? This problem was solved by the polygenists, who believed that human races were separate biological species which had descended from different Adams. This doctrine had its origin primarily in the United States and had proponents such as Louis Agassiz (1807–1873) and Samuel George Morton (1799–1851). As Gould (1996) points out, it is not surprising that **polygenesis** was so prominent in the United States given the times. The United States was still practicing slavery and expanding westward, displacing the native peoples. To view these other races as inferior and a separate species was not an accident. It should also be noted that during the eighteenth century several European writers had presented schemes that would now qualify them as being identified as polygenists (e.g., Henry Home). This movement, however, only really gained strength in the United States during the nineteenth century.

Morton, a Philadelphia physican, provided the raw data for the theory of polygenesis by collecting and examining human skulls. He began his collection in the 1820s, and by the time of his death, he had collected over 1,000 skulls (Gould 1996). Morton's goal was to objectively rank races on the basis of the physical characteristics (primarily size) of the skull. In order to accomplish this task, he measured the cranial capacity of 623 skulls. The "Teutonic Family" within the "Modern Caucasian Group" had the largest skull size (mean of 92 cubic inches), while the "American Group" (79 cubic inches) and "Negro Group" (83 cubic inches) had the smallest skulls. In fact, Morton qualified these numbers by stating that if other groups and more skulls were included, the Caucasian mean would probably drop to 87 while the Negro mean would be reduced to 78 or even 75 cubic inches. See Michael (1988) for an enlightening discussion and "new look" at Morton's data.

In a footnote in his short communication on the size of skulls, Morton (1849, 223) explained the meaning of the word *race*, which is interesting in light of the hierarchical method of classification employed by Stanley Garn (1961) over 100 years later:

> Ethnographic affinities will probably demonstrate that what are now termed the five races of men, would be more appropriately called groups; that each of these groups is again divisible into a greater or smaller number of primary races, each of which has expanded from an aboriginal nucleus or center.

Agassiz, a Swiss naturalist and Harvard professor, provided the theoretical basis to support the "American" concept of polygenesis. Agassiz was a creationist who believed that the story of Adam referred to the origin of Caucasians and that human races were clearly separate species. He even developed a theory of creation centers. Each species radiated from these centers and occupied the surrounding territory (Gould 1996). Modern races occupy these distinctly bounded areas, with some blurring on the edges caused by migration. The differences between the races were permanent, even under the most diversified of climatic forces (Agassiz 1962). Thus, for Agassiz, modern races were nonoverlapping, geographic species.

MORE MEASURING OF SKULLS AND THE IDEAL TYPE

In the middle of the nineteenth century, scientists were quantifying and measuring human bodies, focusing primarily on cranial morphology. Rigorous measurements became reified as being more scientific and accurate than the earlier subjective assumptions and analyses. The coupling of evolution with quantification laid the basis for the first real theory of scientific "racism" (Gould 1996). Statistical methods were applied to the study of human variation. The typological concept of the "average human," the "ideal type," and the "type specimen" frequently appeared in the anthropological literature. A species could be divided into a number of ideal types or races based on numerous, objective, "scientific" measurements. This typological approach espoused that discrete groupings, races if you will, could be created based on an average or an ideal. Today, we recognize that viewing human variation from the perspective of the "average human" or the "ideal type" inhibits and obscures the major focus of study—the actual extent of variation. It does, however, contribute to stereotyping and rankings and makes it easier to lump and classify.

The primary focus of this quantitative research was the skull. In 1842, Anders Retzius (1796–1860) popularized a measurement called the "cranial index." The *cranial index* was simply the maximum breadth of the skull divided by the maximum length. The ratios produced were then grouped and named: (1) long, narrow heads that generated a ratio of 0.75 or less were labeled "dolichocephalic"; (2) short, broad, or round skulls that produced a ratio over 0.8 were called "brachycephalic"; and (3) those between were labeled "mesocephalic." Face angle also became an important measurement, with *prognathic* (jutting out) being much worse (and primitive) than *orthognathic* (less jutting forward or straight). It was thought, incorrectly however, that the shape of the skull was the most resistant to change, making it an excellent feature for tracing a population's (also an individual's) ancestry and origin. So, by comparing cranial indices, scientists could objectively study human variation and delineate human groups. The cranial index not only was purportedly useful in sorting out different human groups but also became the basis for other interpretations and extensions. Paul Broca (1824–1880), a professor of clinical surgery and the founder of the Anthropological Society of Paris, became intrigued with Retzius's work. Broca extended the work in craniology by fitting behaviors and social status to the shape of different skulls (see Gould 1996 for an interesting discussion of Broca's contributions).

TYPOLOGICAL APPROACH

The emphasis on cranial morphology, **anthropometrics**, and anatomy during the late nineteenth century encouraged the continued use of the typological approach in anthropology during the twentieth century. New methods of quantitative analysis were developed, but the typological paradigm continued, changing little in the way that anthropologists studied human variation and viewed or classified races. The metrical and morphological traits used in the analyses and classifications were thought to be stable and environmentally nonadaptive. The traits and classifications were also indistinguishable in many aspects from popular racial stereotypes. Along with intermediate types, the notion of "pure" or primary, unmixed races emerged in the anthropological literature (Hooton 1926, 1931, 1936), helping solidify the image that races were discrete units that were homogenous in their characteristics. This view obscured most of the variability seen within populations. Hooton's polygenist and racial thinking was akin to that of Agassiz, and he was able, through a number of his students, to perpetuate the typological approach to studying human variation (Armelagos et al. 1982). As Hooton (1926, 75) put it, race was

> a great division of mankind, the members of which, though individually varying are characterized as a group by a certain combination of morphological and metrical features, primarily nonadaptive, which have been derived from their common descent.

The idea that races could be identified by a limited number of unique characteristics, presumably transmitted together as a package, helped reify the concept of an ideal type or type specimen. Races were thus discrete, homogeneous entities, lacking variability.

MENDELIAN GENETICS ENTERS THE GAME

In 1900, not only was Mendelian genetics "rediscovered" but Landsteiner discovered the ABO blood group system. It did not take too long before there were studies of the distribution of blood types across the world. Reports appeared that gave details on the frequency of A, B, and O blood in diverse populations in the world. Researchers soon compiled these results and started to analyze them. Hirschfeld and Hirschfeld (1919, 677) suggested that blood groups (A, B, and O) could be used to delineate biochemical races.

> It seemed, therefore, that it would be of interest to make use of the properties of blood . . . to form an anthropological criterion for the discovery of hitherto unknown and anatomically invisible relationships between different races.

Creating what they called a "biochemical race index" (consisting of the ratio of A to B blood in a population), they identified three major racial types: European, Intermediate, and Asio-African. Here again, we see a clear separation of Europeans from the rest of the world. Today, we view these types of "racial index" as biologically meaningless. The article then attempted to trace the origin of the A and B alleles in all races based on two different hypotheses: (1) that A and B were in the same proportions in all races when humans appeared on the earth and (2) that A and B had different origins in different races. They suggested that the latter hypothesis was correct and that India was the cradle for B blood. The origin of A could not be located, but they assumed it arose in north or central Europe and then spread out from there to the rest of the world. Hirschfeld and Hirschfeld (1919, 679) dismissed the first hypothesis by stating that to be correct, it would

> depend on the assumption that for unknown reasons A is more suitable for increased resistance of the organism to disease in a temperate climate, while B is more suitable in a hot climate . . . improbable that the climatic conditions should influence the frequency of A and B.

The idea that these genetic traits were nonadaptive was similar to the reasoning used in suggesting that many anthropometric traits used in racial classifications were also nonadaptive.

Using ABO blood group data and the racial index of Hirschfeld and Hirschfeld, Ottenberg (1925) suggested that there were six main types (races) of human. These groups only partially corresponded to the racial groupings based on other characteristics.

Table 1.2 Examples of two racial classifications based on early genetic data

Ottenberg[1] 1925	Snyder 1926
European type	European
Intermediate type	Intermediate
Hunan type	Hunan type
Indomanchurian type	Indomanchurian
African-South Asiatic type	Africo-Malaysian
Pacific American type	Pacific-American
	Australian

[1]Under each type, Ottenberg (1925) also lists a number of races, such as Gypsies (in Hungary), Germans (in Heidelberg), Javans, and North American Indians.

Another early attempt to use and discuss the usefulness of the newly discovered blood groups to classify humans into races was the work of Snyder (1926). Using similarity in the frequencies of the ABO system, Synder came up with seven types of race that were very similar to those of Ottenberg (Table 1.2).

Snyder also provides subgroups within each of his seven races. At the same time he was making his groupings, he noted that grouping people into races was arbitrary. A few years later, Snyder (1930) argued for the use of blood group data as additional criteria for racial classifications, citing four major advantages: their heritability, stability under varying environments, conscious selection of samples not possible, and the fact that variation in racial groups was striking and correlated with racial affinities.

QUESTIONING THE USEFULNESS OF THE RACE CONCEPT

In the first half of the twentieth century, while racial classifications continued to be generated, a few anthropologists (Ashley Montagu) and biologists (Julian Huxley) began to argue that it was difficult to use zoological nomenclature for classifying humans into groups. Such factors as language, religion, and social institutions helped shape human beings and introduced complications not seen in the zoological world. They argued that the classification of humans into races was simply not a productive endeavor or the correct way to examine human variation. Montagu (1942a,b, 1945, 1950, 1962) was probably the most vocal opponent of the use of the term *race* to classify and study humans. Following T. H. Huxley (1865), Joseph Deniker (1900), and J. S. Huxley and Haddon (1936), Montagu adopted the term *ethnic group* as a replacement for *race* in 1936, arguing that *race* had lost its usefulness for describing human variability. From that time on, he urged others to do the same since the term *race* had taken on too much nonbiological and stereotypic baggage. (See below for Montagu's definition of *ethnic group*.)

Montagu (1942a) did not deny that there were differences between populations, but he noted that there were no clear boundaries in the continuous stream of variation. He also argued that anthropologists should look to Darwinian natural selection to understand the relationships among human groups. Anthropologists should develop a dynamic "genetic theory of race" using such concepts as exogamy, endogamy, migration (gene flow), mutation, selection, isolation, and random events (genetic drift). As

Montagu (1942a, 372) states, "... 'race' is merely an expression of the process of genetic change within a definite ecologic area; that 'race' is a dynamic, not static, condition." He continued by arguing that the goal should not be classification but to discover what factors produce the variation and change gene frequencies. He (1942a, 375) also ventured to put forth a definition of *ethnic group*:

> An ethnic group represents one of a number of populations comprising the single species *Homo sapiens*, which individually maintain their differences, physical and cultural, by means of isolating mechanisms such as geographic and social barriers.

In 1944, Henry Fairchild (1944, 422–423), a social scientist, writing in *Harper's Magazine*, examined seven antiracist arguments that assert race differences are negligible. These arguments are listed here because they provide insight into the questions being asked at the time about human variation, its origins, and the social and biological ramifications.

1. That all men have a common origin.
2. That men of all races are much more alike than they are different.
3. That there are greater differences between extremes of a given race than there are between the average types of different races.
4. That because the extremes of the different races overlap, individuals of a given race may have a particular trait more highly developed than some individuals belonging to some other race of which it is supposed to be characteristic.
5. That there are no pure races today.
6. That all the races of men can interbreed, and such miscegenation is not harmful.
7. That intelligence tests do not reveal simply native ability but are influenced by education and other environmental factors.

Note that many of these same issues persist today, though cloaked in modern terms, theories, and data sets. As an example, geneticists have recently demonstrated that there is more genetic variation within the "so called major geographical races" than between them. Also note that discussion of behavior and intelligence among different races (see Chapter 13) continues to this day.

A GENETIC DEFINITION OF RACE

As early as 1944, Dobzhansky provided a genetic definition of races: "Races are defined as populations differing in the incidence of certain genes, but actually exchanging or potentially able to exchange genes across whatever boundaries separate them" (p. 265). This definition was similar to one offered a bit later: "Races can be defined as populations which differ in the frequencies of some gene or genes" (Dunn and Dobzhansky 1952, 118). Races were dynamic in nature and changed over time by the mixing of groups. These changes could be seen in the fossil record as horizontal and vertical species and races. For Dobzhansky, the traditional morphological races of the anthropologists were inferences of genetic races.

Two books appeared in 1950 that classified humans into discrete races. Coon, Garn, and Birdsell identified 30 races, while Boyd listed only six. These differences were attributed to a number of factors, such as lack of agreement on what was a taxonomic unit in the methods and data used in each of the classifications. Coon, Garn, and Birdsell primarily used morphological data, while Boyd employed blood group (genetic) data.

Even though the data sets were different, the perspective and goal of these two works were the same—to divide the world into races.

In 1950, William C. Boyd, an immunologist, argued for abandoning the traditional anthropometric methods of racial classification in favor of a genetic perspective. He provided five reasons (1950, 21–22) for the unacceptability of skeletal analysis in racial classifications: (1) skeletal morphology is difficult to determine in the living peoples, (2) the skeleton adapts quickly to environmental conditions, (3) skeletal characteristics are controlled by the action of many genes (polygenic), (4) the study of the skeleton is driven by the data alone, and (5) because metric studies were not logically or well conceived, anthropometery and craniometry are obsolete (genetic studies offer more information). Using gene frequencies to define races is both objective and quantitative. As Boyd (1950, 274) stated:

> The genetic classification of races is more objective, and better founded scientifically, than older classifications. The differences we find between races are inherited in a known manner, not influenced by environment, and thus pretty fundamental. But the new criteria differ from some older criteria in an important respect. In certain parts of the world, an individual will be considered "inferior" if he has, for instance, a dark skin, but in no part of the world does the possession of a blood group A gene, or even an Rh negative gene, exclude him from the best society. There are no prejudices against genes.

Boyd then used "nonadaptive traits" in the blood (ABO, Rh, PTC, MN, and secretor systems) and other "nonadaptive" morphological traits to "tentatively" classify humans into six races, noting that they correspond nicely with geography (Boyd 1950, 268–269) (Table 1.3).

Even though Boyd's analysis may have initiated a change in many of the methods of racial analysis, the major questions asked and answered remained virtually unchanged. The analyses remained typologically oriented, with the express goal of classifying human variation into discrete, nonoverlapping groups.

A year after the publication of Boyd's book, T. D. Stewart (1951a,b) noted that the classifications of serologists were not surprisingly different from those of anthropologists

Table 1.3 Boyd's classification 1950 and 1958

1950	1958
Early European group (hypothetical)[1]	Early Europeans
European (Caucasoid) group	Lapps
African (Negroid) group	Northwest Europeans
Asiatic (Mongoloid) group	Eastern and Central Europeans
American Indian group	Mediterraneans
Australoid group	African race
	Asian race
	Indo-Dravidian race
	American Indian race
	Indonesian race
	Melanesian race
	Polynesian race
	Australian (aboriginal) race

[1]"Represented today by their modern descendants, the Basques." (Boyd 1950, 268)

using traditional methods. Stewart suggested that the serologists used existing morphological classifications to draw their samples. Hence, they picked individuals from whom to get blood based on whether they were phenotypically Asiatics, Indians, whites, Africans, etc. They then analyzed the data within this framework, thus manipulating the gene frequencies and obtaining a classification similar or identical to the morphological one. At the same time, Strandskov and Washburn (1951) wrote a short editorial arguing that genetics and anatomy should supplement one another and be used together in racial classifications. For them, races were groups that differed in heredity, and the races should be the same no matter whether one used genetic or anatomical data and methods.

The year 1951 also saw the appearance of a now seminal article titled "The New Physical Anthropology." In this article, Sherwood Washburn argued that physical anthropologists should change their perspective, goals, and approaches. The anthropology of the past was one of technique or the mastery of taking careful measurements, computing indices, and defining type specimens for static classifications. The new physical anthropology should focus on the mechanisms of evolutionary change and adopt a dynamic perspective. The description and speculative methods of the old should be replaced with an emphasis on problems and tests. The concept of a "new physical anthropology" (as defined by Washburn) was controversial from the start. It did, however, reflect and strongly articulate the changing scientific paradigm in anthropology and the shift that was occurring in racial studies and the study of human variation.

As more genetic data accumulated from around the world, Boyd (1958) expanded and updated his classification to 13 races (Table 1.3, note the lack of diversity in Africa and the Americas compared to Europe).

At the same time as these static, typological studies of human variation were appearing, some anthropologists were arguing that the population (breeding unit) should be the basic unit of study of human diversity and adaptation (Thieme 1952). The idea was that each breeding population was subjected to specific environmental constraints and responded through the evolutionary mechanisms of mutation, gene flow, genetic drift, and natural selection. As these populations adapted to these particular environments, they came to manifest traits (measured by gene frequency differences) that were unique. Thus, races could be viewed as episodes in the evolutionary process (Hulse 1962). Races were not static, fixed entities but dynamic units that constantly changed. One could also study the relationship between cultural and biological diversity. This, as Thieme states, is the anthropological perspective of combining cultural and physical anthropology. The concept of race as a breeding unit was, however, not without its problems. For example, what actually constituted a breeding population was not as clear. Also, if races were equated to breeding populations, were all breeding populations automatically separate races (Smedley 1999)?

Expanding upon the suggestion of Rensch (1929) that there are taxonomically broad geographical races and smaller units within local races, Garn and Coon (1955) and Garn (1961) proposed that there were three levels of racial groups: (1) geographical races, (2) local races, and (3) microraces. The *geographical races* corresponded to major continental units and island chains (Fig. 1.5). Garn and Coon (1955) suggest that there are about six or seven geographical races. Finally, Garn (1961) specifically identified nine geographical races: Amerindian, Polynesian, Micronesian, Melanesian-Papuan, Australian, Asiatic, Indian, European, and African. *Local races* were subdivisions within continents (e.g., northwestern Europeans, Bantu, and Iranians), while *microraces* could be equated with breeding units. Garn and Coon (1955, 999) suggested that "if the local race is equated

HUMAN RACES

Figure 1.5. The nine geographical races. (From Garn, *Human Races* 1961. Courtesy of Charles C. Thomas Ltd., Springfield, Illinois.)

with the Mendelian population, then the number of local and micro-geographical races is upwards of thirty." This type of classification system used the older, typological system based on geography and morphology combined with the concept of breeding populations. In a sense, Garn attempted to add a dynamic, evolutionary dimension to the traditional typological classification systems but, in the end, produced a traditional racial classification.

It has been suggested (Marks 1995) that these types of study, the goal of which was to identify and name human groups, came to a crisis and contributed to a paradigm shift in anthropology in 1962 with the publication of Carleton Coon's book *The Origin of Races*. In this work, Coon identified five "tentative" living races: Caucasoid, Mongoloid, Australoid, Congoid, and Capoid. Coon claimed that these five races were also identifiable in the fossil record of the Middle Pleistocene. That is, these five races could be traced to *Homo erectus* specimens throughout the world. *H. erectus* then evolved five separate times in parallel fashion to become modern *H. sapiens*. Interestingly, these races did not become fully sapient at the same time, with the Caucasoids arriving first and the Congoids and Capoids arriving last:

> As far as we know now, the Congoid line started on the same evolutionary level as the Eurasiatic ones in the Early Middle Pleistocene and then stood still for a half million years, after which Negroes and Pygmies appeared as if out of nowhere. (Coon 1962, 658)

Criticism of Coon's approach and conclusions was swift (e.g., Dobzhansky 1963, Montagu 1963) and, according to Marks (1995), helped precipitate a change in anthropological research from the pursuit of racial classifications to the examination and explanation of human biological diversity and adaptation. Brace (1982, 21) noted that many of these studies

> focused on the testable aspects of human biology, but in the end, they generally conclude with a named list of human "races" assigned to various geographic and local regions. The connection between the biology discussed and the races named at the end is never clearly spelled out, and in fact the attentive reader cannot discover, from the information presented, just how the racial classification was constructed—other than the fact that this just seems to be the way anthropologists have always done things.

CLINES AND POPULATIONS

In 1962, Frank Livingstone published an article titled "On the Non-Existence of Human Races." The title succinctly summarizes the arguments put forth in the article. Livingstone pointed out that the static, typological notion of race was simply not compatible with the dynamic concept of natural selection. The continued use of a construct that was based on fixed, nonadaptive traits did not mesh well with studies of the causes of variation in human populations. Livingstone did not deny that there were differences among populations but argued that these differences did not fit into neat little packages called "races." As an alternative to this static approach, he suggested that research should focus on geographical variation of single traits, or what was called "**clinal** variation." In other words, "There are no races, there are only clines" (Livingstone 1962, 279). If the goal of anthropological research was to explain the genetic variation among populations, then the racial approach was simply not adequate. In the same year, Montagu (1962, 919) continued to insist that *race* was an ambiguous, overused, and very loaded term that should be dropped from the scientific literature since it continued to mix biology, culture, intelligence, personality, nations, etc. together:

> Once more, I shall, as irritatingly as the sound of a clanging door heard in the distance in a wind that will not be shut out, raise the question as to whether, with reference to man, it would not be better if the term "race" were altogether abandoned.

Like Livingstone, Montagu (1962) did not deny that there were differences between peoples. He did, however, argue that one should study a population's diversity, ask questions about the observed variation, and then compare it to other populations. For Montagu, it was unproductive to continue using the same nineteenth-century perspective: "In our own time valiant attempts have been made to pour new wine into the old bottles. The shape of the bottle, however, remains the same" (1962, 920).

C. Loring Brace (1964) also advocated for the study of individual traits, stating that races, and even populations, were not adequate for the study of human diversity. The distribution of individual traits and the selective pressures modifying these traits should be the focus of study, not arbitrary entities called races. Thus, clines replaced races as the unit of study for many anthropologists during the 1960s and 1970s.

SO, WHERE ARE WE TODAY?

Some anthropologists feel that racial classifications and the use of the concept of race as a tool for examining human variation are disappearing (Sanjek 1994), while others see a resurgence in their use (Goodman and Armelagos 1996, Lieberman and Jackson 1995). Cartmill (1998) examined articles published in the *American Journal of Physical Anthropology* from 1965 to 1996. He found that there was virtually no change in the way racial taxonomy was used in studying human variation. On average, about 40% of the articles appearing in the journal used racial categories. In a recent survey, Lieberman and Kirk (2000) found evidence for a decline in the use of biological race as an important concept in anthropology.

Subjectivity, classification, and the typological perspective still enter into our current thinking and research designs to some extent. There is clearly no agreement on the number of races or the validity of the classifications. There is not even agreement on a biological definition of race. Anthropologists recognize that confusion stems from numerous sources and problems, among them:

- Human groups are not morphologically homogenous.
- Many polygenic traits are difficult to measure accurately.
- It is difficult, if not impossible, to determine discrete boundaries in continuously varying traits.
- The traits used in a classification may be undergoing different rates of evolutionary change.
- Traits are not linked (i.e., traits are not concordant).
- How many traits should one use? Are three sufficient, or are 25? Would 32 be better? Or, even better yet, one could use 247.
- If there are differences between groups, how much difference is *biologically* significant (10%, 20%, 31%, or 62%)?
- Not everyone can be placed in a category. What does one do with those people who simply do not fit neatly into a group?
- There is actually greater genetic diversity within groups than between major geographical divisions

Given all these problems and issues, many anthropologists now argue that race is biologically meaningless as an explanation or analytical research tool. They further point out that a racial classification does not answer any questions of evolutionary or adaptive significance. Anyone can classify humans into groups if that is the ultimate goal; however, in doing so, he or she has not asked or answered any really interesting questions. Why are there biological differences among peoples? Why one specific trait distribution and not another? Why is there biological variation in the first place, and how much is important from an evolutionary standpoint? How did the variation originate? Can the diversity be lost? How is it maintained in the population over time? How does racism affect the genetic and biological structure of populations? These anthropologists do not deny the fact that there are biological differences among peoples of the world; rather, they suggest that there are more productive ways of examining variation than viewing it simply as racial. As the American Anthropological Association's recent statement on race suggests, "Biophysical diversity has no inherent social meaning except what we humans confer upon it."

CHAPTER SUMMARY

During the past five centuries, most of the research on human diversity resulted in the production of various racial classifications rather than in-depth descriptions of the extent and nature of the variation. Early racial classifications relied primarily on outward, phenotypic traits coupled with cultural or behavioral qualities. These classifications were often ethnocentric and stereotypic in nature. Some questioned whether there were multiple origins (polygenesis) of humans or a single origin (monogenesis). One of the early classifiers was Carolus Linnaeus, who divided the world into four basic varieties (Americans, Europeans, Asians, and Africans) based on geography, skin color, humors, posture, and sociocultural customs. In 1781, Johann Friedrich Blumenbach coined the term *Caucasian* and created a hierarchical racial classification primarily based on skull shape. These early classifications were rigid, linking behavioral traits (morals, customs, values, temperament) to physical traits. The climate of a region continued to be considered the major factor responsible for molding and shaping racial characteristics.

In the mid-nineteenth century, polygenesis and monogenesis became important issues, and the fascination with human skulls become prominent in classification schemes. Samuel George Morton and Louis Agassiz emerged as significant proponents of polygenesis. Morton provided the scientific basis (skull measurements) for this view, while Agassiz become the theoretician. Rigorous measurements of the body, particularly the skull, become reified as being more scientific and accurate than earlier subjective measurements and assumptions. As Gould (1996) suggests, the coupling of rigorous quantification with evolution laid the basis for the first real theories of scientific "racism."

The emphasis on cranial morphology, anthropometrics, and body shape during the late nineteenth century encouraged the continued use of the *typological approach* in anthropology during the twentieth century. New methods and new measurements of the body were devised, but the typological paradigm continued, changing little in the way that anthropologists studied human variation or created racial classifications. In fact, the notion of "pure" races became prominent in the anthropological literature. In addition to measurements of the body, genetic traits started to become important features of early twentieth-century classifications. By the mid-1920s, the ABO blood group system was used to generate racial classifications. By 1944, Dobzhansky provided a genetic definition of race: "Races are defined as populations differing in the incidence of certain genes, but actually exchanging or potentially able to exchange genes across whatever boundaries separate them" (p 265). As these various racial classifications were being generated, a few anthropologists and biologists started to question, in earnest, the scientific validity and utility of the concept of race. Probably the most vocal of these anthropologists was Ashley Montagu, who adopted the use of *ethnic group* instead of *race* in 1936.

In 1950, two books appeared in the anthropological literature that classified humans into races. One of these (Coon, Garn, and Birdsell) relied primarily on morphological data, while Boyd's classification used genetic traits. The data sources were different, but the analyses of human diversity remained typologically oriented, with the express goal of classifying human diversity into discrete, nonoverlapping groups. Races were fixed, static, unchanging units. As Marks (1995) suggests, these types of study, the goal of which was to identify and name human groups, came to a crisis in 1963 with the publication of Carleton Coon's book *The Origin of Races*.

Anthropological research changed from the pursuit of racial classifications to the examination and explanation of human biological diversity and adaptation. Frank

Livingstone (1962) argued that the notion of static, typological races was incompatible with the dynamic concept of natural selection and that one should focus on clines, not races. At this same time, Montagu continued to insist that *race* was ambiguous, overused, and a very loaded term that should be dropped from the scientific literature. Other anthropologists argued that racial classifications were ineffectual since they distorted and obscured the reality of human biological variation and were not compatible with genetics or evolutionary theory. These debates continued into the latter part of the twentieth century, and many anthropologists abandoned the use of *race* and focused their attention on describing and analyzing the genetic and morphological diversity within and between populations. However, as Caspari (2003, 74) notes, "The race *concept* may be rejected by anthropology, but its underlying racial *thinking* persists. Physical anthropologists no longer study races. Populations are now studied, but not all approaches to the study of populations are populational."

This chapter has focused on the history of the race concept for analyzing human biological variation. The remainder of this book takes an evolutionary approach to the study of human biological variation, focusing on the population as the most appropriate unit of analysis and how variation is shaped by evolutionary forces impacting the physical and cultural environments. Further discussion of race as it applies to particular traits is provided throughout the book, such as when considering anthropometrics (Chapter 10) and IQ test scores (Chapter 13).

SUPPLEMENTAL RESOURCES

Augstein H F, ed. (1996) *Race: The Origins of an Idea, 1760–1850.* Bristol, UK: Thoemmes Press.
Cartmill M (1998) The status of the race concept in physical anthropology. *American Anthropologist* 100:651–660.
Caspari R (2003) From types to populations: A century of race, physical anthropology, and the American Anthropological Association. *American Anthropologist* 105:65–76.
Gould S J (1996) *The Mismeasure of Man*, second Edition. New York: W W Norton.
Lieberman L and Kirk R C (2000) "Race" in anthropology in the 20th century: The decline and fall of a core concept, www.chsbs.cmich.edu/rod_kirk/norace/tables.htm.
Marks J (1995) *Human Biodiversity: Genes, Races, and History.* New York: Aldine de Gruyter.
Montagu A (1942) *Man's Most Dangerous Myth: The Fallacy of Race.* New York: Columbia University Press.
Sarich V and Miele F (2003) *Race: The Reality of Human Differences.* Boulder, CO: Westview Press.
Smedley A (1999) *Race in North America: Origin and Evolution of a Worldview.* Boulder, CO: Westview Press.

2

The Genetic Basis of Human Variation

The previous chapter clearly demonstrated the historically shaky grounds human biologists and anthropologists have traversed when studying variation without a clear understanding of basic genetic processes. Our intent in this chapter is to provide a simple understanding of genetic processes as a foundation for the remainder of the book. Like any other aspect of science, our knowledge of even some of the basics is under almost constant revision. However, in this chapter we will focus solely on **transmission genetics** and, to a lesser extent, on the underlying genetic code, areas which have been relatively well understood for at least the last 40 years. We depart from many introductory textbooks on two fronts. First, we do not provide a historical account of the discovery of genetic principles, in the hope of avoiding the tedium of recounting the tale of "Mendel and his peas," with which virtually all students will be familiar. Humans are not entirely like peas, fruit flies, bread molds, or roundworms (some of the favorite organisms for genetic analysis), so while the history of genetic analyses for these organisms may in itself be a gripping yarn, we will forgo its telling here. Second, we limit the terminology we use to the bare minimum necessary to support the remainder of the text. As a consequence, the reader will be deprived of such terms as *anaphase, prophase,* and *metaphase* because knowing the stages of the cell cycle bears little relationship to understanding how the genetic information is passed from parents to children (the subject of transmission genetics, with which we deal here). Some potentially unfamiliar terminology is unavoidable, so we provide these words in the glossary at the end of the text.

A MINIMALIST VIEW OF THE REQUIREMENTS FOR INHERITANCE VIA DNA

It will come as no surprise to the student that **deoxyribonucleic acid (DNA)** is the molecule that transmits genetic information from mothers and fathers to their children. In order for DNA to function in both the development of embryos from single fertilized eggs and sustain genetic transmission from parents to offspring, there are two properties that it must, and indeed does, possess. First, DNA must be able (with a little help) to faithfully reproduce many copies of itself. Second, DNA must be capable of producing maternal copies from the mother and paternal copies from the father that combine to form the initial single offspring cell. Before we turn to describing the structure of DNA, we consider at a broader level how DNA "behaves."

Mitosis Produces Identical Nuclear DNA in Daughter Cells

The requirement that DNA be able to make many identical copies follows from the fact that all humans start life as a single-celled organism known as a **zygote** and end up as a multicellular adult with far more than a trillion cells. The cell division, known as **mitosis**, that creates two identical daughter cells from one parental cell is essential to our growth from a zygote. Mitosis takes the zygote from its single-cell form to a two-cell stage; then, each of the two cells divides to make a total of four cells, and each of these cells divides to make eight and then again to make 16 cells. The process would continue rather like computer memory (32, 64, 128, 256, etc.) except that the synchronicity of mitotic divisions is soon lost, some cell lineages go through fewer mitotic divisions than other lineages, and it becomes too difficult to count the actual number of cells in the human embryo. Clearly, not all cells are identical, so we speak of cellular **differentiation** occurring during growth and development of the embryo. Up until about the 16-cell stage in humans, the mitotically formed daughter cells are identical. Past this point, cells start to commit to differentiated **embryonic stem cell** lines (a topic much in the news these days). However, with only a few exceptions (such as red blood cells, which lose their nuclei as they mature), the DNA contained in each mitotically formed daughter cell is identical to the DNA in all other cells. Differentiation occurs not by modification of the DNA itself but by determination of which parts of the DNA are actually used by the cell. Thus, mitosis is about building many, many identical copies of a "library" of genetic information, which differentiated cells then choose to "check out and read" at will.

We know that the DNA located in the **nucleus** of **diploid** cells (cells that have 23 pairs of **chromosomes** in humans) must be identical across all cells because of experiments in **nuclear transplantation**. In such experiments, the nucleus in an **ovum** (a **haploid gamete** formed by females, also sometimes called an "egg") is inactivated or removed (*enucleation*), and then the nucleus from an adult diploid cell is injected into the egg. The artificially formed zygote is then raised to maturity. In early experiments (Briggs and King 1952), diploid nuclei from frog **blastula** cells (the blastula is an early developmental stage in which the zygote has divided many times to form a hollow ball) were transplanted to frog eggs, which then underwent mitosis to form tadpoles. These early experiments demonstrated that even though the embryonic cells had undergone many mitoses to reach the blastula stage, the nuclei were still *totipotent*, capable of forming all tissues when placed into an enucleated egg. In more recent experiments, nuclei from adult cells have been transplanted in frogs (Gurdon et al. 1975) and sheep (the infamous production of Dolly reported in Wilmut et al. 1997). These transplantation experiments demonstrate that mitosis does not alter the DNA.

Meiosis Produces Haploid Daughter Cells

Because both the mother and father contribute a gamete in the production of the diploid zygote (the mother providing the ovum and the father the **spermatid**), it is necessary for the **germ cells**, precursors to ova and sperm, to undergo a reduction division that takes them from the diploid to the haploid state. This reduction division is achieved by **meiosis**. Unlike mitosis, meiosis does not produce identical copies of the parental DNA. By definition, meiosis cannot produce identical copies because the parental nucleus is diploid, while the daughter nuclei must be haploid. We will see as well when we consider the mechanics of meiosis that there are other reasons that the DNA in daughter cells is not identical to (half of) the parental DNA.

HUMAN CHROMOSOMES

The DNA in human cells is packaged into 46 chromosomes contained within the nucleus and, hence, referred to as "nuclear" DNA. In addition to the nuclear DNA, there is cytoplasmic DNA found within the mitochondria, which we discuss in a subsequent section. In this section, we give a brief description of the chromosomes in cells undergoing mitosis. Prior to division, chromosomes are not visible, so our description of chromosomes must focus on the early stages of mitosis. Following replication of the DNA, 23 pairs of chromosomes are visible, and each chromosome has two identical portions, referred to as **sister chromatids**. It is replication that formed one of the sisters from the other, and as a consequence the chromatids are, barring mutation, identical. The packaging of DNA within the chromatids is quite complex, with the double helix of DNA in each chromatid being wound around various proteins and then compacted and folded back on itself. All chromosomes have a single section called the **centromere** that connects the two sister chromatids. The centromere is primarily composed of many repetitions of a sequence that is about 170 bases long, referred to as *α-satellite DNA*. This sequence can be repeated as many as 29,000 times (Miller and Therman 2001, 49) in a given centromere. Structurally, the centromeres function to align the chromosomes during mitotic division.

Chromosomes can be initially characterized by the overall length of the sister chromatids and the relative position of the centromeres. The centromere may be near the middle of the two chromatids, referred to as **metacentric**; displaced to one end, referred to as **submetacentric**; or near the very end of the chromatids, referred to as **acrocentric**. The chromosomes are usually shown in a very standardized way known as a **karyotype** (see Fig. 2.1), in which the homologous chromosomes (members of a pair) are shown next to each other and the pairs are ordered from longest to shortest chromosome (with the exception of the last two pairs, which are in inverted order). A karyotype can be obtained at any stage in a human life, including prenatally by amniocentesis. The first 22 pairs of chromosomes are referred to as **autosomal** chromosomes or just autosomes. The last pair of chromosomes are the sex chromosomes, which form a homologous pair in females (XX) and a non-homologous pair in males (XY). Because the chromosomes are numbered from longest to shortest, it is possible to refer to a particular chromosome by number so that if we refer to chromosome number four, then we are referring to the fourth longest autosomal pair of chromosomes. The specific arms are "named" relative to the centromeres, with the shorter chromatid arms referred to as "p" and the longer arms as "q" (*p* stands for "petite"). Even for the metacentric chromosomes, it is possible to determine a short and a long arm since the centromere is never exactly in the center of the chromosome. In the case of the fourth chromosome, there is an important disease **locus** (a physical location on a chromosome) for Huntington's disease on the short arm, so we can say that the locus "maps to 4p." When a karyotype is being produced, the chromosomes can also be prepared with various stains that are taken up differentially and lead to patterns of visible bands across the chromatids. On the basis of these bands, regions can be defined within the short and long arms, bands can be defined within the regions, and even sub-bands can be defined within the bands. All of these are numbered moving away

Figure 2.1. A stylized karyotype showing the band labeling used to refer to specific chromosomal locations. This collection of chromosome drawings (each referred to as an "ideogram") was assembled from lower band-resolution images available from the Human Genome Organisation (http://www.hugo-international.org/). These ideograms, as well as the higher resolution 850 band maps, are displayed in "NCBI Map Viewer" (http://www.ncbi.nlm.nih.gov/genome/guide/human/).

from the centromere using single digits and with sub-band position noted with a decimal. For example, the Huntington's disease locus maps to 4p16.3, which means that the locus is on the short arm of the fourth chromosome, in the first region (the one closest to the centromere), in the sixth band within the first region, and within the third sub-band of the sixth band. The ends of the chromosomal arms are sometimes labeled

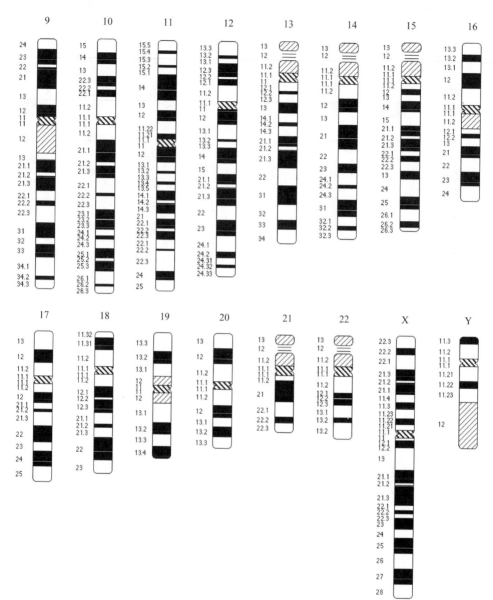

Figure 2.1. (cont'd)

as "ter" for terminus, so the two ends of the fourth chromosome would be "4pter" and "4qter."

THE PROCESSES OF CELL DIVISION

Before turning to some of the specifics of genetic transmission and the functioning of DNA, we need to consider the basics of cell division (mitosis and meiosis).

Mitosis

The process of mitotic cell division is, from a human genetic standpoint, decidedly uninteresting as it does not create variation that can be transmitted from parents to children. Mitosis occurs via replication of the DNA so that the human diploid number of 46 chromosomes (23 pairs) is represented by sister chromatids attached at centromeres (there are consequently 92 sister chromatids, see Fig. 2.2). The 46 chromosomes line up more or less in a row and then tear apart at the centromeres so that one chromatid goes

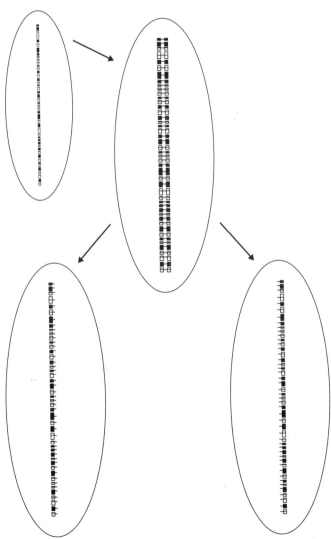

Figure 2.2. A schematic representation of mitosis. The first step shown in the schematic is fictional as the chromosomes are not clearly visible until after replication has produced the sister chromatids. Consequently, the single-chromatid state shown in the first step cannot be observed. Maternally derived chromosomes are shown in white and paternally derived chromosomes are shown in black. Centromeres are shown as horizontal lines.

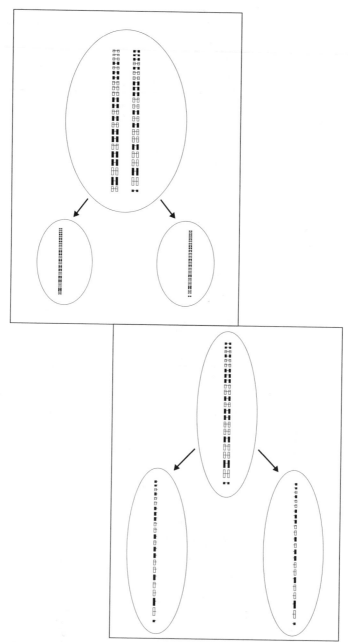

Figure 2.3. A schematic representation of meiosis. The *first panel* shows the first division with pairing of homologous chromosomes (the sex chromosomes are shown at the bottom). The *second panel* shows the second meiotic division.

to one new daughter cell while the other chromatid goes to another new daughter cell. Half of the 92 sister chromatids will go to one of the new daughter cells and half to the other so that each contains the diploid set of 46 chromosomes.

Meiosis

Meiosis begins much like mitosis with DNA replication so that there are 46 chromosomes with 92 sister chromatids (see Fig. 2.3). The cell goes through a division, but unlike in a mitotic division, the homologous chromosomes will pair with one another and the two sex chromosomes will also pair with each other. As a result of this pairing, one member of each homologous pair will go to a new daughter cell so that each cell receives 23 chromosomes (the haploid number), and it is random as to whether a particular cell receives the maternally derived chromosome or the paternally derived chromosome. To complicate matters further, homologous chromosomes may exchange parts of chromatids via **crossing over**, which is a physical swapping between chromatid portions (see Fig. 2.4). Consequently, with the exception of the sex chromosomes (which are not capable of crossing over), each of the 22 autosomes distributed to each of the two daughter cells is typically composed of both maternal and paternal DNA. It is this reassortment of maternal and paternal variation (or really grand-maternal and grand-paternal variation) that leads to additional genetic variation in offspring.

Following the first cell division in meiosis, there is a second division of each of the daughter cells, to ultimately produce four daughter cells. This second division occurs just as in the one division of mitosis so that in each of the two haploid daughter cells the chromosomes align in the middle and then the chromatids tear apart at the 23 centromeres. This produces four haploid cells. If the meiotic division occurs within the ovaries of a female, then two of these four daughter cells will form an ovum (the remaining cells receive very little cytoplasm and are consequently incapable of being fertilized).

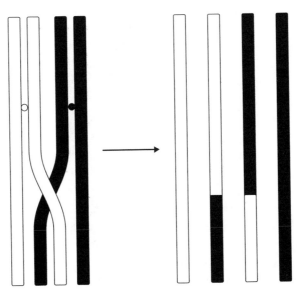

Figure 2.4. Example of a crossover event for homologous chromosomes.

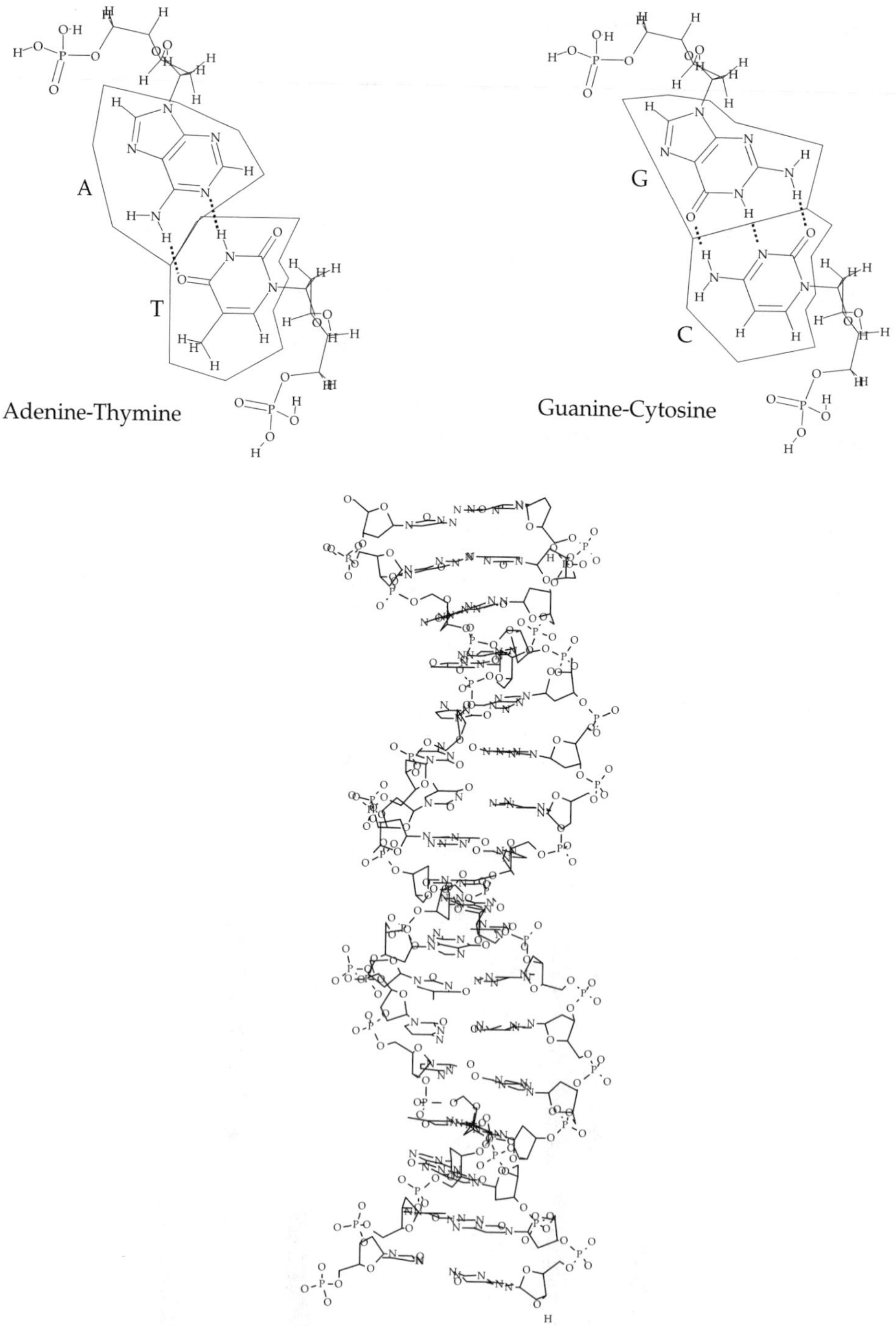

Figure 2.5. Diagram of DNA structure. The *upper left* shows an adenine-thymine base pair, the *upper right* shows a guanine-cytosine base pair, and the *bottom center* shows a DNA molecule with 17 base pairs.

If the meiotic division occurs within the testes of a male, then all four daughter cells will develop into spermatids, which are specialized cells for the delivery of the haploid nuclei from the male. As a consequence, spermatids contain very little cytoplasmic material, a point which is important in discussing the transmission of mitchondrial DNA.

DNA

In this section, we briefly review the properties of DNA as they relate to genetic transmission and, to a lesser extent, the way that DNA functions. First, we will need some basic terminology as well as a description of DNA.

The Molecular Composition of DNA

What can be thought of as the "skeletal structure" of DNA is composed of two sugar–phosphate "backbones" that are twisted in a helix around one another (hence the "double helix" moniker that is often used to describe DNA) (see Fig. 2.5). The specific sugar found repeated many, many times in this backbone is deoxyribose, from which DNA takes the first part of its name. Attached to these backbones are pairs of nitrogenous bases. In a strand of DNA, a single sugar, phosphate group, and nitrogenous base are referred to as a **nucleotide**. The term *acid* is a reference to the phosphate group, which has the potential to form free hydrogen ions and is consequently also known as phosphoric acid. The remainder of the name is a reference to the fact that DNA is found in the nucleus of cells, though it is also found in **mitochondria**, which are cytoplasmic organelles.

The paired nitrogenous bases are the interesting part of the DNA as they can vary (the sugar–phosphate backbone is invariant). There are four bases—**adenine**, **guanine**, **cytosine**, and **thymine**—which as a group are often referred to as "nucleobases". Because of the way that these bases are configured, it is only possible for adenine and thymine to occur as a base pair and similarly for guanine and cytosine (sometimes referred to as the "base pairing principle"). Consequently, if we specify the sequence of bases along one strand of the paired sugar–phosphate backbone, then we will know the sequence on the complementary strand. For example, if the sequence on one strand is A–T–G–T–A–T (for adenine, thymine, guanine, thymine, adenine, and thymine in that order), then the other strand must be T–A–C–A–T–A. Each strand of the double–stranded DNA has a directionality (based on the way that the sugars in the sugar–phosphate backbone are "pointing") so that one end of each strand is referred to as the "3 prime end" (or just 3′) and the other, as the "5 prime end" (or just 5′). The strands are aligned in opposite directions so that the 3′ end of one is paired to the 5′ of the other. Typically, the bases are listed from the 5′ to the 3′ end, so if the short sequence of A–T–G–T–A–T is in "proper" order (5′ to 3′), then the complementary bases would actually be A–T–A–C–A–T and not the T–A–C–A–T–A listed above.

Replication

For the purpose of cell division, the nucleus must be able to make faithful copies of its DNA, a process referred to as "**replication**" (see Fig. 2.6). Replication can occur only on single-stranded DNA, so the first task in replication of a DNA molecule is for the

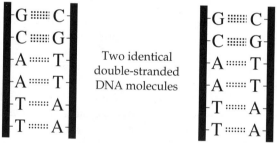

Figure 2.6. Diagram of DNA replication showing a 6 bp segment of DNA. In the *middle panel*, the strands have separated and bases have been added by complementation. In the *bottom panel*, there are now two double-stranded "daughter" DNA molecules that are identical to the original double-stranded molecule.

sugar–phosphate backbones to pull apart, taking with them their associated members of the base pairs. It would be impossible for the double-stranded DNA molecule to come apart into single strands along its entire length at one time as simultaneously unwinding the entire double helix for 46 chromosomes would be extremely disruptive at the micro-cellular level. Instead, each molecule of double-stranded DNA pulls apart into single strands at a number of points along the molecule.

DNA can only grow in the 3′ to 5′ direction along the template strand so that the newly formed strand initiates at its 5′ end and grows to the 3′ end. On one strand, known as the "**leading strand**," the newly formed DNA "grows" continuously from its 5′ to 3′ end. On the complementary strand, known as the "**lagging strand**," the DNA must grow in little segments that are later "stitched" together into a continuous strand by an enzyme known as DNA ligase. For either template strand, the new DNA is formed under the

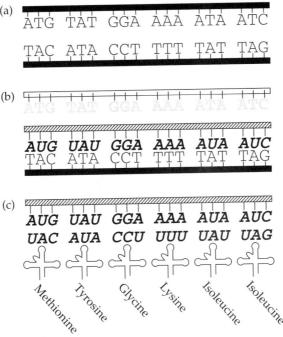

Figure 2.7. Diagram of DNA transcription and translation. (a) The first 18 bp from the coding section of the glycophorin A gene (see Table 2.3). (b) RNA transcription. The gray "backbone" is the forming mRNA. (c) Translation of the mRNA to amino acids.

control of an enzyme known as "**DNA polymerase**." The DNA polymerase moves down the template strands from 3′ to 5′, incorporating nucleotides by complementation. For example, when the polymerase is at an adenine on the template strand, the enzyme will incorporate a thymine on the growing strand. Whether the cell is dividing mitotically or meiotically, replication is the initial step necessary to produce two sister chromatids (each containing a single DNA molecule) from one chromosome (containing one DNA molecule).

Transcription

The ability of DNA to replicate would be of no use unless DNA also had the ability to produce a product. In **transcription** (see Fig. 2.7), just such a product is produced in the form of another nucleic acid, **ribonucleic acid** (or RNA). RNA differs from DNA on three accounts. First, the sugar in the sugar–phosphate backbone has a hydroxyl group that is missing from the deoxyribose found in DNA. Consequently, the sugar is ribose rather than deoxyribose, so we have the acronym RNA instead of DNA. Second, RNA is single-stranded, so it does not form the double helix that DNA forms. Third, RNA has adenine, guanine, and cytosine for its nitrogenous bases; but in place of thymine, it has **uracil**. RNA molecules are built by complementation to one of the strands of the double-stranded DNA, with the DNA strand that serves as the template referred to as the **template strand**. The other strand is referred to as the **nontemplate strand**, which is not transcribed into RNA, and sometimes as the **coding strand** (because with both the messenger RNA [mRNA] and the coding strand complementing the template strand, the mRNA and coding strand will have the same sequence). The bases in an RNA molecule are added by the enzyme RNA polymerase by complementation to the DNA bases of the template strand. The DNA is transcribed from the 3′ to the 5′ end, so the RNA grows from 5′ to 3′. Only a small portion of the long DNA molecule in a chromosome will be transcribed at any one time, with the transcription starting at what is known as a **promotor site** in the DNA. To continue the example from above, the mRNA that would be built off of the template strand would be (from 5′ to 3′) A–U–G–U–A–U, as follows:

5′ A–T–G–T–A–T 3′ nontemplate strand
3′ T–A–C–A–T–A 5′ template strand
5′ A–U–G–U–A–U 3′ mRNA made from template complementation

Note that although the mRNA is built by complementation to the template strand, the mRNA will have an identical sequence to the nontemplate strand (save for substitution of U for T), which is why the nontemplate strand is also called the "coding strand."

There are different kinds of RNA molecules formed from DNA templates, and most are subject to considerable "editing" once they have been initially formed. RNA molecules can travel to cytoplasmic organelles called *ribosomes*, that function in the production of proteins. In this case, the RNA molecules are called **ribosomal RNA (rRNA)**. Other RNA molecules bind to specific amino acids, the building blocks for proteins, and then carry these amino acids to the ribosomes for incorporation into growing protein chains. These RNA molecules are known as **transfer RNA (tRNA)** because they transfer amino acids to the protein chain. Finally, some RNA molecules serve as templates for the production of proteins from individual amino acids, so they bring the "message" from the DNA sequence used to build the protein. These RNA molecules are consequently called **messenger RNA (mRNA)**. mRNA molecules are particularly subject to post-transcription editing. In addition to some repackaging (such as the addition of a string of adenines at one end called the poly-A tail), major portions of the mRNA may be excised. The corresponding DNA sequences for the deleted mRNA sequence are known as **introns**, because they intervene between the functional parts of the DNA sequence, which are known as **exons** because they are expressed. The introns are spliced out of the mRNA molecules at the appropriately named **spliceosomes**, which themselves are composed of RNA and proteins. It is fairly typical for a gene to contain long intron sections that are later edited (spliced) out. For example, phenylalaine hydroxylase is an enzyme that is typically 452 amino acids in length, which would suggest that the gene should be about 1,360 base pairs long. In fact, the gene is a bit in excess of 171,000 base pairs long, with 13 exons (coding sequences). Consequently, about 99.2% of the gene is noncoding.

Translation

Once an mRNA has been appropriately edited, it can be translated into a protein. A protein is just a string of amino acids, of which there are 20 different kinds. Table 2.1 lists the 20 amino acids, along with the single-letter and three-letter abbreviations often used to represent them. Proteins form much of the structure of the human body, so the encoding of proteins by the DNA is the essential connection between genetic information (the **genotype**) and characteristics of individuals (the **phenotype**). That said, it is incorrect to

Table 2.1 **The amino acids and their single-letter codes** (sorted by single-letter code, with three-letter abbreviations in parentheses)

Letter Code	Amino Acid	Letter Code	Amino Acid
A	Alanine (Ala)	M	Methionine (Met)
C	Cysteine (Cys)	N	Aspargine (Asn)
D	Aspartic acid (Asp)	P	Proline (Pro)
E	Glutamic acid (Glu)	Q	Glutamine (Gln)
F	Phenylalanine (Phe)	R	Arginine (Arg)
G	Glycine (Gly)	S	Serine (Ser)
H	Histidine (His)	T	Threonine (Thr)
I	Isoleucine (Ile)	V	Valine (Val)
K	Lysine (Lys)	W	Tryptophan (Trp)
L	Leucine (Leu)	Y	Tyrosine (Tyr)

state that all genes code for proteins, for we have already seen that only sections of DNA are transcribed into the mRNA molecules that actually code for proteins. There are many other genes that code for transfer and ribosomal RNA molecules.

In cells, the tRNA molecules have attached to them specific amino acids, and each tRNA has a specific three-base **anticodon** that complements the mRNA codons. At the ribosomes, mRNA molecules are translated to proteins by the mRNA slipping through the ribosome from 5' to 3', with every mRNA triplet (codon) pairing with a specific tRNA anticodon. When the tRNA pairs, its amino acid is transferred to the growing protein chain. All proteins start with a methionine amino acid (ATG codon), so typically there is a 5' **untranslated region** (UTR) on the mRNA before the beginning of the translation from mRNA to protein. Usually, this initial part of the protein will be cleaved away during the "maturation" of the protein. There are three codons (TAA, TAG, and TGA) that terminate the production of the protein, and these are often referred to as "**stop codons.**" It is quite common for the mRNA to continue far past the stop codon so that there is also a 3' UTR that is generally much longer than the 5' UTR. Table 2.2 shows the complete genetic code for both nuclear and mitochondrial genes, while Table 2.3 shows an actual example of how DNA is ultimately translated into a protein. The protein shown in the table is the precursor for a protein called glycophoran A (GYPA, Kudo et al. 1994), which will be discussed in Chapter 4. We will use this protein through much of the remainder of this chapter to demonstrate a number of points about transmission genetics.

Before we proceed to transmission genetics, we should go through a detailed example of how to read Table 2.2 and show how this rather compactly presented information is more useful than many of the genetic code tables found in other textbooks. As an example of reading Table 2.2, consider the first translated triplet of DNA bases shown in Table 2.3. This triplet is ATG. To find the corresponding amino acid, we start with A in the first row of Table 2.2, then go down to T in the second row, and then end with G in the third row. Underneath this G we read M in the row for nuclear genes, so the ATG triplet codes for methionine (see Table 2.1 for amino acid abbreviations). As another example of reading the genetic code, consider the fifth translated triplet in Table 2.3, which is ATA. If we look this up in Table 2.2, we find that for a nuclear gene the ATA

Table 2.2 The genetic code for nuclear and mitochondrial genes (translation of coding strand DNA triplets to amino acids)

First Base	T				C			
Second Base	T	C	A	G	T	C	A	G
Third Base	TCAG	TCAG	TCAG	TCAG	TCAG	TCAG	TCAG	TCAG
Nuclear	FFLL	SSSS	YY**	CC*W	LLLL	PPPP	HHQQ	RRRR
Mitochondrial	FFLL	SSSS	YY**	CC**WW**	LLLL	PPPP	HHQQ	RRRR

First Base	A				G			
Second Base	T	C	A	G	T	C	A	G
Third Base	TCAG	TCAG	TCAG	TCAG	TCAG	TCAG	TCAG	TCAG
Nuclear	IIIM	TTTT	NNKK	SSRR	VVVV	AAAA	DDEE	GGGG
Mitochondrial	II**MM**	TTTT	NNKK	SS**	VVVV	AAAA	DDEE	GGGG

Amino acid abbreviations are as in Table 2.1. Asterisks represent stop codes, and differences of mitochondrial codes from nuclear are shown in bold.

triplet codes for isoleucine, while for a mitochondrial gene ATA codes for methionine. *GYPA* is a nuclear gene, so the fifth position amino acid is isoleucine, not methionine. What are the advantages of Table 2.2 over other presentations? First, Table 2.2 allows us to show both the nuclear and mitochondrial codes together so that they can be easily compared. These codes are quite similar, but they are not identical. With four possible bases at each of three positions in the triplet, there are $4 \times 4 \times 4 = 64$ possible triplets to code for 20 amino acids and a "stop." Of these 64 possible triplets, 60 (or 93.75%) code for the same thing in nuclear and mitochondrial DNA (mtDNA). Table 2.2 also makes it clear where there are redundancies in the genetic code (different triplets coding for the same thing). If there were doublets instead of triplets for the genetic code, there would be only 16 possible codes, which would be insufficient to code for 20 amino acids and a "stop." However, the triplets provide 64 codes, which is about three times more than are needed. As a consequence, many of the different codes are redundant, and typically it is the third base that is "ignored" in the translation. For example, Table 2.2 shows that TC_ codes for serine, CT_ for leucine, CC_ for proline, CG_ for arginine, AC_ for threonine, GT_ for valine, GC_ for alanine, and GG_ for glycine in both nuclear and mitochondrial genes. Here, the blank represents the third base, which can be any of the four possible bases (A, C, T, or G).

FROM GENOTYPE TO PHENOTYPE

The concepts of genotype and phenotype are already familiar to most students, but now that we have a reasonable handle on how DNA functions, it would be a good time to review them. The *genotype* is just the genetic configuration for an individual. Usually, the

Table 2.3 The glycophorin a (*GYPA*) gene that results in the MM, MN, and NN genotypes (the "M" allele is shown here)

AGTTGTCTTTGGTAGTTTTTTGCACTAACTTCAGGAACCAGCTCATGATCTCAGG

| ATG | TAT | GGA | AAA | ATA | ATC | TTT | GTA | TTA | CTA | TTG | TCA | GCA | ATT |
| M | Y | G | K | I | I | F | V | L | L | L | S | A | I |

| GTG | AGC | ATA | TCA | GCA | **TCA** | AGT | ACC | ACT | **GGT** | GTG | GCA | ATG | CAC |
| V | S | I | S | A | S | S | T | T | G | V | A | M | H |

| ACT | TCA | ACC | TCT | TCT | TCA | GTC | ACA | AAG | AGT | TAC | ATC | TCA | TCA |
| T | S | T | S | S | S | V | T | K | S | Y | I | S | S |

| CAG | ACA | AAT | GAT | ACG | CAC | AAA | CGG | GAC | ACA | TAT | GCA | GCC | ACT |
| Q | T | N | D | T | H | K | R | D | T | Y | A | A | T |

| CCT | AGA | GCT | CAT | GAA | GTT | TCA | GAA | ATT | TCT | GTT | AGA | ACT | GTT |
| P | R | A | H | E | V | S | E | I | S | V | R | T | V |

| TAC | CCT | CCA | GAA | GAG | GAA | ACC | GGA | GAA | AGG | GTA | CAA | CTT | GCC |
| Y | P | P | E | E | E | T | G | E | R | V | Q | L | A |

| CAT | CAT | TTC | TCA | GAA | CCA | GAG | ATA | ACA | CTC | ATT | ATT | TTT | GGG |
| H | H | F | S | E | P | E | I | T | L | I | I | F | G |

| GTG | ATG | GCT | GGT | GTT | ATT | GGA | ACG | ATC | CTC | TTA | ATT | TCT | TAC |
| V | M | A | G | V | I | G | T | I | L | L | I | S | Y |

| GGT | ATT | CGC | CGA | CTG | ATA | AAG | AAA | AGC | CCA | TCT | GAT | GTA | AAA |
| G | I | R | R | L | I | K | K | S | P | S | D | V | K |

| CCT | CTC | CCC | TCA | CCT | GAC | ACA | GAC | GTG | CCT | TTA | AGT | TCT | GTT |
| P | L | P | S | P | D | T | D | V | P | L | S | S | V |

| GAA | ATA | GAA | AAT | CCA | GAG | ACA | AGT | GAT | CAA | TGAG | AATCTGTTCAC |
| E | I | E | N | P | E | T | S | D | Q | Stop | |

The first 55 bases are 5′ untranslated (UTR), then there are 453 bases that translate to 150 amino acids (the last three bases are a stop codon). This 150–amino acid polypeptide is the precursor for GYPA, which is later cut down to 131 amino acids. The 453 DNA bases are underlined and broken into the triplet codons, with the one-letter amino acid code below. The 20th and 24th codons are shown in bold as they are the ones responsible for the differences in the M and N alleles. After the stop codon the beginning of the long 3′ UTR that continues for over 2,000 bases is shown.

genotype under consideration will be at a specific genetic *locus*, which is a physical location along a chromosome, and the variant DNA codes that can occur at the locus are called *alleles*. The specific glycophoran A sequence shown in Table 2.3 is known as the *M allele* at the MN locus and has been mapped to between 4q28 and 4q31. There is another common allele at this locus, the *N allele*, in which the 20th codon, which is TCA for the *M* allele, is instead TTA while the 24th codon, which is GGT for the *M* allele, is instead GAG. These mutational changes cause the 20th amino acid to be leucine in the *N* allele (instead of serine) and the 24th amino acid to be glutamic acid in the *N* allele (instead of glycine). The first 19 codons before the variant 20th code for the "signal" part of the protein. This part of the protein helps direct glycophoran A to where it is heading, which is the cell membranes of red blood cells. The signal is then cleaved off of the mature protein, leaving a 131–amino acid chain (as in Chapter 4). Codons 20–91 (72 amino acids) hang out of the cell membrane of red blood cells (i.e., they are extracellular); codons 92–114 code for 23 *transmembrane* amino acids, which run through the cell membrane; and codons 115–150 code for 36 amino acids within the cell cytoplasm.

Because the GYPA locus is on an autosome, it is possible for an individual to have an *M* allele from one parent and an *M* from the other, an *M* and an *N*, or an *N* and an *N*. The three possible genotypes are therefore *MM*, *MN*, and *NN* at this locus. However, what of the phenotype? *Phenotypes* are just the physical characteristics that can be observed, and ultimately we will see that, for the MN locus, genotype and phenotype mean the same thing because all three genotypes are expressed as different phenotypes. Before we can understand this latter point, though, we will need to delve into the subject of **dominance** at a locus.

Dominance

Dominance refers to how two alleles at a locus are expressed in the final phenotype. When there is complete dominance, the **dominant allele** is expressed in the phenotype, while the **recessive allele** is hidden. This hiding of the recessive allele can occur because the allele contains a disruption that blocks production of a protein or because the allele codes for a nonfunctional protein that is not visible in the phenotype (we will discuss examples of this in Chapter 4). Individuals who are **homozygous** for the dominant allele (i.e., both parental copies of the allele are the dominant allele) are phenotypically indistinguishable from **heterozygous** individuals, with a dominant allele from one parent and a recessive allele from the other; and the homozygous recessive individual is distinguishable from heterozygous and homozygous dominant individuals. We can continue with our example from the GYPA locus to make this point. We will assume that we have obtained some blood from an individual. This individual's genotype is *NN*, meaning that they obtained the *N* allele from both their parents. If we take some of this individual's blood and inject it into a rabbit, then the rabbit's immune system will recognize the red blood cell surface protein (also known as an *antigen*) as foreign and will produce antibodies against the N protein. The antibodies are themselves proteins found circulating in the fluid (serum) portion of the blood, and collectively the serum and antibodies are referred to as an *antiserum*. In this specific case, the antiserum would usually be called by the shorthand name "anti-N." Now, if we take blood from three people who have the *MM*, *MN*, and *NN* genotypes and combine their blood with the anti-N, the blood from the *MN* and *NN* individuals will **agglutinate** (clump together). In other words, the anti-N will recognize the type N proteins sticking out of the red blood cells and will bind to them. Type *MM* individuals lack the type N protein, so their blood will not react with anti-N.

In the above example, we have seen that if the phenotype is defined as reactivity to anti-N, then the *N* allele is completely dominant to the *M* allele. This is true because NN and MN have the same phenotype (they agglutinate when treated with anti-N) while MM has a different phenotype (it does not agglutinate with anti-N). Following our definition from above, the homozygous dominant (NN) phenotype is indistinguishable from the heterozygous (MN) phenotype as both react with anti-N. Now, we can try a slightly different experiment, where we take blood from another individual. This individual's genotype is *MM*, meaning that they obtained the *M* allele from both their father and their mother. If we inject some of this individual's blood into a rabbit, the rabbit's immune system will recognize the M protein as foreign and will start making anti-M. Now, if we combine the anti-M with blood from our original three people who have the *MM*, *MN*, and *NN* genotypes, the MM and MN blood types will agglutinate but the NN will not. In this case, the *M* allele is dominant to the *N* allele as the former "masks" the latter. These two examples show us that we have to be extremely careful to specify exactly what phenotype we are discussing. If the phenotype is "reactivity to anti-N and to anti-M," then the *M* and *N* alleles are **codominant**. For a codominant locus, both homozygotes are phenotypically distinguishable from the heterozygote and from each other. In this example, the *MM* genotype will react with only anti-M, the *MN* genotype will react with anti-M and anti-N, and the *NN* genotype will react with only anti-N.

MUTATION

In the previous sections, we have examined in some detail how variation in the genotype can manifest in variations in the phenotype. However, this begs the question of how alternate alleles arise at a genetic locus. In this section, we briefly consider the types of mutation that can be found in DNA.

Substitutions

In a substitution mutation, one DNA nucleotide (base) is substituted for another. For example, at the 20th codon in the GYPA precursor, the *M* allele is TCA and the *N* allele is TTA. Consequently, at some point in evolutionary history there was a mutation from C to T or T to C at the second base of the codon. This particular mutation leads to a change in the amino acid at the 20th codon and, consequently, changes the reactivity of the protein with the antisera. Other mutations can change a nucleotide while having no effect on the coded amino acid because of redundancy in the genetic code. For example, if the first two nucleotides are thymine and cytosine, then the codon will code for serine, regardless of which of the four nucleotides is present in the third position of the codon. So the mutation of TCT to TCC (or TCA or TCG) will have no phenotypic effect, and such mutations are consequently referred to as **silent mutations**. However, even silent mutations can be detected by determining the nucleotide sequence.

Insertions and Deletions

As opposed to the **point mutations** described above, where a single nucleotide has changed, leading to a **single nucleotide polymorphism**, a stretch of DNA may be lost or gained. If the insertion or deletion occurs within a coding sequence and the number of bases gained or lost is not a multiple of 3, then the entire sequence past the mutation will be mistranslated. This is called a **reading frame shift**. Reading frame shift mutations are exceedingly rare because they usually lead to a nonviable embryo. An interesting example of an insertion/deletion mutation occurs in the first intron of the *CD4* gene (Edwards and Gibbs 1992). CD4 is an antigen involved in T-lymphocyte activity (cells important in mounting an immune response), and the locus for this protein is on the 12th chromosome (mapped to 12pter-p12, meaning that it is on the short arm somewhere between the second region of the first band and the terminus of the short arm). Within the first intron at this locus there has been the insertion of a particular type of sequence referred to as an "**Alu insert**" (we will have much more to say about Alu inserts in Chapter 8). This particular insertion mutation involves the insertion of a 285-base sequence; this alu insert is found in apes (Tishkoff et al. 1996), so it is phylogenetically old. In contrast, there is also a deletion mutation that cuts across the insertion, and this deletion is found only in some humans. The deletion results in a loss of 256 bases, 239 of which are from the alu insert and 17 of which are from the "flanking sequence" toward the second exon. Medical genetics texts are replete with examples of insertion and deletion mutations, but typically they would not mention the *CD4* Alu deletion because it has no phenotypic consequences. In fact, the three authors of this text are in blissful ignorance of the status of their own *CD4* first intron Alu inserts. Certainly, human variation is more than meets the eye.

Recombinations

We have already seen that homologous chromosomes can exchange DNA through crossover events during meiosis. As a consequence, mutations (substitutions, insertions, or deletions) that occurred previously in two separate chromosomes can end up in the same chromosome because of a crossover between the mutational sites. When the crossover occurs between two loci and the exchange of genetic material is equal, this in itself does not constitute a mutational event. However, it is also possible for crossover events to be unequal, which has the result of causing an insertion on one chromosome and a deletion on its homolog. Again, as with insertions and deletions, unequal crossovers are usually found such that they do not disrupt the reading frame, or else they are found within noncoding DNA. In Chapter 8, we will discuss how "replication slippage" can increase or decrease the number of times that a short tandemly repeated piece of DNA is present. Unequal recombination can also alter the number of times that such repeat elements are found.

Translocations and Nondisjunctions

Translocations are the result of crossover events between nonhomologous chromosomes. These mutations are generally so disruptive that they are rarely observed in live births, with the exception being a special type known as a **Robertsonian translocation**. In a *Robertsonian translocation*, two nonhomologous acrocentric chromosomes fuse such that their short arms are lost and their long arms make the p and q arms of a new metacentric chromosome. One example of such a translocation is found in some individuals with trisomy 21 (Down syndrome). Some Down syndrome individuals will have the usual pair of 21st chromosomes but then also have a third 21st chromosome that is fused to another chromosome (typically a member of the 14th pair). Observable **nondisjunctions** are also rare. In a *nondisjunction*, the homologous pairs do not separate from one another at the first meiotic division. After the second meiotic division, this will produce two gametes that are diploid for the chromosome in question and two gametes that lack the chromosome. Nondisjunction, like any other mutational event, can also occur during mitotic division, causing some cell lines to be trisomic and some to be monosomic for the involved chromosome. This leads to mosaicism but is generally of no direct genetic consequence as the mutation cannot be transmitted to an individual's children unless it occurs in the germ line. As examples of genetic mosaics, trisomy 21 can vary across cell lines within an individual and a number of skin lesions can exist in mosaic forms (Paller 2004).

Mutation Frequencies

The previous presentation of possible mutations does not address the frequency with which these mutations might occur, or the *mutation rate*. Mutation rates are difficult to discuss because estimates are given in a number of different ways and can be arrived at from a number of different analytical methods, some of which carry rather strong assumptions. As an example, in a section titled "Why Are There so Many Ways of Expressing Mutation Rates?" Jobling et al. (2004) give the following example. Suppose that we are interested in a 1,500–base pair genetic locus and observe 15 mutational events over a generation time of 25 years. This mutation rate could be described as equal to 0.6 (15/25) mutations per locus per year, 15 mutations per locus per generation, 0.00004 [15/(15,000 × 25)] mutations per nucleotide per year, or 0.001 (15/15,000) mutations per nucleotide per generation. Vogel and Motulsky (1997) give a ballpark rate for mutations per locus per generation of about 2–5 per 100,000. This figure means that on average we would expect that among 100,000 gametes produced by an individual with the "normal gene" at a given locus, two to five mutant gametes would be formed at that locus. Mutation rates can also be given at the base-pair level, in which case the rate is much lower. Nachman and Crowell (2000) have estimated a base-pair mutation rate of about 2.5 per 100,000,000. While this rate seems extremely low, the fact that the human genome is so large means that each individual will, on average, have about 175 base-pair mutations. However, it is unlikely that any of these mutational events would have any phenotypic consequences. The highest mutation rate that Vogel and Motulsky (1997) report is 1 per 10,000 for the Robertsonian translocation causing Down syndrome. Even at this relatively high rate, mutation provides only a source of variation; it does not typically serve as an evolutionary force (see Chapter 3).

BASIC TRANSMISSION GENETICS: THE MENDELIAN LAWS

We will continue our discussion of the MN locus and some similar blood group loci as the alternate alleles for these systems are common, a statement not true if we were examining the results of very rare mutations like nondisjunction events. In this section, we will use the blood group loci to demonstrate some of the basics of transmission genetics. As these basics were first discovered by Mendel, they are usually referred to as Mendelian laws.

Mendel's First Law: Segregation of Alleles

Mendel's first law, in modern terms, states that the two alleles at a locus segregate when gametes are formed. In other words, of the maternally and paternally derived alleles at a locus, only one can be passed from parent to offspring. This is a painfully simple concept (though quite radical in Mendel's day), so it does not bear much further discussion here. Within the context of the MN locus, an individual with the *MM* genotype can produce only gametes with *M* alleles, an individual with the *NN* genotype can produce only gametes with *N* alleles, and an individual with the *MN* genotype will produce about half of his or her gametes with the *M* allele and half with the *N* allele. Given these facts, the unobserved genotype of an individual can sometimes be inferred from the observed genotypes of relatives. For example, if two full siblings have the *MM* and *NN* genotypes and their mother's genotype is *MN*, then their father's genotype must also be *MN*. This first Mendelian law was a radical departure from much of the thought of the day, which viewed genetic inheritance as a blending process. Mendel's deep insight led to the documentation that inheritance was based on the segregation of individual particles (see the web listing for Robert Olby's essay in Supplemental Resources).

Mendel's Second Law: Independent Assortment of Alleles

Mendel's second law, in modern terms, states that alleles at different loci segregate independently of one another, so they follow independent assortment. In a mechanistic sense, if we think of an organism with three chromosome pairs, then at the first meiotic division there will be three maternally derived and three paternally derived chromosomes. We can label these as 1_M, 2_M, and 3_M (for maternal) and 1_P, 2_P, and 3_P (for paternal). When the chromosome pairs align at the first meiotic division, we can think of each member of the pair as being either to the left or to the right of the plane through which the first cell division will occur. For the first pair, we could have 1_M to the left and 1_P to the right, which has no effect on which member of the other two pairs is to the left or the right (thus, the chromosomes sort independently). With three chromosome there are 2^3, or eight, possible ways for the maternally and paternally derived chromosomes to assort; and with 23 chromosomes (the human condition) there are 2^{23}, or 8,388,608, ways for the chromosomes to assort.

To demonstrate independent assortment, we will need to consider an additional locus to the MN locus, for which we use the Colton blood group (see Chapter 4). The Colton blood group is determined by a mutation in the Aquaporin 1 channel-forming integral protein locus (AQP1), which has been mapped to 7p14. The AQP1 locus codes for a 269–amino acid protein that forms a complex helix within the cell membrane of red blood cells, consequently forming a channel for water to move into the cell (hence, the names "Aquaporin" and "channel-forming"). At the 45th codon, the middle base is variable so that the sequence is either GCG (coding for alanine) or GTG (coding for valine). This single nucleotide polymorphism (SNP) causes different reactivity with antisera. By convention, the antiserum that reacts with the alanine variant is called "a," while the antiserum that reacts with the valine variant is called "b." Consequently, an individual who is homozygous for the alanine variant will have the phenotype a+b– (meaning that his or her blood will agglutinate with the a antiserum but not the b antiserum). Heterozygotes will react to both antisera, so the phenotype is a+b+. Individuals who are homozygous for the valine variant have the phenotype a–b+. Finally, there is a small fraction of the population who have major mutations that render the AQP1 protein inactive. Individuals homozygous for these mutations will have the phenotype a–b–. To summarize, there are two major alleles (a+ and b+) as well as a minor allele present at a very low frequency (absence of the protein). There are consequently 3 antigens in this blood group, as shown in Table 4.1, or really two antigens and the absence of an antigen.

Because the MN locus is on the fourth chromosome while the Colton locus is on the seventh chromosome, these loci will segregate independently during meiosis. For example, if an individual is heterozygous at both loci (MN and a+b+), then the gametes he or she produces are equally likely to be Ma+, Mb+, Na+, and Nb+. Knowing which allele at the MN locus was passed to a gamete tells us nothing about which allele was passed at the Colton locus.

There are, however, many examples where Mendel's second law does not apply because the two loci are located near one another on the same chromosome (i.e., they are "linked genes"). We saw previously that the MN locus was determined by the *GYPA* gene. There is a comparable gene called glycophoran B (or *GYPB*) that, like *GYPA*, is mapped to 4q28-q31. *GYPB* codes for a precursor protein with 91 amino acids, of which the first 19 are "signal" that are later removed. The 29th amino acid in the mature protein can be either a threonine (ACG) or a methionine (ATG). Reactivity for the threonine amino acid is represented as an "s," while reactivity for the methionine amino acid is represented as an "S," so the genotypes are *SS*, *Ss*, and *ss*, all of which are phenotypically distinguishable (i.e., it is a codominant system, just like the MN blood group system). Because the *GYPA* and *GYPB* loci are only approximately 80,000 base pairs apart (Lonjou et al. 1999), the alleles do not assort independently. Consequently, if an individual has the *MN* and *Ss* genotypes such that the *M* and *S* alleles are on the same chromosome while the *N* and *s* alleles are on the other (homologous) chromosome, they will produce about half *MS* and half *Ns* gametes, not the quarter *MS*, quarter *Ms*, quarter *NS*, and quarter *Ns* gametes that we would expect if the loci were on different chromosomes. This general lack of **recombination** (i.e., failure to produce the recombinant *Ms* and *NS* gametes) is a very clear indication that the *GYPA* and *GYPB* loci are on the same chromosome. If the loci were farther apart on the chromosome, then we would expect a higher frequency of crossover events and, consequently, more recombinants. The Human Genome Project has used this type of linkage information to produce rough relative maps of chromosomes that can then be tied to physical maps and eventually sequenced.

BEYOND MENDEL

Given that humans are rather different from garden pea plants, there are some additional "wrinkles" on the basic laws espoused by Mendel that we need to consider. First, we need to look at loci on the 23rd pair of chromosomes, the **sex chromosomes**. Second, we need to consider an extranuclear "locus" found in cytoplasmic organelles, the mitochondria.

Sex-Linked Inheritance

The 23rd pair of chromosomes are known as the "sex chromosomes" (or non-autosomal chromosomes) because they determine the sex of the individual. Females have two homologous X chromosomes, one paternally derived and one maternally derived. The X chromosome is a long (only the first seven autosomal pairs are longer) metacentric chromosome, named "X" because when the sister chromatids of the chromosome are spread (but still connected by the centromere) the chromosome forms an X. Males have one X chromosome, which is maternally derived, and one Y chromosome, which is paternally derived. The Y chromosome is a short (only the 21st and 22nd autosomal pairs are shorter) acrocentric chromosome. When the sister chromatids for the long arm are spread, the short arm chromatids appear to form the tail of a Y. The X chromosome contains many important loci, while in contrast the Y chromosome is a "functional wasteland" (Quintana-Murci and Fellous 2001). The Y chromosome contains what are known as "pseudoautosomal" regions at either end (i.e., at the ends of the long and short arms). In males, the pseudoautosomal regions pair with homologous portions of the X chromosome during meiosis and are consequently capable of genetic exchange by crossover events within them. The remainder of the Y chromosome, which makes up about 95% of the length, is referred to as the **nonrecombining Y** (NRY).

The transmission genetics that underlie the sex chromosomes are relatively simple. The actual action of the X chromosomes in females is rather complicated because with two copies of every locus to males' one copy, there is the potential for overproduction of mRNA molecules. Mammals have solved this problem by randomly inactivating one of the paired X chromosomes early in development (Heard et al. 1997). As the X chromosome is not inactivated in the germ line, the process of inactivation does not affect genetic transmission. Consequently, the X chromosome in females follows the same form of genetic transmission as would be found for any autosomal chromosome, in the sense that there is an equal chance of transmitting the paternally and maternally derived X chromosome. In males, the transmission genetics is more complicated. First, males are haploid for both the X and Y chromosomes; and second, the Y chromosome can be passed only from father to son since only males have it. Consequently, NRY variations are specific to males and can be transmitted only paternally to male offspring. The fact that males are haploid for the X chromosome (or **hemizygous**) leads to the interesting effect known as *sex linkage*, in which a recessive trait is much more commonly expressed in males than in females. Sex linkage occurs because a recessive allele on the X chromosome of a male cannot be masked by a dominant allele on the Y (where the X-specific locus is lacking), while females, even in the face of X inactivation, typically will not display the recessive trait unless it is found on both X chromosomes. This is true because the inactivation of X chromosomes does not occur until many mitotic divisions have occurred, and then the inactivation of one of the X chromosomes per cell occurs at random. In about half of the cells, the maternally derived X chromosome will be inactivated, while in about half of the cells the paternally derived X chromosome will be inactivated.

Mitochondrial DNA

mtDNA is a short stretch of DNA (typically 16,569 base pairs) that forms a continuous circle and is found only in the mitochondria. Mitochondria are organelles found outside the nucleus of the cell and are responsible for energy production. Because there are multiple copies of mtDNA within each mitochondrion and there are many mitochondria per cell, there are about 13,000 copies (Tang et al. 2000) of mtDNA per cell (versus the single copy of the 46 chromosomes per cell). Unlike nuclear DNA, mtDNA is almost entirely composed of coding sequences, so there are no introns within genes.

The transmission genetics for mtDNA is similar in some respects to NRY, with one major difference: mtDNA is found in both males and females, while NRY is found only in males. Like NRY, which can be transmitted only from fathers, mtDNA typically can be transmitted only from mothers. There are two reasons for the general lack of paternal inheritance of mtDNA. First, there are far fewer mitochondria in a sperm cell versus an oocyte, so paternal mitochondria would make up less than a tenth of a percent of the total mitochondria in the newly formed zygote. Second, there appears to be some mechanism by which paternal mitochondria that enter the oocyte are inactivated. In point of fact, there has been only one case of human paternal mtDNA transmission reported (Schwartz and Vissing 2002).

To summarize, NRY is transmitted from a father to his sons, while mtDNA is transmitted from a mother to her children (sons and daughters alike). In both cases, it is important to understand that although there are a number of genes on the NRY and mtDNA, each acts as a single locus. This is true because NRY and mtDNA are inherited as single units, with no possibility of recombination (crossover). In fact, in the one documented case of paternal mtDNA transmission, there was recombination between the maternal and paternal mtDNA (Kraytsberg et al. 2004); and there is some statistical evidence that recombination within mtDNA has occurred in the past (Awadall et al. 1999). However, generally recombination can be ignored. We will return to this concept of the NRY and mtDNA acting as single loci in Chapter 8, when we take up again the subject of DNA.

WHAT IS A GENE?

Although we have used the word *gene* with considerable frequency in the first two chapters, we have yet to define it. We can do no better than Hartl's (2000, 2) definition that a *gene* "is a general term meaning, loosely, the physical entity transmitted from parent to offspring during the reproductive process that influences hereditary traits." There is quite a bit packed into this definition, and now that we have a basis for understanding genetic processes, we should unpack it. The definition is particularly interesting as a contrast to cultural inheritance, which does not involve a "physical entity" and need not be "from parent to offspring." Neither does cultural inheritance involve reproduction. What makes a gene a gene is that it is a physical entity (a portion of DNA) that is transmitted only from a father or mother to a child. The stipulation that a gene "influences hereditary traits" is necessary because cytoplasm is a physical entity transmitted from a mother to her children.

CHAPTER SUMMARY

This chapter has described the basic molecular biology underlying transmission genetics, and consequently serves as a foundation for the remaining chapters of this book. Transmission genetics is based on the molecular biology of DNA, which is the material of genetic inheritance. During growth, development, and maintenance of body tissues it is necessary for DNA to be faithfully reproduced. This happens by the process of DNA replication followed by cell mitosis. Thus a single cell gives rise to two identical daughter cells. In meiosis (cell division for sexual reproduction) one cell gives rise to daughter cells that are not identical because they differ as to which member of each pair of homologous chromosomes is transmitted. To keep track of the map location for individual genes on each of the chromosomes, a gene's location is given by the number of the chromosome, the arm of the chromosome (either the short or long arm from the centromere), and a band location. In most cases the gene will form a protein by first serving as a template for messenger RNA production (a process referred to as transcription) and then by the messenger RNA being translated into a sequence of amino acids—the building blocks of a protein.

The production of proteins is based on the genotype that is expressed as a phenotype. But the connection between the two can be quite complicated because of dominance. Ultimately, any difference in alleles at a particular locus must arise from mutation, which takes a number of forms. Mutations can arise by a simple substitution of one DNA base for another, by the insertion or deletion of one or more bases, by translocation (fusion of two chromosomes), or by non-disjunction (lack of separation of homologous chromosome pairs during meiosis). The inheritance of alleles at a locus is governed by the "Mendelian Laws." The first law is that alleles segregate during meiosis, so that a gamete receives either a maternally or a paternally derived allele at a given locus. The second law is that loci that are not linked will sort independently. To these laws can be added additional maxims that postdate Mendel. For example, because Mendel studied pea plants, he did not deal with sex-linked inheritance or with the inheritance of mitochondrial DNA, both of which follow different transmission rules.

SUPPLEMENTAL RESOURCES

http://helios.bto.ed.ac.uk/bto/glossary/ab.htm. Glossary of genetics terms from the University of Edinburgh, School of Biology.

http://www.biology.arizona.edu/. The Biology Project at the University of Arizona. Includes a number of useful tutorials and interactive exercises. See the "DNA Structure" section for an interactive viewer (requires Mac/PCMolecule2, which can be downloaded in a demo version) and the karyotyping activity for some human karyotypes and practice matching chromosomes.

http://www.dnaftb.org/dnaftb/. "DNA from the Beginning" from the Dolan DNA Learning Center at Cold Spring Harbor Laboratory. Features sections such as "Classical Genetics," "Molecules of Genetics," and "Genetic Organization and Control."

http://www.emunix.emich.edu/~rwinning/genetics/. Genetics Online from Dr. Bob Winning at Eastern Michigan University.

http://www.mendelweb.org/MWolby.html. An online essay by Robert C. Olby (Department of the History and Philosophy of Science, University of Pittsburgh), "Mendel, Mendelism, and Genetics."

http://www.umass.edu/microbio/chime/dna/index.htm. An interactive viewer for DNA. Requires "chime" for your browser, which is a free download (see the link at the above URL).

http://www.usask.ca/biology/rank/text/. An online text ("geNETics") on general genetics, from Dr. Gerry Rank in the Department of Biology, University of Saskatchewan.

http://www.viewzone.com/aging.html. A site discussing telomeres and aging.

http://www.weihenstephan.de/~schlind/genglos.html. A hypermedia glossary of genetic terms from Dr. Birgid Schlindwein at Library Weihenstephan, Technische Universität München.

3

Population Genetics and Human Variation

Population genetics is the quantitative study of how genes are distributed within and across populations and how gene distributions pattern against time and space. Consequently, a basic understanding of population genetic theory is at the root of any discussion about human variation. The patterns of genic distribution are affected by evolutionary forces, so population genetic theory is inherently tied to evolutionary theory.

The topic is complex; this chapter will take as gentle an approach as possible, using graphical devices and computer models in place of the underlying mathematics wherever possible. The necessary mathematical tool for this chapter is relatively simple algebra. Probability theory, which is key to understanding many of the evolutionary forces, is presented before dealing directly with population genetic theory.

PROBABILITY THEORY

Probability theory is a way of measuring the frequency with which events occur or stating the likelihood that they will occur and relating the probability of specific events. The use of the word *frequency* here actually provides one of the definitions used for the word *probability*, which is the relative frequency of an event.

If we consider a population containing 100 females, in which 80 have a 6 bp deletion in their mitochondrial genome and 20 do not, we can ask what would be the probability of obtaining an ovum whose mitochondria have the 6 bp deletion if we draw randomly from one of the females. The probability that we will sample an ovum with the 6 bp deletion is 80/100, or 0.8. The word *probability* can also be defined subjectively (Remington and Schork 1970), in which case we do not have to refer to a specific sample, such as the 100 females. A subjective probability is one where we can state the probability of an event without having to assess the relative frequency of events. For example, if we flip a fair coin, we know (subjectively) that the probability of obtaining a "heads" is 0.5 and the probability of obtaining a "tails" is also 0.5. The subjective definition for probability is often used in population genetics. For example, as in the coin toss, we know that the probability of sampling an *A* allele from an *AB* (heterozygous) individual is 0.5 without having to perform this "experiment."

Probability Rules

There are two simple rules that allow us to describe how individual probabilities can be combined into more complex statements. They are referred to here as the "or" rule and the "and" rule, because the words *or* and *and* usually occur in word problems that require applications of these respective rules.

The "Or" Rule

The "or" rule of probability is used to find the probability that one event or the other will occur. For example, what is the probability that an *FG* heterozygote will produce an *F* gamete or a *G* gamete? The "or" rule states that for mutually exclusive events the probability that one or the other event will occur is equal to the sum of the individual events. Here, *mutually exclusive* means that the two events cannot occur together (simultaneously). To continue the example, the probability that an *FG* heterozygote will produce an *F* or a *G* gamete is 1.0, equal to the probability of producing an *F* gamete (0.5) plus the probability of producing a *G* gamete (0.5). As it is not possible for the *FG* heterozygote to produce a gamete that is both an *F* and a *G* allele at the same time, these are mutually exclusive events. The "or" rule can also be used for more than two mutually exclusive events. For example, if we say that the gene pool is composed of *F*, *G*, and *H* alleles with frequencies (probabilities) of 0.7, 0.2, and 0.1, then the probability of sampling an allele that is an *F* or a *G* or an *H* is 0.7 + 0.2 + 0.1 = 1.0. Like the example with two alleles, the "or" rule demonstrates an important point about probabilities, which is that they must be between 0.0 and 1.0 (inclusive). If the only type of alleles that exist are *F*, *G*, and *H*, then the probability that we will sample an *F* or a *G* or an *H* must be 1.0 because we know with certainty that when we sample we must get one (or another) of these three alleles.

The "And" Rule

The "and" rule of probability is used to find the probability that two independent events will both occur. For example, what is the probability that an *FG* heterozygote will produce two *F* gametes (i.e., an *F* and then another *F*)? The "and" rule states that for independent events the probability that both events will occur is equal to the product of the individual events. That is, the probability that an *FG* heterozygote will produce an *F* and then another *F* gamete is 0.25, equal to the probability of first producing an *F* gamete (0.5) times the probability of producing another *F* gamete (0.5). When we refer to *independent* events, we mean that the outcomes of the events do not affect each other, so the fact that we obtain an *F* allele from a heterozygote on the first "draw" does not affect whether we obtain an *F* allele on the second "draw." The "and" rule, like the "or" rule, can also be used to cover more than two events. For example, suppose that an *FG* heterozygote marries a *GG* homozygote, and they produce one offspring, who marries another *GG* homozygote, producing another offspring, who again marries a *GG* homozygote, producing a final child who is a great-grandchild of the original *FG* heterozygote. What is the probability that the *F* allele from the great-grandparent is also present in the great-grandchild? This is equal to the probability that the great-grandparent transmitted the *F* allele to the grandparent and that the grandparent transmitted the *F* allele to the parent of the great-grandchild and that the parent of the great-grandchild transmitted the *F* allele to the great-grandchild. Every time we see the word *and* in the previous sentence, we have the multiplication of 0.5 (the probability that an *FG* heterozgote will transmit an *F* allele), so altogether the "and" rule indicates that the probability the *F* allele will be found in the great-grandchild is $0.5 \times 0.5 \times 0.5 = 0.125$.

HARDY-WEINBERG EQUILIBRIUM

We can use the probability theory we now have in place to understand a very central tenet from population genetics, which is **Hardy-Weinberg equilibrium**. Hardy-Weinberg equilibrium is a statement about expected genotype frequencies in a population, but in order to look at these expected frequencies, we will first need to deal with gene (also known as "allele") frequencies. The mathematics of this equilibrium were published in the same year by Hardy and Weinberg and alluded to in an early publication by Castle (see Li 1976, 9 for a brief history of Castle's original statement of the equilibrium).

Gene (Allele) Frequencies

Gene (allele) frequencies are the proportions of the total number of genes that are in each allelic state. For an autosomal locus, there are 2N genes, where N is the number of individuals in the sample or population. The multiplication by two occurs because everyone has two genes at the locus. Let us suppose that we are dealing with a sample where 400 individuals have been assayed for a genetic locus where there are the codominant alleles F and G, with 100 individuals typed as being FF homozygotes, 80 individuals typed as being FG heterozygotes, and 220 individuals typed as being GG homozygotes. The 400 individuals together have 2×400, or 800, alleles, which constitutes the **gene pool**. Now, we would like to know how frequent is the F allele in this pool. Conventionally, the symbol p is used for the frequency of one of the alleles and q is used for the frequency of the other allele. These frequencies are the proportions (or percentages if we multiply by 100) of the alleles in the gene pool. To find the frequency of the F allele, we just count the number of F alleles and divide by 2N. All FF homozygotes have two F alleles, so the 100 FF homozygotes contribute 2×100, or 200, F alleles. The 80 FG heterozygotes each have one F allele, so the FF homozygotes and the FG heterozygotes together have $(2 \times 100) +$ 80, or 280, F alleles. When we divide 280 by the 800 alleles present in the gene pool, we get that p is equal to 0.35, or the F allele constitutes 35% of the gene pool. Because all of the other alleles in the gene pool must be G alleles, we can find q (the G allele frequency) by subtraction. The G allele frequency is $1 - 0.35$, or 0.65, or on a percentage basis $100\% - 35\%$, which is 65%.

A very small amount of algebra shows that we can directly calculate p and q from the observed genotype frequencies. We let P_{FF} be the genotype frequency for the FF homozygotes, which is just the number of FF homozygotes divided by the number of people, or in our case $P_{FF} = 100/400 = 0.25$. Similarly, P_{FG} is the heterozygote frequency, which is $P_{FG} = 80/400 = 0.20$; and P_{GG} is the GG homozygote frequency, which is $P_{GG} = 220/400 = 0.55$. Using #FF and #FG to mean the observed number of genotypes that are FF homozygotes and FG heterozygotes, respectively, we can (following the discussion above) write the F allele frequency as follows:

$$p = \frac{2 \times (\#FF)}{2 \times N} + \frac{\#FG}{2 \times N} = P_{FF} + \frac{P_{FG}}{2} \tag{3.1}$$

Similarly, the G allele frequency is $q = P_{GG} + P_{FG}/2$.

Expected Genotype Frequencies

Two Alleles

The Hardy-Weinberg relationship is a statement about expected genotype frequencies given the observed allele frequencies. We can use the allele frequencies from above plus the "or" and "and" rules from our discussion of probability to derive the Hardy-Weinberg expectations. To find the expected proportion of *FF* genotypes, we need simply ask what is the probability when we sample two alleles from the gene pool that both alleles will be the *F* allele? This is the probability of sampling an *F* allele "and" an *F* allele. By the "and" rule, this probability is $p \times p = p^2$. Similarly, the probability of getting a *GG* genotype is equal to the probability of sampling a *G* allele and a *G* allele ($q \times q = q^2$). Now, there are two ways to find the expected frequency of *FG* heterozygotes. The first and more explanatory method is to again use our probability rules. Specifically, we need to find the probability that we sample an *F* allele "and" then a *G* allele "or" that we sample a *G* allele "and" then an *F* allele. Applying the "and" and "or" rules gives $(p \times q) + (q \times p) = 2pq$. The less explanatory method is to find the *FG* frequency by subtracting the *FF* and *GG* genotype frequencies from 1.0. The least appealing way to explain the expected Hardy-Weinberg equilibrium genotype frequencies, although this is often the route taken in introductory physical anthropology textbooks, is to note that the Hardy-Weinberg equilibrium frequencies are given by the binomial expansion $(p + q)^2 = p^2 + 2pq + q^2$. This way of deriving the Hardy-Weinberg equilibrium frequencies has little explanatory value to most students, although it does provide a quick way to arrive at the expected proportions.

A laborious way of finding Hardy-Weinberg expectations for genotype frequencies is to use a Punnett square. This device is useful for finding Mendelian expected ratios for particular matings (see http://www.borg.com/~lubehawk/psquprac.htm for some practice problems), but it becomes very cumbersome when we try to apply the device to a population. For example, let us suppose that we have a small population composed of two *FF* homozygotes, two *FG* heterozygotes, and one *GG* homozygote. This means that the gene pool contains six *F* alleles (two *F* alleles in each of the two *FF* homozygous individuals plus one *F* allele in each of the two FG heterozygotes) and four *G* alleles. Table 3.1 shows the Punnett square for this small population. If we count the genotypes in each "box" of the square, we find 36 *FF* genotypes, 48 *FG* genotypes, and 16 *GG* genotypes, for a total of 100 genotypes in the 4 boxes of the Punnett square. The expected genotype frequencies are consequently 36/100, 48/100, and 16/100, or 0.36, 0.48, and 0.16, respectively. Alternatively, we could have found these frequencies from the "and"

Table 3.1 Punnett square for finding Hardy-Weinberg expected genotype proportions when the F allele has a frequency of 0.6 and the G allele has a frequency of 0.4

First Allele	Second Allele									
	F	*F*	*F*	*F*	*F*	*F*	*G*	*G*	*G*	*G*
F	FF	FF	FF	FF	FF	FF	FG	FG	FG	FG
F	FF	FF	FF	FF	FF	FF	FG	FG	FG	FG
F	FF	FF	FF	FF	FF	FF	FG	FG	FG	FG
F	FF	FF	FF	FF	FF	FF	FG	FG	FG	FG
F	FF	FF	FF	FF	FF	FF	FG	FG	FG	FG
F	FF	FF	FF	FF	FF	FF	FG	FG	FG	FG
G	FG	FG	FG	FG	FG	FG	GG	GG	GG	GG
G	FG	FG	FG	FG	FG	FG	GG	GG	GG	GG
G	FG	FG	FG	FG	FG	FG	GG	GG	GG	GG
G	FG	FG	FG	FG	FG	FG	GG	GG	GG	GG

The square gives 36 *FF* homozygotes, 48 *FG* heterozygotes, and 16 *GG* homozygotes out of a total of 100, so the genotype proportions are 0.36, 0.48, and 0.16.

and "or" rules as $0.6^2 = 0.36$, $(0.6 \times 0.4) + (0.4 \times 0.6) = 0.48$, and $0.4^2 = 0.16$. For larger populations, the Punnett square becomes a completely impractical approach.

Application of Hardy-Weinberg Equilibrium for Two Alleles

As an example of applying the Hardy-Weinberg equilibrium expectations, we can calculate the allele frequency for Tay-Sachs disease among North American Jews. Peterson et al. (1983) found in a study of 46,304 North American Jews that 1,500 were heterozygous for the Tay-Sachs allele. Tay-Sachs is a lethal recessive disease in which the homozygote for the Tay-Sachs allele dies usually by the third year of life. If we use a capital T for the non-Tay-Sachs allele and a lower case t for the Tay-Sachs allele, then the TT homozygote does not have Tay-Sachs, the Tt heterozygote carries the Tay-Sachs allele but does not express it phenotypically, while the tt homozygote has Tay-Sachs disease. The 1,500 individuals in the Peterson et al. (1983) study were carriers of the Tay-Sachs allele (Tt heterozygotes), and they constituted 3.24% of the sample of 46,304 individuals. The remaining 96.76% of the individuals were all TT homozygotes as any tt homozygotes would have died of Tay-Sachs prior to reaching an age suitable for the study. If the genotypes were present in Hardy-Weinberg equilibrium proportions, then the proportion of the sample who should be carriers is as follows:

$$\frac{f(Tt)}{f(Tt) + f(TT)} = \frac{2pq}{2pq + p^2} = 0.0324 \tag{3.2}$$

If we rewrite $\dfrac{f(Tt)}{f(Tt) + f(TT)}$ as $f(C)$ (for frequency of carriers), then we can solve equation (3.2) for q to find the frequency of the Tay-Sachs allele as follows:

$$q = \frac{f(C)}{2 - f(C)} = \frac{0.0324}{2 - 0.0324} = 0.01647 \tag{3.3}$$

This gives the expected genotype frequencies as $f(TT) = p^2 = 0.967331$, $f(Tt) = 2pq = 0.032397$, and $f(tt) = q^2 = 0.000271$, implying that there were 13 tt homozygotes who died and therefore were not counted in the sample of 46,304 individuals.

Three or More Alleles

The Hardy-Weinberg relationship extends beyond the two-allele example we considered above. For example, if we consider three alleles, F, G, and H, at a single locus, we can again apply the probability rules to find the expected genotype frequencies. Letting the frequencies of the F, G, and H alleles be p, q, and r, respectively, the frequencies of homozygotes are given by the squares of the allele frequencies [$f(FF) = p^2$, $f(GG) = q^2$, and $f(HH) = r^2$]. For example, the frequency of the HH homozygote is the probability of sampling an H allele and an H allele, or $r \times r = r^2$. The heterozygote frequencies are all two times the allele frequencies for the two alleles included in the particular heterozygote [$f(FG) = 2pq$, $f(FH) = 2pr$, $f(GH) = 2qr$]. For example, the frequency of the GH heterozygote is equal to the probability of sampling a G allele and then an H allele or an H allele and then a G allele, which is $(q \times r) + (r \times q) = 2qr$. The frequencies can also be found using the trinomial expansion, $(p + q + r)^2 = p^2 + 2pq + q^2 + 2pr + 2qr + r^2$, where the order of genotype frequencies is $f(FF)$, $f(FG)$, $f(GG)$, $f(FH)$, $f(GH)$, and $f(HH)$. Parenthetically, we should note from the Hardy-Weinberg proportions that the larger the number of alleles at a

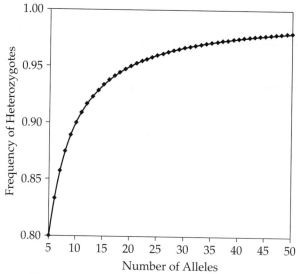

Figure 3.1. Frequency of heterozygotes when there are 5–50 equally frequent alleles at a locus.

given locus, the more heterozygotes we should expect (provided that the alleles are equally frequent). For example, with only two equally frequent alleles, the expected proportion of heterozygotes in the population is 0.5; with three equally frequent alleles, the expected proportion is 0.67; with four alleles, the expected proportion is 0.75; and with 5 alleles, the expected proportion is 0.80. Figure 3.1 shows the expected proportion of heterozygotes when there are 5 equally frequent alleles, 6 alleles, and so on up to 50.

Application of Hardy-Weinberg Equilibrium for Three Alleles

For an example of the Hardy-Weinberg equilibrium expected proportions with three alleles, we look at the ABO blood group system (see Chapter 4 for an explanation of this system). We will assume that there are only three alleles, although in reality there are many variants within the main alleles. The ABO system presents an interesting challenge in that some of the genotypes are not directly distinguishable. The A and B alleles are codominant to one another, while both are dominant to the O allele. This means that individuals who phenotypically have type A blood may be the AA genotype or the AO genotype, and similarly individuals with type B blood may have the BB or the BO genotype. Only in individuals with type AB or type O blood is the genotype known (AB in the first case, OO in the second). As a consequence, we will need to use Hardy-Weinberg equilibrium expectations in order to estimate the allele frequency from phenotypic data. Taylor and Prior (1938) studied the ABO system in England based on 422 individuals, 179 of whom had type A blood, 35 of whom had type B blood, 6 of whom had type AB blood, and 202 of whom had type O blood.

We can use a method known as the expectation-maximization (EM) algorithm (Dempster et al. 1977) to calculate the allele and genotype frequencies from Taylor and Prior's phenotypic data. The EM is a very simple method that involves repeated application of the Hardy-Weinberg equilibrium law and calculation of the allele frequencies

Table 3.2 Expectation-maximization (EM) algorithm calculation of allele and genotype frequencies from phenotypes in the ABO blood system

A (n = 179)		B (n = 35)		AB (n = 6)	OO (n = 202)	p (0.15407)	q (0.15407)	r (0.69186)
AA	AO	BB	BO					
17.93372	161.06628	3.50659	31.49341	6	202	0.24044	0.05273	0.70682
26.01988	152.98012	1.25864	33.74136	6	202	0.25002	0.05007	0.69991
27.12644	151.87356	1.20867	33.79133	6	202	0.25133	0.05001	0.69866
27.28841	151.71159	1.20938	33.79062	6	202	0.25153	0.05001	0.69846
27.31246	151.68754	1.20972	33.79028	6	202	0.25156	0.05001	0.69843
27.31604	151.68396	1.20978	33.79022	6	202	0.25156	0.05001	0.69843

using equation (3.1). Table 3.2 shows the results from these repeated calculations, which can be easily done in a computer spreadsheet by copying a row a number of times, until the values change little between each row. The method has to start by assuming values for the three allele frequencies. We start by assuming that r equals the square root of 202/422 because we have the following:

$$f(OO) = r^2 = \frac{202}{422}$$

$$\sqrt{r^2} = r = \sqrt{\frac{202}{422}} = 0.69186 \tag{3.4}$$

We also assume that $p = q$ to begin with. By the end of the table we have found allele frequencies of $p = 0.25156$, $q = 0.05001$, and $r = 0.69843$, from which we can find the expected counts in phenotypes as 175.0 type A, 30.5 type B, 10.6 type AB, and 205.9 type O. These expected counts are found by assuming Hardy-Weinberg equilibrium, as follows:

$$[0.25156^2 + (2 \times 0.25156 \times 0.69843)] \times 422 = 175.0 \text{ Type A}$$

$$[0.05001^2 + (2 \times 0.05001 \times 0.69843)] \times 422 = 30.5 \quad \text{Type B} \tag{3.5}$$

$$(2 \times 0.25156 \times 0.05001) \times 422 = 10.6 \qquad \text{Type AB}$$

$$0.69843^2 \times 422 = 205.9 \qquad \text{Type O}$$

What Is Hardy-Weinberg Equilibrium, and Why Is It Important?

We have expended a considerable amount of effort in describing the Hardy-Weinberg expectations for genotype frequencies but have not yet addressed the Hardy-Weinberg equilibrium or its central importance in population genetics. In population genetics, *equilibrium* means that some aspect describing the population remains unchanged through time. In the case of Hardy-Weinberg equilibrium, the allele frequencies do not change through time; and after the genotypes reach their Hardy-Weinberg expectations (which takes a single generation of random mating), the genotype frequencies also remain unchanged. The fact that allele frequencies do not change if the population is in Hardy-Weinberg equilibrium is very important because this provides the static case against which we can view evolving (changing) allele frequencies. In this section, we examine the nature of Hardy-Weinberg equilibrium.

If a population starts with genotype frequencies that are not in the expected Hardy-Weinberg proportions, then it takes only a single generation of random mating to produce the expected Hardy-Weinberg proportions. For example, if $P_{FF} = P_{FG} = P_{GG} = 1/3$, then the population clearly is not in the expected Hardy-Weinberg genotype frequencies because the F allele frequency is 0.5. The Hardy-Weinberg expected genotype frequencies are $P_{FF} = 0.25$, $P_{FG} = 0.5$, and $P_{GG} = 0.25$, which is what the population would achieve after one generation of random mating. However, in the initial generation and the next generation, the allele frequency has remained at 0.5.

It is deviation from Hardy-Weinberg equilibrium that is truly interesting because such deviations may suggest the presence of evolutionary forces at play. The deviations can take one of two forms. In the first case, the allele frequencies do not change across time, but the genotype frequencies differ from their Hardy-Weinberg expectations. Such cases arise when the population is not mating randomly, and indeed, random mating was an implicit assumption used in deriving the Hardy-Weinberg genotype frequencies. The "and" rule of probability starts with the assumption that the events are independent, which is the case for the union of gametes only if mating is random. The second and more interesting case of deviation from Hardy-Weinberg expectations can lead to evolution (change) in allele frequencies across time, a topic we will take up later in this chapter when we discuss the four evolutionary forces.

How Can We Tell If a Population Is in Hardy–Weinberg Equilibrium?

Provided the genotypes are distinguishable in a genetic system, we can tell whether or not the sample is in Hardy-Weinberg equilibrium by calculating the allele frequencies, calculating the expected Hardy-Weinberg genotype proportions from the allele frequencies, and comparing these to the observed proportions. For example, Mitchell et al. (1999) reported genotype frequencies among a number of different populations for the COL1A2 gene. This autosomal gene, which codes for a type of **collagen** (a structural protein found in many connective tissues), contains a 38 bp insertion in one of its introns, such that the DNA sequence is either 603 or 641 bp long. For the Kets, an aboriginal group from central Siberia, Mitchell et al. (1999) typed 9 individuals, 5 of whom were homozygous for the 641 bp sequence, 2 of whom were heterozygous, and 2 of whom were homozygous for the 603 bp sequence. By counting the alleles, we find that the frequency of the 641 bp allele is $(5 \times 2) + 2/(2 \times 9)$, or two-thirds. The expected number of 641 bp homozygotes under Hardy-Weinberg equilibrium is then $(2/3)^2 \times 9 = 4$, the expected number of heterozygotes is $2 \times (2/3) \times (1/3) = 4$, and the expected number of 603 bp homozygotes is $(1/3)^2 \times 9 = 1$. So while the observed genotype counts were 5, 2, and 2, the expected counts under Hardy-Weinberg equilibrium are 4, 4, and 1. Is this a meaningful departure from Hardy-Weinberg expectations?

To answer the above question, we need to find the probability of getting the observed genotype counts or any more extreme counts under Hardy-Weinberg expectations. The use of the word *or* again tells us that we need to sum the probability of getting the observed count with the probabilities of getting more extreme counts. In the current example, there is only one more extreme genotype count, which is 6, 0, and 3 (note that this preserves the 641 bp allele frequency, which is a requirement for this type of analysis). We will skip over the method for calculating the probability of getting a particular genotype count as this is rather involved (see Evett and Weir 1998, 150–154). The probability of getting the observed 5, 2, and 2 genotype count is 0.162896, while the probability of getting the more extreme count (6, 0, and 3) is 0.004525. Summed together, this gives a probability of 0.167421, which is the probability of getting the observed genotype counts (or more extreme counts) under Hardy-Weinberg equilibrium. As it seems fairly likely that we would get this observed count (or the more extreme one) under Hardy-Weinberg equilibrium, we would be inclined to believe that the sample is in Hardy-Weinberg equilibrium. However, if Mitchell et al. (1999) had observed a 6, 0, and 3 genotype count, there would only be a 0.004525 probability of getting this count under Hardy-Weinberg equilibrium. In this case, we would be inclined to reject the idea that the sample is in Hardy-Weinberg equilibrium. As a rule of thumb, statisticians use a probability value of 0.05, below which they believe that the possibility of obtaining the observed data under the presumed hypothesis is remote. Consequently, they would suggest that the hypothesis should be rejected.

In addition to the data from the Kets, Mitchell et al. (1999) collected genotypic counts for 9 other populations from around the world. Considering these 10 populations together, their genotypic count was 136 641/641 bp homozygotes, 66 641/603 bp heterozygotes, and 17 603/603 bp homozygotes. Is the total sample in Hardy-Weinberg equilibrium? The expected counts are 130.4, 77.2, and 11.4 versus the observed counts of 136, 66, and 17. From this comparison we can see that there are fewer heterozygotes and more homozygotes than expected under Hardy-Weinberg equilibrium. Later, we will see that there is a theoretical reason for this depletion of heterozygotes (it is due to genetic drift), but for now we simply want to know if the departure from Hardy-Weinberg equilibrium is greater than we would expect by chance. Table 3.3 contains all the non-zero probabilities for genotype counts that preserve the allele counts. The sum of these probabilities is 1.0 as they are all the possible counts. The sum of the probabilities that are equal to or less than that for the observed count is 0.035094, indicating that the data (or any more extreme departures from Hardy-Weinberg equilibrium) are unlikely to have arisen if the total sample was in Hardy-Weinberg equilibrium.

DEPARTURES FROM HARDY-WEINBERG EQUILIBRIUM

There are a number of reasons that genotypic counts may deviate from the expected Hardy-Weinberg proportions. In some cases, the deviations are also a signal that the allele frequencies are changing across time and, hence, that genetic evolution is occurring. In the following section, we will discuss these evolutionary forces that alter allele frequencies; but in this section, we will only be concerned with non-evolutionary departures from Hardy-Weinberg equilibrium.

Multiple-Locus Hardy-Weinberg Equilibrium

In theory, Hardy-Weinberg equilibrium extends to more than one locus, so that if p_A is the allele frequency at the A locus (with alleles A and a) and p_B is the allele frequency

Table 3.3 Calculation of probability of getting various genotype counts if the sample is in Hardy-Weinberg equilibrium

641/641	641/603	603/603	Prob.	P(obs)≥
120	98	1	0.000007	0.000007
121	96	2	0.000073	0.000073
122	94	3	0.000452	0.000452
123	92	4	0.002009	0.002009
124	90	5	0.006782	0.006782
125	88	6	0.018107	
126	86	7	0.039294	
127	84	8	0.070680	
128	82	9	0.106940	
129	80	10	0.137654	
130	78	11	0.152093	
131	76	12	0.145272	
132	74	13	0.120637	
133	72	14	0.087497	
134	70	15	0.055633	
135	68	16	0.031100	
136	**66**	**17**	**0.015321**	**0.015321**
137	64	18	0.006664	0.006664
138	62	19	0.002562	0.002562
139	60	20	0.000871	0.000871
140	58	21	0.000262	0.000262
141	56	22	0.000070	0.000070
142	54	23	0.000016	0.000016
143	52	24	0.000003	0.000003
144	50	25	0.000001	0.000001
			1.000000	0.035094

The line in bold gives the observed genotype counts (136 of the 641 bp homozygotes for the *COL1A2* gene, 66 heterozygotes, and 17 of the 603 bp homozygotes, data from Mitchell et al. 1999). The column labeled "Prob." is the probability of the genotype counts under Hardy-Weinberg equilibrium, while the column labeled "P(obs)≥" only lists the probability if it is less than or equal to the observed count. The sums are given at the end of the last two columns.

at the B locus (with alleles B and b), the frequency of the $AaBb$ genotype should be $2p_Aq_A \times 2p_Bq_B$ (the probability of getting an Aa genotype and a Bb genotype). Here, unlike in Table 3.2, we use the alleles A, a, B, and b to represent the fictional alleles at an A and a B locus, not to represent ABO blood type alleles at a single locus. In practice, it is quite possible for individual loci to be in Hardy-Weinberg equilibrium while pairs of loci are not. As an example, we can consider two equally sized populations where everyone in the first population has an $aaBB$ genotype and everyone in the second population has an $AAbb$ genotype. Now, we assume that whatever barrier existed between the two populations is broken down so that they become one randomly mating population. In the initial generation, the only gametes that exist are aB and Ab, which are present in equal frequency. The probability of the $aaBB$ genotype being formed from these two types

Table 3.4 Approach to two-locus Hardy-Weinberg equilibrium for two unlinked loci (recombination rate $\rho = 0.5$)

	AABB	AABb	AAbb	AaBB	AaBb	Aabb	aaBB	aaBb	aabb	AB	aB	Ab	ab	ρ
0	0.0000	0.0000	0.5000	0.0000	0.0000	0.0000	0.5000	0.0000	0.0000	0.0000	0.5000	0.5000	0.0000	−1.0000
1	0.0000	0.0000	0.2500	0.0000	0.5000	0.0000	0.2500	0.0000	0.0000	0.1250	0.3750	0.3750	0.1250	−0.5000
2	0.0156	0.0938	0.1406	0.0938	0.3125	0.0938	0.1406	0.0938	0.0156	0.1875	0.3125	0.3125	0.1875	−0.2500
3	0.0352	0.1172	0.0977	0.1172	0.2656	0.1172	0.0977	0.1172	0.0352	0.2188	0.2813	0.2813	0.2188	−0.1250
4	0.0479	0.1230	0.0791	0.1230	0.2539	0.1230	0.0791	0.1230	0.0479	0.2344	0.2656	0.2656	0.2344	−0.0625
5	0.0549	0.1245	0.0706	0.1245	0.2510	0.1245	0.0706	0.1245	0.0549	0.2422	0.2578	0.2578	0.2422	−0.0313
6	0.0587	0.1249	0.0665	0.1249	0.2502	0.1249	0.0665	0.1249	0.0587	0.2461	0.2539	0.2539	0.2461	−0.0156
7	0.0606	0.1250	0.0645	0.1250	0.2501	0.1250	0.0645	0.1250	0.0606	0.2480	0.2520	0.2520	0.2480	−0.0078
8	0.0615	0.1250	0.0635	0.1250	0.2500	0.1250	0.0635	0.1250	0.0615	0.2490	0.2510	0.2510	0.2490	−0.0039
9	0.0620	0.1250	0.0630	0.1250	0.2500	0.1250	0.0630	0.1250	0.0620	0.2495	0.2505	0.2505	0.2495	−0.0020
10	0.0623	0.1250	0.0627	0.1250	0.2500	0.1250	0.0627	0.1250	0.0623	0.2498	0.2502	0.2502	0.2498	−0.0010
11	0.0624	0.1250	0.0626	0.1250	0.2500	0.1250	0.0626	0.1250	0.0624	0.2499	0.2501	0.2501	0.2499	−0.0005
12	0.0624	0.1250	0.0626	0.1250	0.2500	0.1250	0.0626	0.1250	0.0624	0.2499	0.2501	0.2501	0.2499	−0.0002
13	0.0625	0.1250	0.0625	0.1250	0.2500	0.1250	0.0625	0.1250	0.0625	0.2500	0.2500	0.2500	0.2500	−0.0001
14	0.0625	0.1250	0.0625	0.1250	0.2500	0.1250	0.0625	0.1250	0.0625	0.2500	0.2500	0.2500	0.2500	−0.0001
15	0.0625	0.1250	0.0625	0.1250	0.2500	0.1250	0.0625	0.1250	0.0625	0.2500	0.2500	0.2500	0.2500	0.0000

of gamete is 0.5^2, or 0.25 (the probability of sampling an aB and an aB gamete), the probability of the $AAbb$ genotype forming is also 0.25, and the probability of the $AaBb$ genotype is $2 \times 0.5 \times 0.5$, or 0.5 (the probability of sampling an aB and then an Ab gamete or sampling an Ab and then an aB). There are, however, six other two-locus genotypes ($AABB$, $AABb$, $AaBB$, $Aabb$, $aaBb$, and $aabb$) which still are not represented. We show this in Table 3.4. In the current generation, the AB gamete frequency that will go into the next generation's zygotes is one-fourth the $AaBb$ genotype frequency (the $AaBb$ genotype produces AB, Ab, aB, and ab gametes in equal frequency). This gives 0.25×0.50, or 0.125.

The record keeping for gamete and genotype frequencies becomes a fairly onerous task, so we handle this in a spreadsheet, which is how Table 3.4 was produced. We do this by writing the relevant Hardy-Weinberg equations in the first row and then copying this row into the subsequent rows. From Table 3.4 it is clear that the genotype frequencies are changing across generations until they reach fixed values by the thirteenth generation. Thus, it took 13 generations to reach multiple-locus Hardy-Weinberg equilibrium ($AaBb$ has a frequency of 0.25, equal to $2p_Aq_A \times 2p_Bq_B$). In order to show this approach to two-locus Hardy-Weinberg equilibrium, the last column of Table 3.4 lists the **correlation** between alleles within gametes. A correlation coefficient is just a measure of association which equals 0 when there is no association, −1 when there is complete negative association, and +1 when there is complete positive association. Table 3.4 starts with the correlation being −1 because there is complete negative association (A alleles are found only with b alleles and a alleles with B alleles within gametes). By generation 15 the correlation is 0 so that there is no association between alleles at the two loci within gametes. The 0 correlation between alleles is the hallmark of multiple-locus Hardy-Weinberg equilibrium.

The example we have just given made the assumption that the A and B loci were on separate chromosomes or that they were far enough apart on the same chromosome that Mendel's law of independent assortment held. If the two loci are on the same chromosome and very close together, then the AB and ab gametes will never be produced and

Table 3.5 Approach to two-locus Hardy–Weinberg equilibrium for two linked loci (recombination rate = 0.1)

	AABB	AABb	AAbb	AaBB	AaBb	Aabb	aaBB	aaBb	aabb	AB	aB	Ab	ab	ρ
0	0.0000	0.0000	0.5000	0.0000	0.0000	0.0000	0.5000	0.0000	0.0000	0.0000	0.5000	0.5000	0.0000	−1.0000
1	0.0000	0.0000	0.2500	0.0000	0.5000	0.0000	0.2500	0.0000	0.0000	0.0250	0.4750	0.4750	0.0250	−0.9000
2	0.0006	0.0238	0.2256	0.0238	0.4525	0.0238	0.2256	0.0238	0.0006	0.0475	0.4525	0.4525	0.0475	−0.8100
3	0.0023	0.0430	0.2048	0.0430	0.4140	0.0430	0.2048	0.0430	0.0023	0.0678	0.4323	0.4323	0.0678	−0.7290
4	0.0046	0.0586	0.1868	0.0586	0.3829	0.0586	0.1868	0.0586	0.0046	0.0860	0.4140	0.4140	0.0860	−0.6561
5	0.0074	0.0712	0.1714	0.0712	0.3576	0.0712	0.1714	0.0712	0.0074	0.1024	0.3976	0.3976	0.1024	−0.5905
6	0.0105	0.0814	0.1581	0.0814	0.3372	0.0814	0.1581	0.0814	0.0105	0.1171	0.3829	0.3829	0.1171	−0.5314
7	0.0137	0.0897	0.1466	0.0897	0.3206	0.0897	0.1466	0.0897	0.0137	0.1304	0.3696	0.3696	0.1304	−0.4783
8	0.0170	0.0964	0.1366	0.0964	0.3072	0.0964	0.1366	0.0964	0.0170	0.1424	0.3576	0.3576	0.1424	−0.4305
9	0.0203	0.1018	0.1279	0.1018	0.2963	0.1018	0.1279	0.1018	0.0203	0.1531	0.3469	0.3469	0.1531	−0.3874
10	0.0235	0.1062	0.1203	0.1062	0.2875	0.1062	0.1203	0.1062	0.0235	0.1628	0.3372	0.3372	0.1628	−0.3487
11	0.0265	0.1098	0.1137	0.1098	0.2804	0.1098	0.1137	0.1098	0.0265	0.1715	0.3285	0.3285	0.1715	−0.3138
12	0.0294	0.1127	0.1079	0.1127	0.2746	0.1127	0.1079	0.1127	0.0294	0.1794	0.3206	0.3206	0.1794	−0.2824
13	0.0322	0.1150	0.1028	0.1150	0.2699	0.1150	0.1028	0.1150	0.0322	0.1865	0.3135	0.3135	0.1865	−0.2542
14	0.0348	0.1169	0.0983	0.1169	0.2662	0.1169	0.0983	0.1169	0.0348	0.1928	0.3072	0.3072	0.1928	−0.2288
15	0.0372	0.1185	0.0944	0.1185	0.2631	0.1185	0.0944	0.1185	0.0372	0.1985	0.3015	0.3015	0.1985	−0.2059
16	0.0394	0.1197	0.0909	0.1197	0.2606	0.1197	0.0909	0.1197	0.0394	0.2037	0.2963	0.2963	0.2037	−0.1853
17	0.0415	0.1207	0.0878	0.1207	0.2586	0.1207	0.0878	0.1207	0.0415	0.2083	0.2917	0.2917	0.2083	−0.1668
18	0.0434	0.1215	0.0851	0.1215	0.2570	0.1215	0.0851	0.1215	0.0434	0.2125	0.2875	0.2875	0.2125	−0.1501
19	0.0451	0.1222	0.0827	0.1222	0.2556	0.1222	0.0827	0.1222	0.0451	0.2162	0.2838	0.2838	0.2162	−0.1351
20	0.0468	0.1227	0.0805	0.1227	0.2546	0.1227	0.0805	0.1227	0.0468	0.2196	0.2804	0.2804	0.2196	−0.1216
21	0.0482	0.1232	0.0786	0.1232	0.2537	0.1232	0.0786	0.1232	0.0482	0.2226	0.2774	0.2774	0.2226	−0.1094
22	0.0496	0.1235	0.0769	0.1235	0.2530	0.1235	0.0769	0.1235	0.0496	0.2254	0.2746	0.2746	0.2254	−0.0985
23	0.0508	0.1238	0.0754	0.1238	0.2524	0.1238	0.0754	0.1238	0.0508	0.2278	0.2722	0.2722	0.2278	−0.0886
24	0.0519	0.1240	0.0741	0.1240	0.2520	0.1240	0.0741	0.1240	0.0519	0.2301	0.2699	0.2699	0.2301	−0.0798
25	0.0529	0.1242	0.0729	0.1242	0.2516	0.1242	0.0729	0.1242	0.0529	0.2321	0.2679	0.2679	0.2321	−0.0718
26	0.0538	0.1244	0.0718	0.1244	0.2513	0.1244	0.0718	0.1244	0.0538	0.2338	0.2662	0.2662	0.2338	−0.0646
27	0.0547	0.1245	0.0708	0.1245	0.2510	0.1245	0.0708	0.1245	0.0547	0.2355	0.2645	0.2645	0.2355	−0.0581
28	0.0554	0.1246	0.0700	0.1246	0.2508	0.1246	0.0700	0.1246	0.0554	0.2369	0.2631	0.2631	0.2369	−0.0523
29	0.0561	0.1247	0.0692	0.1247	0.2507	0.1247	0.0692	0.1247	0.0561	0.2382	0.2618	0.2618	0.2382	−0.0471
30	0.0568	0.1247	0.0685	0.1247	0.2506	0.1247	0.0685	0.1247	0.0568	0.2394	0.2606	0.2606	0.2394	−0.0424

the genotype frequencies will always be 0.25 *AAbb*, 0.5 *AaBb*, and 0.25 *aaBB* so that the population will never reach two-locus Hardy–Weinberg equilibrium. If the loci are on the same chromosome but far enough apart that recombination can occur at rate r, then Table 3.4 needs to be redone. The recombination rate r has a maximum of 0.5, in which case the loci are said to be unlinked, and a minimum of 0.0, in which case they are complete linked. We have redone Table 3.4 as Table 3.5, where the recombination rate is $r = 0.1$. The record keeping for this example is more complicated than for the previous example of unlinked loci, but the results are readily interpretable. Now, because of the linkage between the loci, the approach to two-locus Hardy–Weinberg equilibrium is much slower. By 30 generations the correlation between alleles at the two loci is still negative, indicating that the population is not in two-locus Hardy–Weinberg equilibrium.

Our discussion of multiple-locus Hardy–Weinberg equilibrium may appear to be highly esoteric, but *linkage analysis* (i.e., the measurement of the correlation between alleles at two or more loci) is an extremely important tool in human genetics. Linkage analysis has been used extensively to map loci to specific locations on specific chromosomes (Lander and Green 1987, Ott 1991, Jorde 1995, Xiong and Guo 1997, Jorde et al. 1994, De la Chapelle and Wright 1998), and many genetic tests to screen for disease alleles are based on tight linkage to known markers.

Assortative Mating

In deriving the Hardy-Weinberg expected genotypic proportions, we made explicit use of the "and" rule, which states that the probability of two or more independent events occurring is equal to the product of the individual probabilities. In **assortative mating**, like genotypes are more (or less) likely to mate than we would expect at random; and as a consequence, the combination of gametes into zygotes cannot be thought of as independent. Assortative mating can take two forms. The first, called *positive assortative mating*, occurs when like genotypes are more likely to mate with each other than we would expect at random. The second, called *negative assortative mating*, occurs when like genotypes are less likely to mate with each other than at random. To simplify the result, we will look only at complete positive assortative mating (only like genotypes mate with each other) and complete negative assortative mating (like genotypes never mate with each other).

For a *biallelic* locus (a locus with two alleles), where we label the alleles as A and B, if there is complete positive assortative mating, the only matings are AA with AA, AB with AB, and BB with BB. We will start with a population where the A allele frequency is 0.6 and the initial population (at generation 0) is in Hardy-Weinberg equilibrium so that the genotype frequencies are 0.36, 0.48, and 0.16 (see Table 3.6 and Table 13-1 from Levitan and Montagu, 1971). Now, we apply Hardy-Weinberg expectations but based on the restriction that only like genotypes will mate with each other. It is a simple matter to set

Table 3.6 Genotype frequencies and the Fixation index (F) in positive assortative mating

Generation	AA	AB	BB	F
0	0.3600	0.4800	0.1600	0.0000
1	0.4800	0.2400	0.2800	0.5000
2	0.5400	0.1200	0.3400	0.7500
3	0.5700	0.0600	0.3700	0.8750
4	0.5850	0.0300	0.3850	0.9375
5	0.5925	0.0150	0.3925	0.9688
6	0.5963	0.0075	0.3963	0.9844
7	0.5981	0.0038	0.3981	0.9922
8	0.5991	0.0019	0.3991	0.9961
9	0.5995	0.0009	0.3995	0.9980
10	0.5998	0.0005	0.3998	0.9990
11	0.5999	0.0002	0.3999	0.9995
12	0.5999	0.0001	0.3999	0.9998
13	0.6000	0.0001	0.4000	0.9999
14	0.6000	0.0000	0.4000	0.9999
15	0.6000	0.0000	0.4000	1.0000

up formulae for these matings in a spreadsheet and then copy them for each generation, which is how Table 3.6 was produced.

Table 3.6 shows that with positive assortative mating the frequency of homozygotes increases and the frequency of heterozygotes decreases. Ultimately, a population that undergoes complete assortative mating will reach a state where the AA genotype frequency is equal to p, the BB genotype frequency is equal to q, and the AB genotype frequency is 0. Importantly, though, the allele frequencies never change, which is why positive assortative mating is not considered an evolutionary force per se. To measure the extent to which heterozygotes are depleted when positive assortative mating occurs (relative to Hardy-Weinberg expectations), we can define a coefficient, F, as follows:

$$F = \frac{2pq - f(AB)}{2pq} = \frac{H_e - H_o}{H_e} \tag{3.6}$$

where H_e and H_o mean the proportion of heterozygotes expected under Hardy-Weinberg equilibrium and the proportion observed, respectively. These terms are often referred to as "heterozygosity" because they indicate the proportions of the population expected and observed to be heterozygous. Decreased heterozygosity can increase the opportunity for selection so that natural selection operates more effectively. However, without any selection, increased or decreased heterozygosity in a population has no effect on allele frequencies (only genotype frequencies). This again is why assortative mating in and of itself is not considered an evolutionary force. Table 3.6 shows that complete positive assortative mating will eventually bring the index F up to 1.0, at which point there are no observed heterozygotes.

The effects of negative assortative mating are more complicated than is true for positive assortative mating (Falk and Li 1969 give a detailed derivation). While positive assortative mating led to decreased heterozygosity, negative assortative mating leads to increased heterozygosity. Assortative mating does to some extent occur for quantitative traits such as stature and intelligence quotient (IQ) (Spuhler 1968, Vandenberg 1972), but the evidence for assortative mating for simple genetic loci is less substantial. Indeed, Morton (1982, 56) wrote that "for most phenotypes of interest to genetic epidemiology, the possibility of assortative mating is remote." His rationale for stating this was that "potential mates do not know one another's genotypes," but to some extent this statement is now out of date. With the increased ability for genetic testing, carriers (heterozygotes) for rare lethal diseases may begin to avoid mating with each other, which would constitute a form of negative assortative mating. Ober et al. (1997) have also presented evidence that **human leukocyte antigen (HLA)** genotypes for spouses among the Hutterites, a religious isolate in the northern United States and southern Canada, match less frequently than one would expect by chance. This is an indication that there is some negative assortative mating by HLA genotype within the Hutterites. The exact mechanism for avoidance of like HLA genotypes among mates in humans remains obscure. There is ample evidence that mice can detect **major histocompatibility complex (MHC)** types by olfaction, and there is some evidence from humans that similar cues may be operative (see Beauchamp and Yamazaki 1997). An interesting genetic consequence of negative assortative mating for HLA (if it is indeed occurring) is that heterozygosity for HLA would be increased over Hardy-Weinberg expectations. In the Hutterites, where "remote" inbreeding as a result of small population size can deplete heterozygosity (see below), negative assortative mating may counterbalance the depletion of heterozygotes.

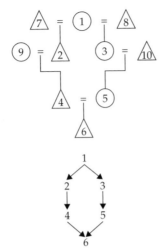

Figure 3.2. Kinship diagram and pedigree diagram for calculating the inbreeding coefficient in the son of a half-first cousin mating.

Inbreeding

Inbreeding is the production of offspring by mates who are consanguineous ("blood") relatives. Like positive assortative mating, inbreeding decreases heterozygosity at the population level but leaves allele frequencies unchanged. Indeed, inbreeding can be thought of in the same general framework as positive assortative mating because consanguineous kin are more likely to be similar in genotype (due to common descent). In order to measure the effects of inbreeding, we need some way of quantifying the degree of consanguinity between mates, which is done using a **kinship coefficient**. The coefficient of kinship can also be referred to as an **inbreeding coefficient**, in which case it is defined for the offspring of the consanguineous mating. The formal definition of the *inbreeding coefficient* is the probability of **identity by descent (IBD)** for the two alleles at an autosomal locus. IBD means that the two alleles at the locus are identical because both came from a common ancestor of the two mates.

Calculation of the inbreeding coefficient is easiest to understand if we use an example to explain. Figure 3.2 shows a fictitious pedigree where two half-first cousins (individuals 4 and 5) produce a son (individual 6). The common ancestor (individual 1) is the grandmother of individuals 4 and 5. This pedigree is drawn using a fairly standard format from cultural anthropology. Below, the pedigree is drawn in diagrammatic form, with only the gene transfers that come (ultimately) from the common ancestor shown as arrows. For the moment, we will presume that individual 1 (the great-grandmother of individual 6) is an AB heterozygote. To find the inbreeding coefficient for individual 6 (i.e., the probability of IBD), we can consider the probability that either the *A* or the *B* allele is passed from individual 1 through both lines (i.e., from individual 1 to 2, from 2 to 4, and from 4 to 6 and from individual 1 to 3, 3 to 5, and 5 to 6). The "ands" and "ors" in the previous sentence are our clues on how to calculate the probability of IBD. The probability of IBD is the sum of the probability that individual 6 is AA (with both alleles coming from the great-grandmother) with the probability that they are BB (again with both alleles coming from the great-grandmother). Every time that an allele passes "down" an arrow, there is half a chance that one allele, versus the other allele at the locus, will be transmitted. Consequently, for individual 6 to have two copies of his great-grandmother's A allele, the allele would had to have been transmitted from individual 1 to 2 and from 2 to 4 and from 4 to 6 and from 1 to 3 and from 3 to 5 and from 5 to 6. Using the "and" rule, we find that the probability of this occurring is $(1/2)^6$. Similarly, the probability that individual 6 will have two copies of his great-grandmother's B allele is $(1/2)^6$, so the probability that individual 6 has inherited two copies of his great-grandmother's A allele or two copies of her B allele is $(1/2)^6 + (1/2)^6 = 2(1/2)^6 = (1/2)^5$.

We have just seen how to calculate the inbreeding coefficient, which for the case of a half-first cousin mating produces a child with an inbreeding coefficient of $(1/2)^5$, or $1/32$. The usual symbol for the inbreeding coefficient is F, the same symbol we previously used for the decrease in heterozygosity relative to the expected heterozygosity in positive assortative mating. As in positive assortative mating, inbreeding also leads to decreased heterozygosity. We can see this by continuing our example of the half-first cousin mating. In the above paragraph, we assumed that the great-grandmother was an AB heterozygote, but this was simply done so that we would have a way of distinguishing her two alleles. If we had no knowledge of her genotype, our best guess (from Hardy-Weinberg proportions) would be that the probabilities she was AA, AB, or BB were p^2, $2pq$, and q^2, respectively. Similarly, any "marry-ins" (individuals 7, 8, 9, and 10) would be expected to have genotypes in the Hardy-Weinberg proportions. Now, what is the probability that individual 6 is a heterozygote? By definition, individual 6 cannot be a heterozygote if he inherited two copies of an allele from his great-grandmother. The probability that he did not inherit two copies of the same allele is $1 - F$, so the probability that he did not inherit two copies of the same allele from his great-grandmother and he is a heterozygote is $2pq(1 - F)$. We can also find the probability that he is an AA homozygote. In order to be an AA homozygote, he must have inherited the A allele from two different ancestors, or he must have inherited them both from his great-grandmother. The probability that he would inherit the A allele from two different ancestors is just the Hardy-Weinberg expectation p^2 times the probability that he did not get two copies of the same allele from his great-grandmother (i.e., $1 - F$). The probability that he inherited two copies of the A allele from his great-grandmother is pF as F is the probability that he inherited two copies of the A or B allele from his grandmother and p is the probability that the allele he did inherit as two copies is an A allele. Putting this together, we have the probability that individual 6 is an AA homozygote is $p^2(1 - F) + pF$; similarly, the probability he is a BB homozygote is $q^2(1 - F) + qF$.

There are two more issues we need to address concerning inbreeding. The first is the generalization of our example pedigree so that the inbreeding coefficient can be calculated for any pedigree. From the bottom part of Figure 3.2 we can see that there are five individuals above individual 6, in what is usually referred to as an "inbreeding loop." To form a loop, we must be able to go "against the flow" of the arrows up from ego to a common ancestor, then turn at the common ancestor and go down the other side of the pedigree to reach ego again. This loop has five individuals, from which we can infer that the inbreeding coefficient for an individual is $(1/2)^n$, where n is the number of individuals in the inbreeding loop. There are two other complications we need to consider when calculating inbreeding coefficients from more complex pedigrees. First, it may be the case that the common ancestor at the top of the inbreeding loop is inbred him- or herself. If F_c is the inbreeding coefficient for the common ancestor, then we need to adjust up our estimate for ego to account for the increased probability of IBD. This adjustment

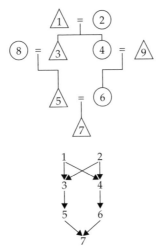

Figure 3.3. Kinship diagram and pedigree diagram for calculating the inbreeding coefficient in the son of a first cousin mating.

gives $F = (1/2)^n(1 + F_c)$, where we find F_c first so that it can be incorporated in the inbreeding calculation for ego. The second complication is that there may be more than one inbreeding loop in the pedigree, as shown in Figure 3.3 for a mating between first cousins. In this case, ego (individual 7) may have two copies of an allele from his great-grandfather (individual 1) or great-grandmother (individual 2). Using the "or" rule, we can write a general equation for calculating the inbreeding coefficient:

$$F = \sum_{i=1}^{\#loops} \left(\frac{1}{2}\right)^{n_i} (1 + F_{ci}) \qquad (3.7)$$

where the summation is across the number of loops and the subscript i indicates which loop we are working on. In Figure 3.3, the inbreeding coefficient is $(1/2)^5 + (1/2)^5 = 2(1/2)^5 = (1/2)^4 = 1/16$.

The second issue we need to address is the population-level effect of inbreeding. Just as we can define an inbreeding coefficient for an individual, we can also define an inbreeding coefficient for a population, which is the average inbreeding coefficient. For example, if a quarter of the individuals in a population are the offspring of first cousin mating and three-quarters are from nonconsanguineous mating, then the (average) inbreeding coefficient for the population is $(1/4) \times (1/16) + (3/4) \times 0 = 1/(4 \times 16) = 0.015625$. The usual symbol for average inbreeding within a population is F_{IS}, where the subscript IS refers to the probability of IBD within individuals relative to subpopulations. The "sub" part of *subpopulation* reminds us that most populations are hierarchically organized within larger populations. The relationships we saw regarding the probability of getting a heterozygote in a consanguineous mating apply as well to the probability of getting a heterozygote in a population once we know the allele frequency and F_{IS}. Specifically, the expected proportion of heterozygotes is $2pq(1 - F_{IS})$, while the proportions of *AA* and *BB* genotypes are $p^2 + F_{IS}p$ and $q^2 + F_{IS}q$. Consequently, inbreeding decreases the proportion of heterozygotes expected under Hardy-Weinberg equilibrium and increases the proportion of homozygotes. If there is a recessive allele that causes a disease, then inbred populations would be expected to express the disease more frequently than Hardy-Weinberg equilibrium would predict. This said, a little algebra shows that inbreeding in and of itself is not an evolutionary force because the allele frequencies are unaffected. We can find the allele frequency in the next generation as

$$p^2(1 - F_{IS}) + F_{IS}p + 2pq(1 - F_{IS})/2 = p[p(1 - F_{IS}) + F_{IS} + q(1 - F_{IS})] = p[(p + q)(1 - F_{IS}) + F_{IS}] = p.$$

EVOLUTIONARY FORCES

While positive assortative mating and inbreeding alter genotype frequencies so that they are not in Hardy-Weinberg equilibrium, these processes do not alter allele frequencies. Consequently, positive assortative mating and inbreeding are not evolutionary forces. In this section, we consider the four evolutionary processes (mutation, selection, drift, and migration) that can change allele frequencies across generations.

Mutation

Mutation is the source of all new alleles, and as a consequence, none of the other evolutionary forces would be operative in the absence of mutation. However, the frequency of mutational events is generally so low that allele frequencies cannot change appreciably simply by the force of mutation. In order to gain an appreciation for the slow rate of allele frequency change that would obtain by mutation alone, we first need to have an estimate of the **mutation rate**. The *mutation rate* is the probability that a gene will mutate and is given as the probability that a gamete will have a mutation at the specified locus in one generation. The simplest way to estimate a mutation rate is to use the spontaneous appearance of a dominant allele expressed in a child whose parents are both homozygous recessive.

As an example for estimating mutation rates, we will use mutations to achondroplasia. **Achondroplasia** is a nonproportional dwarfism (it primarily affects the limbs) caused by mutation to a dominant allele in the fibroblast growth factor receptor-3 gene (4p16.3). Specifically, the mutation usually arises by a base substitution of an A for a G at the 1,138th nucleotide of the *FGFR3* gene. Nelson and Holmes (1989) counted two achondroplastic children born out of 69,277 children with parents who were not achondroplastic. The 69,277 children represent $2 \times 69,277$ gametes, of which two had mutated to the dominant achondroplasia allele. This gives a mutation rate of $2/(2 \times 69,277)$, or 1.44×10^{-5}. Stoll et al. (1989) report a slightly higher mutation rate of 3.3×10^{-5}, which they attribute to the fact that their study included prenatal diagnoses via ultrasound. Other estimates of mutation rates for achondroplasia (Orioli et al. 1986, Oberklaid et al. 1979, Gardner 1977) are generally in the range of 2 to 6 mutations per 10^5 gametes.

Using the higher mutation rate (probability) of $\mu = 6 \times 10^{-5}$, we can write what is known as a **recurrence relationship**, which gives the allele frequency after one generation of mutation. The recurrence relationship gives the allele frequency in the first generation (which we write as p_1) as a function of the allele frequency at time 0 (p_0):

$$p_1 = p_0(1 - \mu) \tag{3.8}$$

where p is the allele frequency for the recessive allele and q is the allele frequency for the dominant (achondroplastic) allele. We assume that at time 0 there are no achondroplastic alleles, so $p_0 = 1.0$ and $q_0 = 0.0$. We can apply equation (3.8) again to obtain the allele frequency in the second generation as follows:

$$p_2 = p_1(1 - \mu) \tag{3.9}$$

We substitute equation (3.8) into equation (3.9) to obtain the following:

$$p_2 = p_0(1 - \mu)(1 - \mu) = p_0(1 - \mu)^2 \tag{3.10}$$

or in general:

$$p_t = p_0(1 - \mu)^t \tag{3.11}$$

Equation (3.11) can be solved for t to find the number of generations of mutation necessary in order for p to decrease from 1.0 to 0.99 (i.e., the number of generations necessary for the achondroplastic allele frequency to increase from 0.0 to 0.01). It would take about 167.5 generations of mutation to reach an allele frequency of 1%. Using a generation length of 25 years, this comes to 4,188 years, which shows that evolution by mutation (alone) is quite laborious.

Even though mutation is a remarkably slow evolutionary force, given enough time, the achondroplasia allele frequency would continue to increase to an upper value of 1.0. For example, after about 2 million years (76,751 generations), the achondroplasia allele would reach a frequency of 99%. There is consequently no equilibrium if mutation occurs only in one direction (from non-achondroplastic to achondroplastic alleles).

Selection

Because of the Darwinian paradigm, natural selection is a concept with which most students are at least broadly familiar. The effects that selective regimes have on allele frequencies are actually quite easy to discern using a computer spreadsheet, while the mathematics of selection can mostly be dealt with by some algebra. We will, as far as possible, avoid the algebra. Selection must operate on genetic variation, so mutation is an essential precursor to evolution by natural selection.

An Example of Selection at the β-Hemoglobin Locus

It is simplest to examine selection by using an actual example, for which we use some data on polymorphism in the β-hemoglobin gene. Thompson (1962) provided genotype count data for 6 genotypes (*AA, AS, SS, AC, SC, CC*) in 840 Ghanaian policemen and in 1,222 Ghanaian policemen's children. To simplify our presentation, we include the *AC* and *CC* genotype counts with the *AA* genotype and the *SC* genotype with the *AS* count. Table 3.7 lists the genotype counts for the adults and for the children. Because there were

Table 3.7 Viability information on the β-hemoglobin locus

| | Children | | Adults | | | |
	Count	Percent	Count	Percent	Adult%/Child%	Relative Fitness
AA	701	83.45%	222	79.86%	0.9569	0.7635
AS	135	16.07%	56	20.14%	1.2534	1.0000
SS	4	0.48%	0	0.00%	0.0000	0.0000
Total	840		278			

slightly more than three children genotyped for every one adult, in Table 3.7 we give the percentage of genotypes among children and among adults, to standardize the sample sizes. From these percentages we can get an idea of the genotype's effects on survival. For example, the SS genotype comprises only 0.48% of the child sample and 0.00% of the adult sample, suggesting that none of the children with the SS genotype is likely to survive to adulthood. The AA genotype comprises 83.45% of the child sample and 79.86% of the adult sample, so the adult sample's AA frequency is only about 96% of that found in the children. Only the AS genotype increases in relative frequency from children (16.07%) to adults (20.14%). We divide the adult percentages by the child percentages (see Table 3.4) and then divide that value by the maximum (1.2534 for the AS genotype). The final column in Table 3.7 is labeled **relative fitness**, which is the contribution of each genotype to the next generation, standardized so that the maximum relative fitness is 1.0. Ideally, relative fitness should be measured based on *absolute fitness*, which is the contribution by genotype to the next generation. The literal definition of *contribution to the next generation* is the number of offspring who survive to the same age as the individuals (parents) for whom we are measuring fitness. For example, the fitness of a 41-year-old would be the number of his or her children who survive until 41 years. The data in Table 3.7 are consequently not sufficient to calculate real absolute fitness, but as an example they suffice.

The usual symbol for relative fitness is w, which we subscript to indicate the genotypes. If we think of relative fitness as survivorship, then the expected frequencies of the genotypes are $f(AA) = p^2w_{AA}, f(AS) = 2pqw_{AS}$, and $f(SS) = q^2w_{SS}$. In the Hardy-Weinberg model, all survivorships are 1.0, so the expected genotype frequencies are p^2, $2pq$, and q^2. When one or more of the genotype's relative fitness values are less than 1.0, then $f(AA) + f(AS) + f(SS)$ is less than 1.0. To adjust the frequencies so that they sum to 1.0, we divide $f(AA)$, $f(AS)$, and $f(SS)$ by $p^2w_{AA} + 2pqw_{AS} + q^2w_{SS}$. The quantity $p^2w_{AA} + 2pqw_{AS} + q^2w_{SS}$ is called the **average fitness** because it weights each genotype's relative fitness by the Hardy-Weinberg expected genotype frequency. The usual symbol for average fitness is $\bar{w}$. After one generation of selection, we can find the new A allele frequency as $(p^2w_{AA} + pqw_{AS})/\bar{w}$. It is a simple matter to copy the relevant formulae through a spreadsheet, which is how Table 3.8 was prepared, where we start the initial allele frequencies at 0.01 and 0.99. It is interesting to note that when we start at the low allele frequency the frequency increases, while when we start at the high allele frequency the frequency decreases. This type of selection pattern, where the heterozygote has the highest fitness, produces what is known as a **balanced polymorphism** because both alleles are maintained no matter what the initial allele frequency is. Figure 3.4 shows a graph of the allele frequencies starting at initial frequencies near 0 and 1. Both lines come to a **stable equilibrium** value at an allele frequency of about 0.808734.

Figure 3.4 is one of three different graphs we can use to understand the dynamics of evolution by natural selection. Another useful graph is one that plots the change in allele frequency with one generation of selection on the y axis and the allele frequency on the x axis. The change in allele frequency is usually written as Δp and is equal to $p' - p$, where the prime sign means the allele frequency in the next generation. Figure 3.5 contains the graph of Δp against p. This graph shows that if the allele frequency is less than the equilibrium value (marked with a vertical dashed line), then Δp will be positive. Consequently, the allele frequency will increase to the equilibrium point. Conversely, if the allele frequency is above the equilibrium value, then Δp will be negative and the allele frequency will decrease to the equilibrium. Figure 3.6 shows these dynamics in the vicinity

Table 3.8 Example of selection at the β-hemoglobin Locus, where the relative fitnesses are $w_{AA} = 0.7635$, $w_{AS} = 1.0$, and $w_{SS} = 0.0$

f(AA)	f(AS)	$\bar{w}$	p 0.01	f(AA)	f(AS)	$\bar{w}$	p 0.99
0.0001	0.0198	0.0199	0.5019	0.7483	0.0198	0.7681	0.9871
0.1923	0.5000	0.6923	0.6389	0.7439	0.0254	0.7694	0.9835
0.3117	0.4614	0.7731	0.7016	0.7385	0.0325	0.7710	0.9789
0.3758	0.4187	0.7945	0.7365	0.7316	0.0413	0.7729	0.9733
0.4141	0.3881	0.8023	0.7581	0.7233	0.0520	0.7753	0.9665
0.4388	0.3668	0.8056	0.7724	0.7131	0.0648	0.7780	0.9583
0.4554	0.3516	0.8071	0.7822	0.7012	0.0799	0.7811	0.9489
0.4671	0.3408	0.8079	0.7891	0.6874	0.0970	0.7844	0.9382
0.4754	0.3329	0.8083	0.7941	0.6720	0.1160	0.7880	0.9264
0.4814	0.3270	0.8085	0.7978	0.6552	0.1364	0.7916	0.9138
0.4859	0.3227	0.8086	0.8005	0.6376	0.1575	0.7951	0.9010
0.4892	0.3194	0.8086	0.8025	0.6198	0.1784	0.7982	0.8882
0.4917	0.3170	0.8087	0.8040	0.6024	0.1986	0.8009	0.8760
0.4935	0.3152	0.8087	0.8051	0.5860	0.2172	0.8031	0.8648
0.4949	0.3138	0.8087	0.8060	0.5710	0.2339	0.8048	0.8547
0.4960	0.3127	0.8087	0.8067	0.5578	0.2483	0.8061	0.8460
0.4968	0.3119	0.8087	0.8072	0.5464	0.2606	0.8070	0.8385
0.4974	0.3113	0.8087	0.8075	0.5368	0.2708	0.8076	0.8324
0.4979	0.3109	0.8087	0.8078	0.5290	0.2791	0.8080	0.8273
0.4982	0.3105	0.8087	0.8080	0.5226	0.2857	0.8083	0.8233

The example to the left of the vertical line starts with an allele frequency for the *A* allele of 0.01, while to the right the example starts at an allele frequency of 0.99. In each example, the first two columns give the *AA* and *AS* genotype frequencies after selection (the *SS* genotype frequency after selection is always 0, so it is not shown). The third and fourth columns give the average fitness and allele frequency after selection.

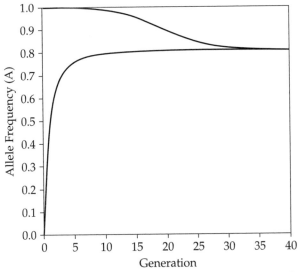

Figure 3.4. Change in allele frequency across time when selection favors the heterozygote.

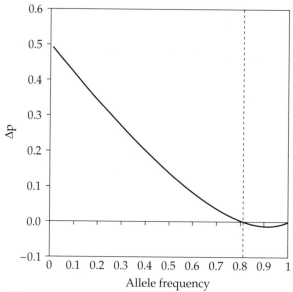

Figure 3.5. Graph of the change in allele frequency between generations as a function of the allele frequency. The *vertical dotted line* shows the equilibrium point where the change in allele frequency is 0.

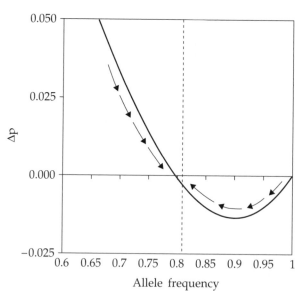

Figure 3.6. Close-up of Figure 3.5 in the vicinity of the equilibrium. The *arrows* indicate the direction of evolution.

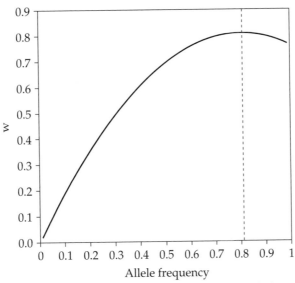

Figure 3.7. Graph of average fitness against allele frequency for the balanced polymorphism shown in Figures 3.4–3.6. The *vertical dotted line* shows the equilibrium point.

of the equilibrium allele frequency, where the arrows indicate the direction of change for the allele frequency.

The third graph we can examine is one that plots average fitness against allele frequency, as shown in Figure 3.7. It is possible to show that the direction of evolution is determined by the slope of the line tangent to the line in Figure 3.7 at the current allele frequency. A *tangent line* is a line that touches the curve in Figure 3.7 at a single point. Figure 3.8 contains the plot of average fitness against allele frequency in the vicinity of the equilibrium, and it also shows the slope of the tangent lines at allele frequencies of 0.75, 0.9, and the equilibrium. The slope of the tangent line is positive at an allele frequency of 0.75, negative at an allele frequency of 0.9, and 0 at the equilibrium. At the equilibrium, the tangent line has a slope of 0, so Δp is also 0 and there is no evolution. At allele frequencies below the equilibrium, the slope of the tangent to the curve is always positive, so Δp is also positive and the allele frequency will be increasing. Above the equilibrium, the slope is negative, so Δp is negative and the allele frequency will be decreasing. As a consequence, when the allele frequency is below the equilibrium, the allele frequency will increase by natural selection, while when it is above, it will decrease. The arrows in Figure 3.8 indicate the direction of evolution. From this graph it is clear that evolution by natural selection can only lead to increased average fitness. Selection cannot move a population to lower fitness. In essence, evolution by natural selection is about climbing to the top of local fitness "peaks." Indeed, the graph shown in Figures 3.7 and 3.8 is often known as an "adaptive topography," after an analogy to a cross-sectional view through a mountain. So even though the mathematics underlying the process may be obscure to some, the important message is that selection can only take populations "uphill" in the adaptive topography.

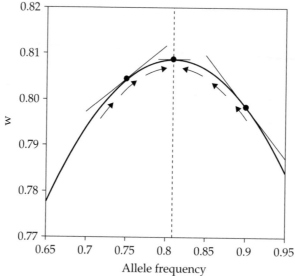

Figure 3.8. Close-up of Figure 3.7 in the vicinity of the equilibrium. The *arrows* show the direction of evolution, while the two *straight lines* give the first derivative (slope of the tangent) at lower average fitnesses.

Other Patterns of Selection

We can use the three graphs (*p* against generation, Δ*p* against *p*, and *w̄* against *p*) to look at other selection regimes. Figures 3.9–3.11 show the graphs for a lethal recessive trait, where $w_{AA} = w_{AB} = 1$ and $w_{BB} = 0$. The graphs show that, unlike the balanced polymorphism, there is no intermediate allele frequency that selection will maintain. Instead,

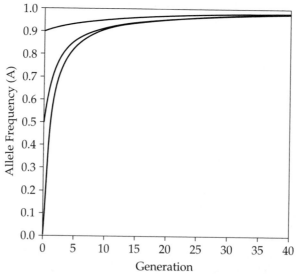

Figure 3.9. Change in allele frequency against time when the recessive state (BB) is lethal.

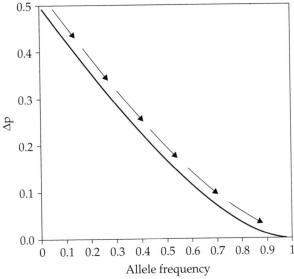

Figure 3.10. Graph of the change in allele frequency against allele frequency when the recessive state is lethal. Note that the change is always positive, so that the *A* allele increases to a frequency of 1.0. The *arrows* give the direction of evolution.

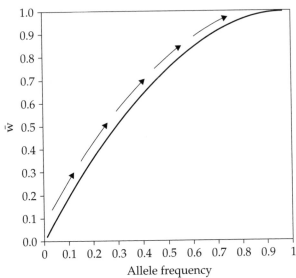

Figure 3.11. Graph of average fitness against allele frequency when the recessive state is lethal. The *arrows* give the direction of evolution.

selection against the recessive genotype will eventually eliminate the allele in the recessive genotype. The rate of evolution does, however, slow down as the dominant allele frequency increases. This is, however, completely different from the evolutionary pattern for a lethal dominant trait. For a lethal dominant trait, all of the dominant alleles are removed by one generation of selection. This is true because both the homozygous dominant individuals (if there are any) and the heterozygous individuals die, while only the homozygous recessive individuals survive. Consequently, evolution by natural selection proceeds very rapidly if there is selection against dominant alleles, while it proceeds more slowly if the selection is against recessive alleles (because they are "hidden" from selection in heterozygous individuals).

Drift

Random *genetic drift* is change in allele frequencies within finite populations across time. The simplest way to understand drift is to start with a simple (though humanly impossible) example. We will assume that there are an infinite number of subpopulations, each composed of one individual who "selfs" each generation (i.e., mates with him- or herself) to produce the next generation composed of one individual. Within each population, we might assume that an infinite number of gametes are produced but that only two gametes will ultimately combine to form a zygote. We call this the "boundless bucket" model because we presume that there are an infinite number of gametes in each subpopulation's "bucket" but that we will only choose two to form the zygote. We start this "thought experiment" by presuming that every subpopulation consists of one *AB* heterozygote. When the *AB* individual "selfs," Mendelian expectations tell us that there is a quarter-chance the offspring will be *AA*, a half-chance it will be *AB*, and a quarter-chance it will be *BB*. However, we select only one individual, so we must "pick" either the *AA*, *AB*, or *BB*. If we pick an *AA* individual, then the allele frequency will have drifted up from 0.5 to 1.0, while if we pick a *BB* individual, the allele frequency will have drifted down to 0.0. When a subpopulation has gone to an allele frequency of 1.0, we talk about the allele as having reached **fixation** in that subpopulation because in the absence of migration or mutation the subpopulation has lost all but the one allele. The subpopulation is consequently "fixed" for the one present allele.

In the above example, we can calculate the probability that a subpopulation will go to fixation in one generation of drift. If the subpopulation size is one and the genotype for the one individual is *AB*, then the probability of an *AA* or *BB* genotype in the next generation is $1/4 + 1/4$, or $1/2$. Consequently, the probability of not going to fixation is $1 - 1/2$, or $1/2$. If the population size is two and the allele frequency is 0.5, then the probability of obtaining two *AA* homozygotes in the next generation is 0.5^4 and the probability of obtaining two *BB* homozygotes is also 0.5^4. Taken together, the probability of going to fixation is 0.125, considerably lower than the 0.5 for one individual. For three individuals and an allele frequency of 0.5, the probability of going to fixation drops even more, to 0.03125. In general, the probability of going to fixation in one generation is $p^{2N} + q^{2N}$. Clearly, drift depends on subpopulation size: the larger the subpopulation, the less important drift becomes. To demonstrate this, Figures 3.12–3.14 show the results of computer simulations of random genetic drift. Each simulation shows allele frequencies for 100 subpopulations across time, where the subpopulations all start with an allele frequency of 0.5. The first simulation is for a small population size of 10 individuals, the next is for 100, and finally 1,000. Clearly, for larger subpopulations the fluctuation of allele

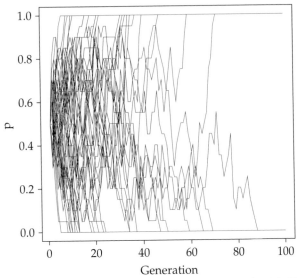

Figure 3.12. Simulation of genetic drift for 100 subpopulations each with 10 individuals.

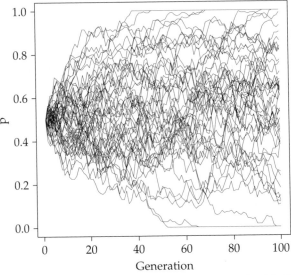

Figure 3.13. Simulation of genetic drift for 100 subpopulations each with 100 individuals.

frequencies across the generations (due to random genetic drift) becomes ever less extreme. For small subpopulations the fluctuation in allele frequencies is so extreme that all subpopulations reach fixation fairly rapidly.

The computer simulation results in Figures 3.12–3.14 are from what are known as stochastic simulations. In *stochastic simulations*, we specify a probability model and then let the computer do the work of randomly simulating from the specified model. We can

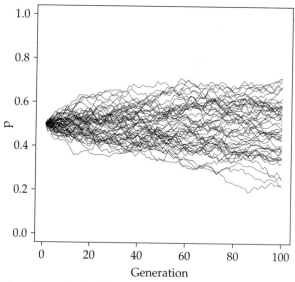

Figure 3.14. Simulation of genetic drift for 100 subpopulations each with 1,000 individuals.

also use what are known as deterministic simulations to look at the effects of random genetic drift on allele frequencies. *Deterministic simulations* use rather complicated math in order to draw the expected histogram of subpopulation allele frequencies after some specified number of generations. Figure 3.15 contains histograms of the expected proportion of subpopulations at each allele count across generations, where all subpopulations consist of 5 individuals. We start this deterministic simulation with every subpopulation having an allele frequency of 0.5; or, in other words, of the 10 alleles sampled for each founding subpopulation, every subpopulation will have 5 A alleles (and 5 B alleles). Consequently, the first histogram in Figure 3.15 shows that all subpopulations have 5 of the A alleles. After one generation, the subpopulation allele frequencies have drifted such that a portion of them (almost 25%) still have 5 A alleles but many have gone to a lesser or greater number of A alleles. Figure 3.15 also shows the expected distribution of subpopulation A allele counts after 10, 20, and 30 generations of drift. By 30 generations, about 47% of the populations have gone to an A allele count of 0 (i.e., those populations are fixed for the B allele) and an equivalent 47% have gone to an allele count of 10 (fixation for the A allele). Only about 6% of the subpopulations at the 30th generation are still polymorphic. Figure 3.16 shows an identical deterministic simulation but with subpopulation sizes of 20 individuals, while Figure 3.17 shows drift with subpopulation sizes of 40 individuals. Comparing the three figures, we can see that the rate at which subpopulations are expected to reach fixation depends on subpopulation size. Small subpopulations drift more rapidly than do large subpopulations and, consequently, reach fixation more rapidly. Interestingly, the proportion of subpopulations that reach fixation for the A allele is equal to the initial allele frequency (p) for the subpopulations. As a consequence, the average (across subpopulations) allele frequencies do not change, while the allele frequencies within subpopulations do change until fixation is reached.

For comparison to the previous figures, we show the case where the initial allele frequency is 0.75 and there are 24 individuals within each subpopulation (see Fig. 3.18).

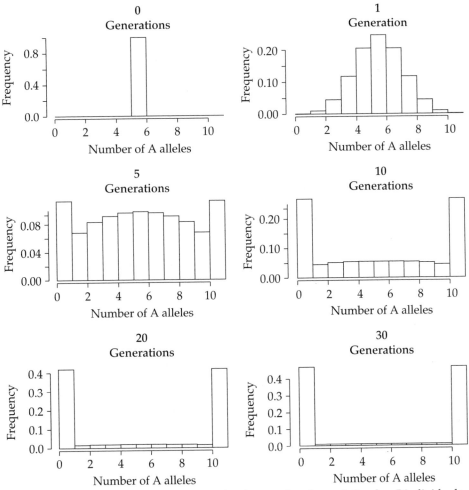

Figure 3.15. Deterministic simulation of drift, where each subpopulation has 5 individuals and starts with 5 (of 10) alleles being *A*.

By 400 generations of drift, all subpopulations have reached fixation, with 75% having gone to fixation for the *A* allele and 25% having gone to fixation for the *B* allele. Again, the overall allele frequency has not changed, though evolution has clearly occurred within subpopulations (as all have gone from being polymorphic for the A and B alleles to being in a state of fixation). What has been the effect on Hardy-Weinberg equilibrium? Within each subpopulation there has been no departure from Hardy-Weinberg equilibrium. For example, when a population has reached fixation for the *A* allele ($p = 1$), the expected genotype frequencies are 1 *AA*:0 *AB*:0 *BB*, which are equal to the observed genotype frequencies. If, however, we consider the expected Hardy-Weinberg proportions for the total population (ignoring subpopulation division), then we would expect genotype frequencies of 0.75^2, $2 \times 0.75 \times 0.25$, and 0.25^2, while we obtain genotype frequencies of 0.75 *AA*, 0.0 *AB*, and 0.25 *BB*. Clearly, drift has led to a complete depletion of heterozygotes because all subpopulations are composed of homozygotes (for the *A* allele in *p* of

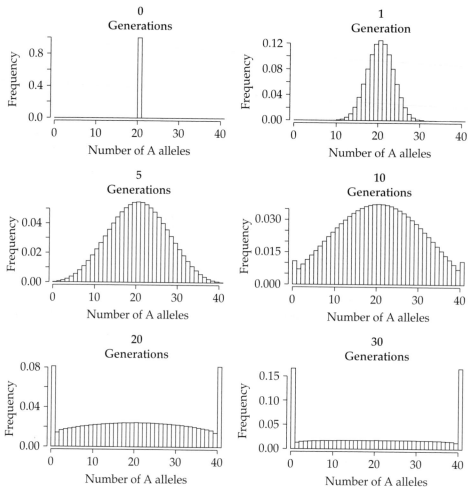

Figure 3.16. Deterministic simulation of drift, where each subpopulation has 20 individuals and starts with 20 (of 40) alleles being A.

the subpopulations and for the B allele in q of the subpopulations). We can consequently use a slightly different form of equation (3.6) to measure the extent of genetic drift:

$$F_{ST} = \frac{2\bar{p}\bar{q} - f(AB)}{2\bar{p}\bar{q}} = \frac{H_e - H_o}{H_e} \tag{3.12}$$

where F_{ST} is a fixation index (we have seen that drift leads to fixation) and the subscript ST means fixation of subpopulation allele frequencies relative to the expected heterozygosity for the total population (the S and T refer specifically to subpopulation relative to total population). F_{ST} can range between 0 (in which case there has been no drift) and 1, in which case the division of the population into finite subpopulations has led to complete fixation within subpopulations (so that H_o, the observed heterozygosity, is 0). Ultimately, drift starts with subpopulations that are genetically heterogenous and

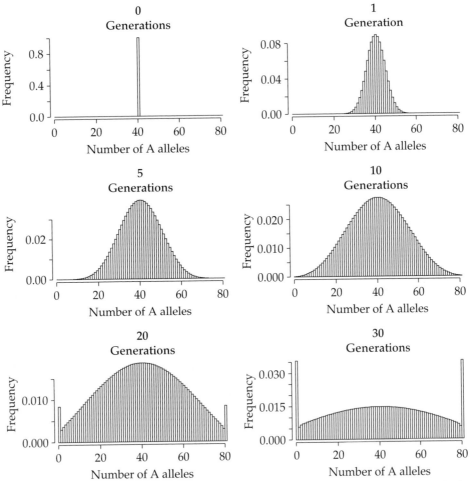

Figure 3.17. Deterministic simulation of drift, where each subpopulation has 40 individuals and starts with 40 (of 80) alleles being *A*.

"transfers" this heterogeneity to between-group differences. We will see in the next section that migration has the opposite effect of homogenizing across subpopulations and heterogenizing within subpopulations.

While F_{ST} can be defined as a "fixation index," or the decrease in total heterozygosity relative to expected heterozygosity, it can also be defined as one-fourth of the genetic distance between a pair of populations or as a probability of IBD. The relationship between F_{ST} and genetic distance gives the F_{ST} a very intuitive feel because we can see that as the F_{ST} for the two populations increases, so does the genetic distance between the two populations. In Chapter 12, we will discuss in depth genetic distance analysis, but it is important to keep in mind that distance analysis and F_{ST} have a very intimate relationship and that both are fundamentally related to genetic kinship.

F_{ST} can also be defined as a probability of IBD. In the section on inbreeding, we saw that the inbreeding coefficient can be defined as the probability of IBD for the two alleles

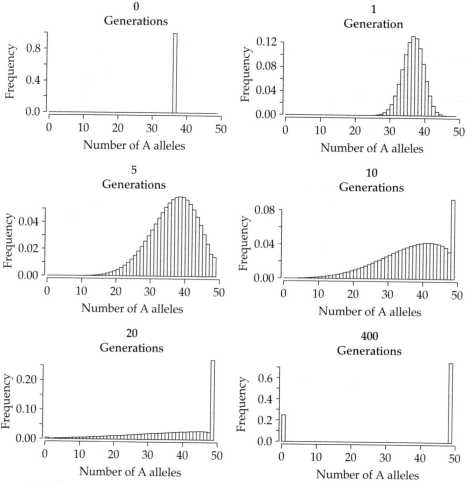

Figure 3.18. Deterministic simulation of drift, where each subpopulation has 24 individuals and starts with 36 (of 48) alleles being A.

at an autosomal locus in a given individual. Similarly, F_{ST} can be defined as the probability of sampling two alleles that are IBD within a subpopulation. This definition of F_{ST} allows us to write a recurrence relationship so that we can find the value of F_{ST} in the next generation as a function of F_{ST} in the current generation. When we sample two alleles (with replacement) from within a subpopulation, the chance that we will sample the same allele twice is $1/(2N)$. This can be confirmed using the "and" and "or" rules since the probability of drawing the same allele twice is the probability of drawing the first allele in the subpopulation and then drawing it again or drawing the second and then the second or the third and then third, etc.:

$$\sum_{i=1}^{2N}\left(\frac{1}{2N}\right)^2 = 2N\left(\frac{1}{2N}\right)^2 = \frac{1}{2N} \qquad (3.13)$$

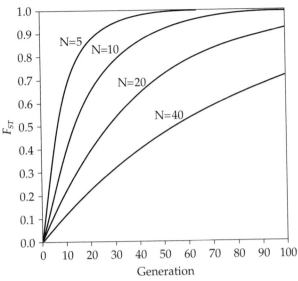

Figure 3.19. Graph of the increase in F_{ST} at different subpopulation sizes.

With probability $1/(2N)$, we would draw the same allele twice, which by definition must be IBD with itself. The probability that we draw two distinct alleles is then $1 - 1/(2N)$, but the probability these alleles are IBD from a previous generation is F_{ST}; thus, the probability that we draw two distinct alleles and they are IBD from a previous generation is $[1 - 1/(2N)]F_{ST}$. Putting this together we have the following:

$$F'_{ST} = \frac{1}{2N} + \left(1 - \frac{1}{2N}\right)F_{ST} \tag{3.14}$$

In the initial generation, we assume that F_{ST} is 0 since there is no descent yet, so none of the alleles can be IBD. In the first generation, F_{ST} will be $1/(2N)$ and will continue to increase across generations (by drift) until it reaches 1.0 We use equation (3.14) repeatedly in Figure 3.19 to draw the increase in F_{ST} across time for subpopulation sizes of 5, 10, 20, and 40 individuals. As we have already seen, the rate of drift (here measured by the increase in the probability of IBD) depends on subpopulation size. In the example with subpopulation sizes of 5 individuals each, by 100 generations F_{ST} is virtually at its maximum value of 1.0, while in the example with subpopulation sizes of 40, F_{ST} only reaches about 0.7 by the 100th generation.

We have seen that F_{ST} can be defined as a fixation index (reduction in total heterozygosity relative to expected heterozygosity) or as the probability of IBD for alleles within subpopulations. There is a final definition for F_{ST} that gives it primacy in any discussion of human population variation. F_{ST} can be defined as the proportion of total variation that is due to population subdivision. As a consequence of this definition, if we say that F_{ST} is 0.11 for subpopulations (where we might define subpopulations as continental human populations within the worldwide human population), then this implies that 11% of the total variation is between subpopulations (and the remaining 89% is within subpopulations). Table 3.9 gives a fictitious example of four subpopulations, each with

Table 3.9 Example calculation of F_{ST} using four subpopulations and a fictitious RFLP (+/+ means homozygous for the restriction site, +/− means heterozygous, and −/− means homozygous for the loss of the restriction site)

Subpopulation	+/+	+/−	−/−	p	Between	Total
1	10	180	810	0.1	0.030625	241.25
2	40	320	640	0.2	0.005625	331.25
3	90	420	490	0.3	0.000625	421.25
4	250	500	250	0.5	0.050625	601.25
Totals	390	1,420	2,190	1.1	0.0875	1,595
			N =	4	4	8,000
			Ave	0.275	0.021875	0.199375

H_o 1,420/(390 + 1,420 +2,190)
= 0.355

H_e 2 × 0.275 × (1 − 0.275)
= 0.39875

            0.021875

$(H_e − H_o)/H_e$ (0.39875 − 0.355)/0.39875 V_b/V_t 0.199375
= 0.1097179   = 0.1097179

1000 individuals who have been typed for a locus with two alleles (referred to as + and −). The allele frequencies for the + allele are 0.1, 0.2, 0.3, and 0.5, giving an average (i.e., frequency in the total population) of 0.275. The expected heterozygosity for the total population is therefore 2 × 0.275 × 0.725 = 0.39875, while the observed heterozygosity is 0.355. This gives a fixation index of 0.1097179, as shown at the bottom left of the table. Now, we can calculate F_{ST} using its definition as the proportion of between-subpopulation variation out of the total population variation. First, we will need a formal definition of *variation*. *Variation* is just a gloss for the statistical concept of **variance**, which is the average squared deviation around the average. For example, the variance of the numbers 1, 2, and 3 is 2/3 because their average is 2 and the average squared deviation around this mean is $(1 − 2)^2/3 + (2 − 2)^2/3 + (3 − 2)^2/3$. In Table 3.9, we first find the variance of the allele frequency across the four subpopulations. The column labeled "Between" gives the squared deviations around the average allele frequency of 0.275, e.g., $(0.1 − 0.275)^2 = 0.030625$, which is averaged at the bottom to give the between-subpopulation allele frequency variance of 0.021875. Next, we find the total allelic variance (ignoring subpopulations). To find the total allelic variance, we assign a score of 1.0 if the allele is a + allele and 0.0 if it is not. In the first subpopulation, there are 2 × 10 + 180 = 200 of the + alleles and 2 × 810 + 180 = 1,800 of the − alleles. In the total population, the average of the allelic values is just the average allele frequency of 0.275. Using our scoring method, subpopulation 1 contributes $200 × (1 − 0.275)^2 + 1,800 × (0 − 0.275)^2 = 241.25$ to the total sum of squared deviations around the average. The numbers in the "Total" column are then summed and divided by 8,000, the number of alleles in the total population, to give the allelic variance of 0.199375. The between-subpopulation variance divided by the total (i.e., F_{ST}) is 0.109719, agreeing precisely with the calculation of the fixation index. This example indicates that about 11% of the total variation is due to differences between the four subpopulations. We should make one final comment before moving on to migration, which is that the F_{ST} calculated from Table 3.9 is a bit lower than the true estimate should be. This is because of the way we defined variance as the average squared deviation around the mean, when we should have defined it as the sum of squared deviations around the mean divided by $n - 1$. When we make this type of adjustment in Table 3.9 the correctly calculated F_{ST} is 0.140699, indicating that about 14% (rather than 11%) of the variation is between subpopulations. These calculations are available in computer programs such as GDA and Arlequin (see the Supplemental Resources section).

Migration

Genetic migration, which is sometimes also called "gene flow", is the movement of alleles between subpopulations. Alleles move as they are carried by individuals who migrate, but we refer to migration of the alleles themselves to contrast demographic with genetic migration. In demographic migration, people may move to a new locale, but unless they transmit alleles to the next generation they do not migrate in a genetic sense.

We will first consider the simplest case, where there are two subpopulations that exchange migrants at rate m per generation. If p_1 and p_2 are the allele frequencies in the two subpopulations, then after one generation of migration the allele frequencies in subpopulations 1 and 2 will be as follows:

$$p'_1 = p_1 (1 - m) + p_2 m$$

$$p'_2 = p_2 (1 - m) + p_1 m \tag{3.15}$$

As an example of applying equation (3.15) we can start with $p_1 = 1.0$, $p_2 = 0.0$, and a migration rate of 0.1 (i.e., 10% of the two subpopulations migrate per generation). The allele frequency in the first subpopulation after one generation of migration will then be (1.0×0.9) + $(0.0 \times 0.1) = 0.9$, while the allele frequency in the second subpopulation after one generation of migration will be $(0.0 \times 0.9) + (1.0 \times 0.1) = 0.1$. Now we can use equation (3.15) again with the allele frequencies of 0.9 and 0.1, and after another generation of migration, the new allele frequencies will be $p_1 = 0.82$ and $p_2 = 0.18$. We can copy these formulae many times through a spreadsheet in order to draw a graph of the evolution of the allele frequencies within the subpopulations. Figure 3.20 shows this graph both for $m = 0.1$ and for a lower migration rate of $m = 0.05$. From this figure we can see that, given enough generations of migration, the two subpopulations will come to the same allele frequency halfway between their starting values, $(0.0 + 1.0)/2 = 0.5$. The speed with which the two populations come to this common value depends on the migration rate. The higher the migration rate, the more quickly the two populations will come to the midpoint allele frequency.

The effect of migration is opposite to that of drift, for drift removes heterozygosity at the total population level, whereas migration restores heterozygosity. Another way to think of the two processes is in terms of genetic variation within and between subpopulations. Figures 3.12–3.19 show that drift starts with all subpopulations having the same allele frequency so that there is no variation between subpopulations. In generation 0, the locus is polymorphic within all subpopulations, so all variation exists within subpopulations and none exists between. After many generations of drift, p of the populations have

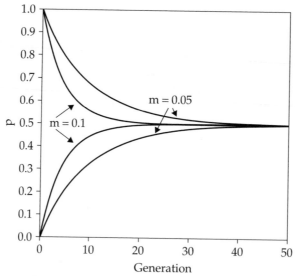

Figure 3.20. Change in allele frequencies when two subpopulations exchange migrants. One population starts with an allele frequency of 1.0, the other with a frequency of 0.0. Both come to an equilibrium value of 0.5. The speed with which this equilibrium is reached depends on the migration rate (shown here for $m = 0.1$ and 0.05).

gone to fixation for the A allele and q have gone to fixation for the B allele. Now there is no variation within subpopulations, and all of the variation is between subpopulations. This can be contrasted with Figure 3.20, where prior to migration there was no genetic variation within subpopulations and all variation was between subpopulations. After many generations of migration, all of the genetic variation is within subpopulations and none is between.

A slightly different form of equation (3.15) can be used to estimate the proportion of ancestry from two populations that are ancestral to a descendant population. We give examples of such applications when we discuss population structure analysis in Chapter 12.

Combining Forces

We have considered the four evolutionary forces (and the nonevolutionary forces of assortative mating and inbreeding) separately. In point of fact, the forces often occur in tandem, with two or more forces balancing each other or with one or more forces setting the stage for a subsequent force. Sewall Wright (1977, 454–455) provided a succinct description of a model for evolution that incorporates all four of the evolutionary forces, and Templeton (1982) has given a good summary. Wright's model, which is referred to as the **shifting balance theory,** posits that mutation will create new alleles and that selection then operates to increase the average fitness of the population. In Figure 3.7, we looked at only two alleles at a single locus, but Wright envisioned many loci affecting fitness. For two loci the curve in Figure 3.7 becomes a surface, while for more than two loci the curve becomes a "hypersurface" that we cannot draw. Selection will take the population "uphill" to the highest local fitness. Because selection cannot take the population "downhill," the population cannot move to a higher fitness peak if this involves crossing a fitness "valley." Random genetic drift is, however, blind to the fitness surface and so can move the population across the "valley" to the beginning of a new upward fitness slope. At this point, selection will carry the population to the top of the new fitness peak. Once the population has "arrived," it will have a higher average fitness, which translates into population growth. The population growth exerts pressure such that some individuals will have to migrate away, and when they do so they carry the new set of alleles with them to other populations. The new alleles may be enough to "nudge" these other populations across the fitness "valley" and start them on the selective path to the higher fitness peak. The end result of this combination of mutation, selection, drift, and migration is a very fluid pattern of evolution with instances of "shifting balance."

The mathematics of three or more forces working at the same time can be rather complex, so in the remainder of this chapter we look at a few examples of pairs of forces interacting. Our first example is of inbreeding and drift occurring at the same time. Equation (3.7) gave individual estimates of inbreeding, which could be averaged within a subpopulation to obtain F_{IS}. Similarly, equation (3.14) gave the probability of IBD of alleles within a subpopulation (F_{ST}) as a function of the probability of IBD in the previous generation. This later probability (F_{ST}) is a measure of drift. To put inbreeding and drift together, we calculate the probability of IBD within individuals. This probability is defined relative to the total population, so it is written as F_{IT}. F_{IT} is the probability of getting IBD alleles in an individual due to inbreeding or due to drift. We need to be careful in applying the "or" rule here because it is possible for alleles to be IBD because of both inbreeding and drift—i.e., these are not mutually exclusive events. Consequently, we need to subtract out the "intersection," giving $F_{IT} = F_{IS} + F_{ST} - F_{IS}F_{ST}$. Usually, F_{IS} is so small relative to F_{ST} that we choose to ignore the effect of inbreeding and take F_{IT} as approximately equal to F_{ST}.

Another interesting example of two forces interacting is when there is drift because of population subdivision but also migration between the subpopulations. In this case, we need to adjust equation (3.14), which gives the probability of IBD for two alleles within a subpopulation, to account for the fact that neither allele can have migrated in from another subpopulation. The probability that neither allele is a migrant is the probability that the first is not a migrant and the second is not, or $(1 - m)^2$. Putting this together with equation (3.14) gives the following:

$$F'_{ST} = \left[\frac{1}{2N} + \left(1 - \frac{1}{2N} \right) F_{ST} \right] (1 - m)^2 \qquad (3.16)$$

In Figure 3.21, we graph F_{ST} across time for the case where $N = 100$ and $m = 0.02$ (i.e., 2 out of the 100 individuals in a subpopulation migrate per generation). Drift acts to decrease total heterozygosity, but migration acts to restore total heterozygosity. As a result, F_{ST} does not increase to 1.0 but instead reaches a lower equilibrium at about 0.11. In this model, we have assumed an infinite number of subpopulations, and we have also assumed that a migrant is as likely to migrate into one subpopulation as another. In other words, in the model there is no spatial structure to migration, while in actuality people typically do not migrate great distances to find mates. In the chapter on population structure (Chapter 12), we consider more realistic models for human migration.

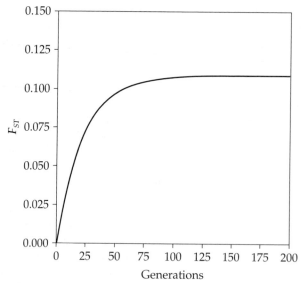

Figure 3.21. Balance between drift and migration leading to an equilibrium F_{ST} of 0.11. The example shown is for subpopulations of size 100 that exchange 2 migrants per generation. There is no spatial structure to the migration in this model.

CHAPTER SUMMARY

This chapter has presented population genetics models that will be useful throughout this book. Population genetic models can be derived using a modicum of probability theory, and in particular what we refer to as the "or" rule and the "and" rule. The "or" rule says that the probability of one "*or*" another event occurring is equal to the sum of individual probabilities. The "and" rule says that the probability of one "*and*" another event occurring is equal to the product of the individual probabilities, provided that the events are independent. From these rules we can rather easily derive the Hard-Weinberg expectations, which state the expected genotype frequencies given known allele frequencies. Although most introductory texts only consider loci with two alleles, the "and" rule and "or" rule can be easily applied to any number of alleles at a locus. It can also be shown (either by working an example or using some algebra) that it takes a single generation for the genotypes to come to the expected Hardy-Weinberg expectations if they were not already there, and that the allele frequencies will not change. Because neither the allele frequencies nor the genotypes change, we can refer to a population as being in Hardy-Weinberg "equilibrium." This state of evolutionary stasis is the backdrop against which we can view deviations from Hardy-Weinberg equilibrium, and hence the process of genetic evolution (change in allele frequencies).

Before looking at processes that alter allele frequencies, we need to summarize some special cases where genotypic frequencies are not in Hardy-Weinberg equilibrium but the allele frequencies are unchanging. In (positive) assortative mating, there are fewer heterozygotes than one would predict from Hardy-Weinberg, but the population maintains constant allele frequencies. Similarly, with inbreeding there is a reduction in the number of heterozygotes, but again the allele frequencies remain unchanged. Assortative mating and inbreeding differ from the true four evolutionary forces, which not only cause departures from Hardy-Weinberg expectations but also lead to allele frequency change. The first force to consider is mutation, which clearly can increase or decrease the frequency of allele. Typically, mutation occurs at too low of a rate to be a driving force in evolution, but mutation is critical to all subsequent evolution, providing the raw source for new alleles. Natural selection, the second force to consider, can increase or decrease allele frequencies, and when combined with mutation can be a powerful evolutionary force. Genetic drift and migration are the third and fourth evolutionary forces to consider. Neither force has a predictable effect as far as whether it will increase or decrease a particular allele frequency, but both affect the structure of genetic variation and lead to departures from Hardy-Weinberg equilibrium.

SUPPLEMENTAL RESOURCES

Lewis P O and Zaykin D (2001) Genetic data analysis: Computer program for the analysis of allelic data. Version 1.0 (d16c). Free program distributed by the authors over the internet, http://lewis.eeb.uconn.edu/lewishome/software.html. GDA is a program for analyzing population genetics data.

Schneider S, Roessli D, and Excoffier L (2000) Arlequin ver. 2.000: A software for population genetics analysis. Geneva: Genetics and Biometry Laboratory, University of Geneva, Switzerland, http://anthro.unige.ch/arlequin. Arlequin is a general program for analyzing population genetics data.

http://konig.la.utk.edu/relethsoft.html. Site for downloading Micro, a computer simulation of evolutionary processes.

http://ucsu.colorado.edu/~lemmon/EvoTutor/EvoGen/EG1A.html. An online Java-based simulation of evolutionary processes.

http://www.modares.ac.ir/elearning/mnaderi/Genetic%20Engineering%20course%20II/Pages/history_of_genetics2.htm. A site with a very brief history of Reginald Crundall Punnett of Punnett square fame, which also points out his contribution to the development of the Hardy-Weinberg law (he brought the problem to G. H. Hardy's attention).

Section 2

Variation in Genes, Simple Genetic Traits, and DNA Markers

4

Blood Group Polymorphisms

As we walk down the street or sit in an airport, we notice that people are biologically very diverse. Their skin colors vary, they have straight or wavy hair, they have blue or brown eyes, and they vary in height and weight. We often think about these differences in terms of race or ethnic group: that person who just walked by must be from one race because she has white skin, wavy hair, and black eyes and is about five and a half feet tall. Since we are such visual creatures, seldom do we think about the biological variation that exists under the skin of the people we see or the variation we see in terms of its evolutionary origin or adaptive significance. Most anthropologists, however, view biological differences and variation in just this manner, asking such questions as, How did that variation arise? How much diversity exists in the population? How is that specific trait maintained in a population? Is that trait adaptive, and if so, what are the selective factors that mold it? In this chapter, we will explore human biological diversity as it is expressed in the blood. We will examine how these blood group traits vary over geographic space and if we have evidence of their adaptive nature. Later, in Chapters 5 and 12, we will show how these genetic traits can be used to study human diversity.

GENETIC POLYMORPHISMS IN THE BLOOD

The blood group variations that we will examine in this chapter are known as genetic **polymorphisms**. A *polymorphism* is a discrete genetic trait (monogenic, or single-gene, trait) that exists in a population in at least two forms. That is, there is more than one allele at a gene locus in the population. By convention, but also somewhat arbitrarily, the frequency of the rarest allele must be no less than 1%. Thus, a locus with allele frequencies of $A = 0.9$ and $a = 0.1$ would be polymorphic. However, a population with allele frequencies of $A = 0.999$ and $a = 0.001$ would not be polymorphic. As we saw in Chapter 3, mutation rates are low and do not lead, by themselves, to major changes in allele frequencies in a population. Thus, allele frequencies above 1% are explained by the other mechanisms of microevolution: natural selection, genetic drift, or gene flow.

Sometimes the term **genetic marker** is used to describe the genetic variation (polymorphisms) that we will examine in this chapter and in Chapters 5–7 (e.g., plasma proteins, red cell enzymes, and human leukocyte antigen [HLA]). In some cases, these Mendelian traits are also called "classical" genetic markers, to contrast them with the newer molecular polymorphisms and DNA markers (Chapter 8). Please also be aware that the term *genetic marker* will be used very differently in molecular genetics compared to this context.

Genetic polymorphisms can be found in body fluids and almost every cell of the human body. These polymorphisms are detected by various means. This chapter will detail the genetic diversity known as the blood groups (**antigens**). Chapter 5 will examine other genetic polymorphisms in the blood, plasma proteins, and enzymes.

The first human polymorphism discovered was the ABO blood group (Landsteiner 1900, 1901). This initial discovery opened an unseen world of biological diversity for anthropologists and geneticists to study (Table 4.1). Landsteiner's discovery was accomplished

Table 4.1 Blood group systems

Name	Symbol	Date of Discovery	Number of Antigens	Chromosome Location
ABO	ABO	1900	4	9q34.1-q34.2
A₁A₂		1911		
H	H	1952	1	19q13.1-qter
MN	MN	1927		
MNS	MNS	1947	40	4q28-q31
Uu		1953		
P	P1	1927	1	22q11.2-qter
Rhesus	RH	1939	45	1p36.13-p34.3
Lutheran	LU	1945	18	19q12-q13
Lewis	LE	1946	3	19p13.3
Kell	KEL	1946	22	7q33
Kp		1956		
Sutter (Js)		1959		
Karhula (Ul)		1968		
Duffy	FY	1950	6	1q22-q23
Kidd	JK	1951	3	18q11-q12
Vel		1952		—
Diego	DI	1955	7	17q12-q21
Cartwright	YT	1956	2	7q22
Auberger		1961		—
Xg	XG	1962	1	Xp22.32
Dombrock	DO	1965	5	—
Stoltzfus		1969		4q28-q31
Colton	CO	1967	3	7p14
Landsteiner-Wiener	LW	1940	3	19p13.3
Chido/Rogers	CH/RG	1962/1967	9	6p21.3
Scianna	SC	1962	3	1p35-p32
Kx	XK	1961	1	Xp21.1
Gerbich	GE	1960	7	2q14-q21
Cromer	CROMER	1965	10	1q32
Knops	KN	1970	5	1q32
Indian	IN	1973	2	11p13
Secretor system[1]		1932	—	19

[1]Not a blood group system but related to ABO(H) and Lewis (see text).

Adapted from Crawford 1973, Le Pennec and Rouger 1995, Reid and Lomas-Francis 1997.

Blood Group	Antigens in Red Cell	Antibodies in Serum	Reaction patterns (agglutination) of anti-A and anti-B serum with A, B, AB, and O blood			
			O	A	B	AB
O	O	Anti-A Anti-B				
A	A	Anti-B				
B	B	Anti-A				
AB	AB	None				

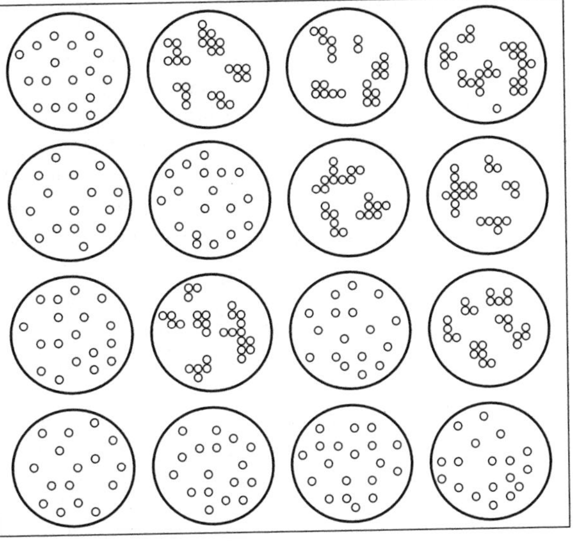

Figure 4.1. Determination of ABO blood types. (Adapted after Bodmer and Cavalli-Sforza, 1976, 215.)

by observing the cross-agglutination patterns when red blood cells of one individual were mixed with the sera of others. The reaction or clumping (called agglutination) that was observed occurs when antibody molecules attach to antigens that are located on the surface of the red blood cells. This attachment causes the red blood cells to form clumps, as depicted in Figure 4.1. The factors that hold the antigen and antibody together are both physical (complementary shape, sort of like a lock and key) and chemical (a relatively weak factor that is not well understood).

Antibodies are protein molecules that are produced in reaction to a foreign substance (antigen) entering the body. These foreign substances can be viruses, bacteria, another person's blood cells, or certain chemicals. The body responds to these invaders by producing antibodies (immune response). Thus, antibodies enhance the body's ability to fight infectious diseases by helping render viruses, bacteria, and other pathogenic organisms harmless. Sometimes antibodies neutralize toxins by eliminating them from the system or by blocking their actions (after binding to them). Often, antibodies serve as markers that attract other immune cells, such as phagocytes, which then render the invaders harmless and aid in their elimination from the body. This antigen–antibody reaction (*recognition*) can be used to detect the existence of human polymorphisms. Some of the antibodies used for detecting red cell blood groups come from people who have had reactions to blood transfusions or from women who have made antibodies in reaction to the antigens of their own fetus. Others are obtained by placing blood of one species into another (e.g., by immunizing rabbits with human group O cells).

ABO, SECRETOR (*FUT2*), Hh (*FUT1*), AND LEWIS (*FUT3*) SYSTEMS

The ABO Blood Group System

The ABO blood group system is the most studied and well known of the simple genetic traits in humans. This genetic system is located on chromosome 9, where it is closely linked to another genetic polymorphism, the red cell enzyme adenylate kinase (AK_1). There are three major alleles (*A*, *B*, and *O*) in the system: *A* and *B* behave as codominant alleles, and *O* is recessive. There are two basic antigens (A and B) and two basic antibodies (anti-A and anti-B). Unlike other blood group systems that will be discussed, the antibodies in the ABO system are referred to as "naturally occurring," even though this is really a misnomer. The antibodies of the ABO system are present in the blood before birth. It was initially thought that their expression was due to the action of a pair of linked genes (one for the antigen, the other for the antibody). It is now thought that the antigens are the result of gene action, while the immune system is responsible for producing the antibodies. The very early manifestation of these antibodies is thus due to exposure to various exogenous antigens (microorganisms, parasites), present in utero, that mimic the A and B antigens. This early exposure to antigens then produces the "natural" antibodies that can be detected in the blood (Springer and Horton 1969, Smith et al. 1983, Pittiglio 1986).

Since antibodies usually react only with specific antigens, it is possible to determine the blood type of individuals by combining their blood with anti-A and anti-B antibodies. For example, if we take blood from a person and expose it to both anti-A and anti-B antibodies and it clumps, we know the person is blood type *AB*. However, if the blood of the individual clumps (agglutinates) on exposure to anti-A antibodies but not anti-B antibodies, the person has A antigens and type A blood (see Table 4.2 and Fig. 4.1).

In actuality, the ABO system is more complex than depicted above. Blood group A has two principal subgroups: A_1 and A_2. Both of these subtypes agglutinate in the presence of anti-A antibodies, but only A_1 red blood cells will react with anti-A_1 reagents. Recent studies have shown that these differences are both qualitative and quantitative in nature (Clausen and Hakomori 1989). Hence, there are six possible ABO phenotypes: A_1, A_2, A_1B, A_2B, B, and O. The inheritance of A_1 and A_2 depends upon two alleles at the ABO locus, with A_1 being dominant to A_2 (Table 4.3).

There are also rare alleles or weak variants of both *A* and *B*. Some subgroups of *A* include A_3, A_x, A_{bantu}, A_{el}, A_{end}, A_{finn}, A_h, A_m, and A_y, and some subgroups of *B* include B_3, B_x, B_h, B_m, and B_{el} (Oriol 1995, Blancher and Socha 1997a). These subgroups primarily result

Table 4.2 ABO blood group phenotypes, genotypes, antigens, and antibodies

Phenotype	Genotypes[1]	Red Cell Antigens	Serum Antibodies
O	$I^O I^O$	None[2]	Anti-A, Anti-B
A	$I^A I^A$ or $I^A I^O$	A	Anti-B
B	$I^B I^B$ or $I^B I^O$	B	Anti-A
AB	$I^A I^B$	A and B	None

[1]Genotypes may be written either as $I^A I^A$ or as simply *AA*. The *I* stands for isoagglutinin.
[2]Individuals with O phenotype have H antigen.

Table 4.3 ABO blood group

Blood Group or Phenotype	Genotype(s)	Reaction with Anti-A Antibodies	Reaction with Anti-A$_1$ Antibodies	Reaction with Anti-B Antibodies
O	OO	No	No	No
A$_1$	A$_1$A$_1$, A$_1$A$_2$, A$_1$O	Yes	Yes	No
A$_2$	A$_2$A$_2$, A$_2$O	Yes	No	No
B	BB, BO	No	No	Yes
A$_1$B	A$_1$B	Yes	Yes	Yes
A$_2$B	A$_2$B	Yes	No	Yes

from single–amino acid substitutions. When examining human diversity, adaptation, and origins, anthropologists do not often study these rare variants but focus on the more common alleles or those that are polymorphic in populations.

Hh and Bombay

Red cells of people who belong to blood group O lack both antigen A and antigen B but possess an antigen known as H. This antigen depends on a separate gene, *H* (also known as *FUT1*), which is located on chromosome 19. Even though individuals with A and B blood have this gene, their red cells do not usually react to the presence of anti-H antibodies (A$_2$ individuals do, however, react to anti-H). Individuals who lack the *H* allele are very rare. These individuals, designated *hh* or O$_h$ (called Bombay), do not have any detectable A, B, or H antigen activity on the red cells. Their serum does, however, contain anti-A, anti-B, and anti-H antibodies. The Bombay phenotype apparently blocks the synthesis of A, B, and H antigens. The H antigen is an obligate precursor to both A and B antigens (see Fig. 4.2). Therefore, A and B antigens are not found in Bombay

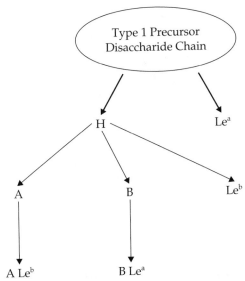

Figure 4.2. Basic biosynthesis pathways for ABH and Lewis antigens. (Adapted from Oriol 1995.)

Table 4.4 The secretor system

Phenotype or Secretor Status	Genotype
Secretor	*SeSe*
Secretor	*Sese*
Nonsecretor	*sese*

individuals (Pittiglio 1986, Gahmberg 1992, Oriol 1995). The Bombay variant reaches its highest frequencies around the city of Bombay, India (1/10,000 individuals), and on Reunion Island, just east of Madagascar (Gerard et al. 1982), where it is estimated that one out of every 100 individuals in the area of Cilaos has the Bombay variant.

An extremely rare type, called para-Bombay, is known in individuals who do not express A, B, or H on their red cells but do have H in their saliva. These individuals are genotypically *hh* and *Sese* or *SeSe* (see The Secretor System, below). O_h and para-Bombay are outcomes of point mutations (Reid and Lomas-Francis 1997).

The Secretor System

Red cell antigens A, B, and H can be detected in bodily secretions (saliva, urine, tears, seminal fluid, gastric juice, amniotic fluid) of individuals who are designated as "secretors." Thus, these body fluids can be tested for the presence of the A, B, and H substances. The secretor system segregates independently from this blood group system (i.e., the systems are not linked). Secretor status depends on the inheritance of at least one *Se* allele (also called the *FUT2* locus) (Table 4.4). Secretor and H are closely linked on chromosome 19 (Ball et al. 1991). The *Se* gene is considered a regulatory gene by most researchers; however, Oriol et al. (1981) have proposed a structural gene model in which the *Sese* and *Hh* genes are closely linked.

The Lewis System

The Lewis system (Lele or *FUT3*), even though located on chromosome 19, is distant from the secretor and H loci (Ball et al. 1991). The Lewis system contains two main antigens, Le[a] and Le[b], which give rise to three main phenotypes: Le(a+b−), Le(a−b+), and Le(a−b−) (Table 4.5). These phenotypes are produced by the epistatic interaction of two

Table 4.5 Genotypes, phenotypes, enzymes, and antigens of the lewis and secretor systems

Genotype		Enzymes Found		Phenotype		
Secretor	Lewis	Secretor	Lewis	Saliva	Plasma	Antigen
se/se	*Le/Le, Le/le*	−	+	Nonsecretor	Le(a+b−)	Le[a]
Se/Se, Se/se	*Le/Le, Le/le*	+	+	Secretor	Le(a−b+)	Le[b]
se/se	*le/le*	−	−	Nonsecretor	Le(a−b−)	Precursor
Se/Se, Se/se	*le/le*	+	−	Secretor	Le(a−b−)	H type 1

[1]A fourth Lewis phenotype, Le(a+b+), exists. It is possibly associated with the weak secretor (Se[w]). Adapted from Oriol 1995.

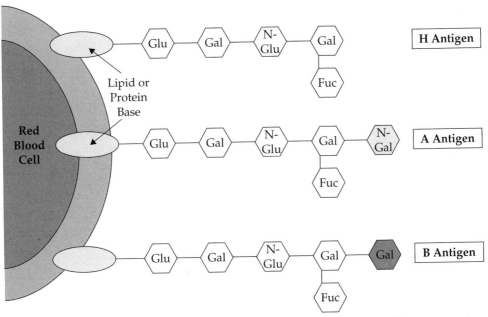

Figure 4.3. Chemical structure of the ABO blood group antigens. The *A* and *B* genes encode enzymes that add different terminal sugars to a common precursor substance, H antigen. Glu, glucose; Gal, galactose; N-Glu, *N*-acetylglucosamine; N-Gal, *N*-acetylgalactosamine; Fuc, fucose.

independent loci, the Le-le and the Se-se (secretor). That is, the products of two or more different genes interact to synthesize a more complex molecule (e.g., an antigen). The Lewis antigens are secreted in the saliva and plasma and subsequently absorbed onto the red cell membrane. Since the Lewis antigens are soluble, their production is also dependent upon the Hh and secretor systems. The Lewis types are closely related to the ABO blood group and the A, B, and H secretor types in a complex and interesting manner (Figs. 4.3 and 4.4).

The ABH and Lewis antigens are members of the same family of oligosaccharides (sugars). Their biosynthesis is intimately related to the Hh and Lewis blood group systems and the secretor system. The gene products of the ABO system transfer sugars to a precursor structure, which results in the synthesis of specific antigens (Oriol 1995). Prior to the action of the *A* and *B* genes, the Lewis, secretor, and Hh systems code for enzymes that add sugars to a common precursor substrate. In order to produce antigen H, the *H* gene codes for an enzyme (α-2-L-fucosyltransferase), whose function is to add a sugar (α-L-fucose) to the precursor substrate. In turn, the H antigen serves as a substrate for the action of the A- and B-produced enzymes, α-3-N-acetylgalactosaminyltransferase and α-3-D-galactosyltransferase, respectively (Fig. 4.3). The *h* gene does not produce an enzyme. Therefore, A and B antigens are not found in persons with the Bombay phenotype. Individuals who are phenotypically O do not show any A or B glycosyltransferase activity and express the unmodified H structure on their red cells. Hence, the H antigens present can be transformed into Le^b or Le^y by the addition of a second sugar (fucose) onto type 1 or type 2 precursor. As an added note, the term *histo-blood group antigens* is more

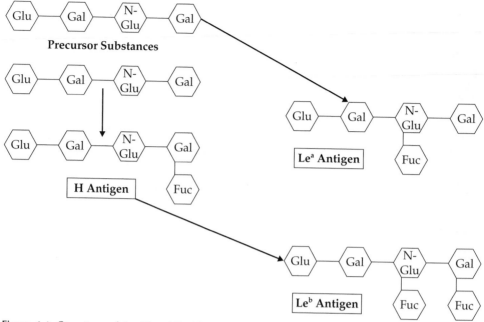

Figure 4.4. Structure of the H and Le blood group antigens. Two similar precursors can be converted into H antigen. If the *Le* gene is present, fucose is added to the top precursor to form Le^a. By adding another fucose to the H antigen, *Le^b* is formed.

accurate than simply *blood group antigens* since ABH and Lewis antigens are found in all organs of the human body (Oriol 1995).

GEOGRAPHICAL DISTRIBUTION AND NATURAL SELECTION OF THE ABO HISTO-BLOOD GROUP SYSTEM

The worldwide variation and clines for the *A*, *B*, and *O* alleles suggest that mechanisms of evolution (gene flow, genetic drift, and natural selection) have played a role in their distribution and frequency (Fig. 4.5). Worldwide, the *O* allele is the most common (about 63%), while *A* is next at about 21%, and *B* at 16%. There is, however, much variation, with the frequency of *O* ranging from about 40% to 100% in populations across the world. The other alleles do not reach 100% but are restricted in their frequencies: allele *A* ranges from 0.0 to 0.55 and *B*, from 0.0 to 0.4 (Brues 1977, Cavalli-Sforza et al. 1994). Since these frequencies are too high to be explained by recurrent mutation alone, other mechanisms of evolution must be shaping these distributions. In 1954, Alice Brues plotted the gene frequencies of *A* and *B* in 215 populations and found that they clustered in about one-fifth of the possible range, with neither exceeding 55% (Fig. 4.6). She argued that such a distribution could be explained by the action of balancing selection (heterozygote, *AO*, advantage). If the evolution of the ABO system were dependent upon gene flow and genetic drift, there would probably be a wide range of allele frequencies for *A* and *B*.

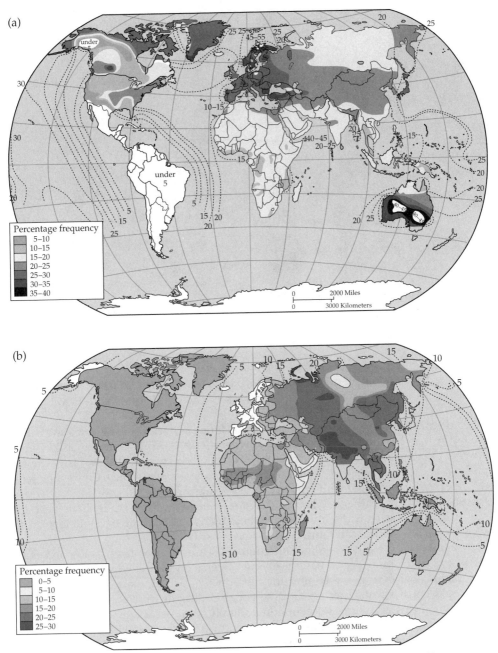

Figure 4.5. (a) Worldwide distribution of A blood. Note the high frequencies in Europe, Australia, and one region of North America. Also note low frequencies in the Americas and Africa (Mourant et al. 1976, Plate 1). (b) Worldwide distribution of B blood. Note the high frequencies in Asia and the low frequencies in Australia and the Americas (Mourant et al. 1976, Plate 2). (c) Worldwide distribution of O blood. Note the high frequencies in the Americas and low frequencies in Asia. (From Mourant et al. 1976a. Plate 3.)

(c)

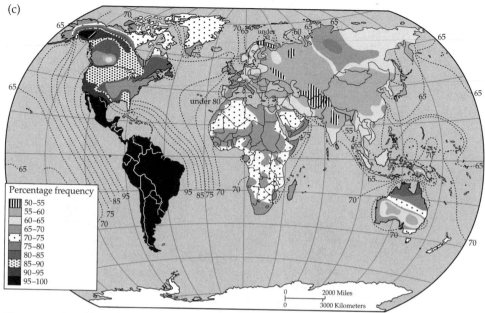

Figure 4.5. (cont'd)

Haldane (1949) had earlier suggested that many genetic polymorphisms were probably maintained by the action of natural selection, particularly disease selection. Soon, investigators began searching for adaptations and associations of diseases with blood group systems (Mourant et al. 1978).

Aird et al. (1953) reported the first association found between blood groups and a disease (other than **hemolytic disease of the newborn [HDN]**). They found an association between carcinoma of the stomach and A. Soon, other chronic disease associations appeared in the literature (see Table 4.6 for some of these associations). Cancers in general appear to be associated with group A, with a weaker association with group B, while O is associated with gastric and duodenal ulcers. Some researchers argue that cancerous tissue exhibits A-like antigens, thus making those with A blood more susceptible to these cancers since they do not carry anti-A antibodies.

A number of these chronic (noninfectious) diseases appear later in life and may not exert as much influence as natural selective agents on the gene frequency distributions. (Remember that mortality after the reproductive period, usually 15–45 years of age, would not have much effect on differential fertility or differential mortality, the major components of natural selection.) In addition, some associations (e.g., rheumatic fever, peptic ulcers) may be due to differences in the immune response and cell surface receptors (**glycoproteins**). For example, in the 1950s and 1960s, it was found that individuals with O blood were at an increased risk of getting duodenal and peptic ulcers. The mechanism for this association was elusive until recently, when it was found that the bacterium *Helicobacter pylori* is responsible, in part, for duodenal ulcers (and possibly gastric ulcers). In addition to this discovery, Borén et al. (1993) found that Lewis[b] (Le[b]) acts as a mediator for the attachment of *H. pylori* to the surface of mucosa cells. That is, *H. pylori* did not bind to Le[b] antigen in the presence of blood group A, suggesting that

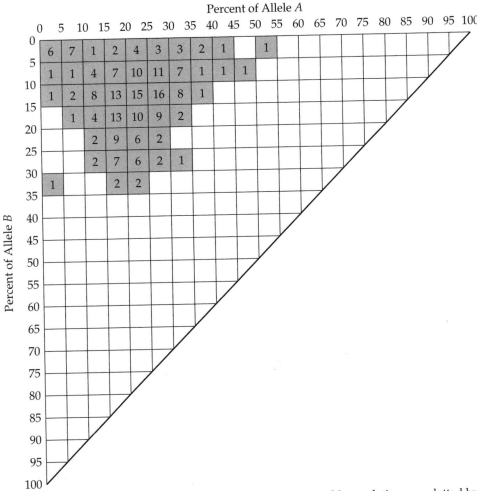

Figure 4.6. Limitations of *A* and *B* allele frequencies in 215 world populations, as plotted by Alice Brues in 1954. Note that the potential for both allele *A* and allele *B* is 100%, yet this frequency is not realized. Note also that most populations cluster within the ranges of 10%–35% for allele *A* and 5%–20% for allele *B*. (Adapted from Brues 1954, 560.)

type O individuals have more receptor sites for bacterial attachment and are thus more susceptible to gastric ulcers. Since Le[b] individuals are also secretors, secretor status may be involved in the susceptibility to *H. pylori* infection. However, Alkout et al. (2000) report that H type 2 antigen (the blood group O antigen) is a more efficient receptor for *H. pylori* than either Lewis[a] or Lewis[b]. They also found an increased density of colonization and greater inflammatory response to *H. pylori* infection in persons with blood group O. These associations are complex, and more work needs to be done on microorganism attachment and colonization of cells and on the interaction of various etiological factors.

Infectious diseases may present a different picture since many select individuals out of the population before they reach reproductive age. Differential mortality (natural

Table 4.6 Some significant (χ^2, $p < 0.05$) associations between blood groups and noninfectious diseases

Disease	Comparison	Relative Incidence[1]
Stomach cancer	A:O	1.22
Cancer of cervix	A:O	1.13
Malignant tumors of salivary glands	A:O	1.64
Nonmalignant tumors of salivary glands	A:O	2.02
Duodenal ulcers	O:A	1.35
	O:A+B+AB	1.33
Gastric ulcers	O:A	1.17
	O:A+B+AB	1.18
Rheumatic disease	A:O	1.23
Diabetes mellitus	A:O	1.07
	A+B+AB:O	1.07
Ischemic heart disease	A:O	1.18
	A+B+AB:O	1.17
Thromboembolic disease	A:O	1.61
	A+B+AB:O	1.60

[1]Relative incidence of $\text{A:O} = \dfrac{\text{A (Patients)} \times \text{O (Controls)}}{\text{O (Patients)} \times \text{A (Controls)}}$

Adapted from Vogel and Motulsky 1997.

selection) based on genetic variation in the blood types would be expected to influence genetic polymorphisms. Thus, recurrent epidemics of diseases such as smallpox, cholera, plague, and measles, which swept through continents, may have been influential in shaping the genetic landscape. We need to keep in mind, however, that these epidemic diseases (acute crowd diseases) are of recent origin (<10,000 years) and were not prevalent until after the development of a sedentary way of life and an increase in population densities greater than those found in hunting and foraging societies.

Organisms that cause some of these infections have been found to exhibit, on their outer surface, antigens that are similar in structure to the A, B, and H antigens on the surface of red blood cells. Hence, it is argued that our defense system would not recognize and produce antibodies against disease-causing organisms (antigens) that are similar to our own antigens. Thus, we would be more or less susceptible to various infections based on our blood type.

Recently, research has focused on cell receptor sites for pathogens. The capacity of pathogens to attach or adhere to the surface of cells of a host has been shown to be important both in contracting disease (microbial colonization) and in its virulence. Attachment results from an interaction of bacterial surface structures called adhesins (exact structures unknown) and host cell receptors. Since receptors consist of surface carbohydrates (either glycoproteins or glycolipids), blood groups probably play an important role in infection. Pathogens unable to colonize are removed by the mucous layer and by shedding of surface cells. Thus, it has been suggested that blood groups influence the susceptibility, severity, and mortality rates for many diseases. Some of these associations are discussed below.

Cholera

Two studies (Baru and Paguio 1977, Chaudhuri 1977) found that in areas where cholera was endemic patients who were hospitalized were more likely to have O blood than any of the other blood types. Research in Bangladesh showed a significant association between infections of *Vibrio cholerae* 01 and blood group O (Glass et al. 1985). In addition to a greater risk of infection, individuals with blood group O had increased severity of disease and risk of developing severe diarrhea. It is suggested that the low frequency of O and the high frequency of B in the Ganges Delta area (Fig. 4.5B, C) may be the result of the endemic presence of cholera adversely affecting these individuals. Why this association occurs is not known.

Syphilis (Treponemal Diseases)

In the immune response to syphilis, those individuals with O blood apparently have an advantage over those with A, B, or AB. Before the use of penicillin, tertiary syphilis was almost 1.7 times as frequent in (A+B+AB) individuals than O individuals. When neosalvarsan was used to treat syphilis, O group individuals became seronegative more often than individuals with other blood types (Vogel et al. 1960). Vogel and Motulsky (1997) suggest that the high frequency of O blood among native peoples of the Americas (Fig. 4.5C) may be related to natural selection by syphilis and other treponemal infections (pinta and yaws). If syphilis originated in the Americas, as many researchers contend, this is an interesting hypothesis. Even if syphilis originated in the Old World, there is ample evidence that treponemal diseases (pinta and yaws) were present in pre-Columbian America. How these diseases affected the survival of young children and the fertility of women is still not certain.

Plague

The frequency of O blood is relatively low in many parts of Europe and Asia (Fig. 4.5C). Is this distribution a reflection of the devastation wrought on this part of the world by plague epidemics in the past? Because humans and many bacteria share ABH-like antigens, researchers examined *Yersinia pestis*, the pathogen that causes plague (Vogel et al. 1960, Pettenkofer et al. 1962). An antigen very similar to H antigen was detected on the surface of *Y. pestis*, suggesting that an immune response would not occur among group O individuals because they possess H antigen and do not produce any anti-H antibodies. Thus, the plague pathogen, even though it is a "foreign substance," would not be recognized as such. As a consequence, individuals with O blood may have been at a selective disadvantage and died at higher rates than individuals with the other blood types. The distribution of O in central Asia and India and the existence of an H-like antigen on the surface of the plague bacterium are only indirect evidence for this hypothesis and should be viewed with caution.

Smallpox

The association between smallpox and A blood is controversial. Pettenkofer et al. (1962) report that smallpox scarring was much greater in individuals with A and AB blood types than other blood types. Experiments looking for ABH antigen activity showed strong A activity on the surface of vaccinia virus, a closely related virus to variola, the cause of smallpox (Pettenkofer and Bicherich 1960, Harris et al. 1963, Vogel and Motulsky 1997). It was then suggested that since A and AB individuals do not produce anti-A antibodies, they would be at a selective disadvantage where smallpox was present since they would be more susceptible to the virus. This finding had implications for smallpox susceptibility, severity, and mortality. Also, since smallpox primarily affects children, any differential selection according to blood type would shape the genetic structure of a region. It would be expected that the allele frequencies of *A* would be lower in regions experiencing recurrent smallpox epidemics. If we couple this with the actions of the plague, which should decrease the frequency of *O*, we should see a high frequency of *B* in the region where both diseases are present. In fact, this situation occurs along major river systems in India and China (Fig. 4.5B), suggesting that smallpox and plague have acted as selective agents molding the ABO blood group frequencies over the centuries.

The results of these studies were, however, criticized, suggesting that the detected A antigen was actually derived from the medium (egg) in which the virus was grown and not the virus itself (Springer and Wiener 1962, Harris et al. 1963). In addition, Krieger and Vicente (1969) were unable to demonstrate any association between smallpox and ABO in Brazil. However, two studies in India (Chakravartti et al. 1966, Vogel and Helmbold 1972), where smallpox was prevalent for ages, showed that individuals with A and AB were at a distinct disadvantage. A and AB individuals got smallpox more frequently, had much more severe cases, and died at a higher rate than those with O and B blood (Fig. 4.7). The attack rate, or incidence, of smallpox was four to six times higher among individuals with (A+AB) than those individual with (B+O) blood. Since smallpox has been eradicated, we will not be able to further test this possible relationship and settle the controversy.

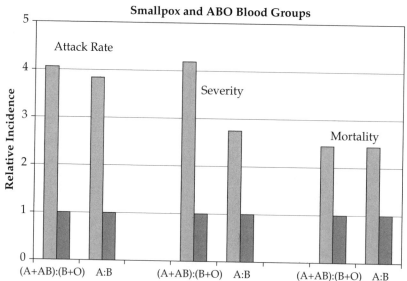

Figure 4.7. Smallpox attack rate, severity, and mortality. Comparisons of A blood type to B and O. (Adapted from Vogel and Motulsky 1997, 541.)

Escherichia coli and Infectious Diarrhea

The finding that plasma antibody levels to the pathogen *E. coli* 086B7 are lowest in blood group O individuals and highest in AB individuals suggests that individuals with O blood type are less able to produce antibodies to *E. coli*. In fact, it has been suggested that *E. coli* may carry an H-like antigen on its surface (Eichner et al. 1963). ABH substances and blood group P have been shown to act as cell surface receptor (attachment) sites for *E. coli*. Binding is essential, as has been shown for urogenital infections (Lomberg et al. 1986, Linstedt et al. 1991). Associations between diarrhea and blood group O may vary depending upon the strain of *E. coli* since some association studies have shown negative results (e.g., Van Loon et al. 1991).

Blood Group Associations

Associations between blood groups and diseases (or diets) are suggestive that some type of relationship between the two variables exists. Once a statistical association (X^2) has been established, a search for the cause of the association should begin. One should not accept these associations without some caution and further study. As early as 1967, Otten voiced criticisms and concerns about these studies and their implications. Potential biases may also affect these types of association study. For example, appropriate and unbiased controls must be used in all comparisons. There is a slight chance that only those researchers who found significant and "positive" results published their findings, while those who did not find any association did not publish. However, these possible biases are not very influential, especially given the consistency in negative association reports for many diseases using the same methodology and selection criteria. As an example of negative results, Vogel and Motulsky (1997) point out that even though 4,762 patients and 156,716 controls were studied, no associations were found between blood types and hydrocephalus, harelip/cleft palate, congenital heart disease, and malformation of the kidney and urinary tact.

Other Possible Selective Factors and the ABO System

Other selective forces maintaining the ABO polymorphism have been suggested: (1) preferential feeding patterns of mosquitoes, (2) differences in diet, and (3) variation in morphology and body dimensions or measurements.

Mosquitoes preferentially fed on individuals with O blood type over those with A or B blood (Wood 1974). The research suggests that in areas where malaria (or other mosquito-borne diseases) is endemic, this preference could contribute to selection against O individuals. More research needs to be conducted to clarify these findings since skin color, temperature, and body chemicals may influence the mosquito's choice.

Differences in diet and body shape (morphology) have been associated with the ABO blood group system. Kelso (1962) initially proposed that a high dietary intake of meat was associated with high frequencies of A blood, while high-carbohydrate diets were associated with B blood. Later, Kelso and Armelagos (1963) suggested that high A blood group frequencies may be linked in some manner to increased fat consumption. Variation in body morphology has also been associated with the ABO system. Beals et al. (1992) found that the B phenotype was the best predictor of weight and "ponderality" (or relative weight which is surface area/mass ratio). The reason for this relationship is unknown. These studies suggest that the ABO polymorphism may play a role in digestion, metabolism, and tissue recognition.

We cannot eliminate the role that other mechanisms of evolution, gene flow and genetic drift, may have had in shaping the genetic landscape. Many human populations are relatively small, and the continuous action of genetic drift or the intermittent influences of founder effects and bottlenecks have clearly been instrumental in increasing and decreasing the *ABO* alleles. As part of this exploration, we must also address the question of the longevity of the genetic structure of an area. The gene frequency clines that we observe today are the result of past events that shaped the **gene pool**. Gene flow, genetic drift, and natural selection all interact to mold the genetic structure of a region. These features can also be modified or blocked by human cultural behavior. In some cases, these evolutionary mechanisms work to increase gene frequencies and, at other times, to decrease frequencies. Selective agents may intensify or disappear over time. Is the genetic signature of a region wiped out by a few years of intense natural selection only to be reshaped later by drift or migration? Just how deep is the genetic structure of an area? These are important questions to ask when one is studying genetic polymorphisms, adaptation, and the genetic structure of an area.

THE Rh SYSTEM

The discovery of the Rh blood group system was an important scientific breakthrough because it finally explained some unexpected transfusion reactions and hemolytic disease of the newborn (HDN). In 1939, Levine and Stetson found a new antibody in a woman's serum that had crossed the placenta and caused her fetus's death by lysing the red cells and depriving the fetus of oxygen. A year later, Landsteiner and Wiener (1940) produced antibodies by injecting rabbits with rhesus monkey red cells. The immune serum that was produced agglutinated the red cells of 39 of 45 (~85%) humans. These 39 individuals were designated *Rhesus-positive*, or *Rh+*. The individuals who did not react were called *Rhesus-negative*, or *Rh−*. The antigens and specificities of the antibodies discovered by Levine and Stetson and those found by Landsteiner and Wiener were later compared and found to be similar but not identical. The clinically important antibody discovered by Levine and Stetson is now recognized as the Rh system, while the antibody discovered by Landsteiner and Wiener is now designated *LW* in their honor. Even though the Rh system is located on chromosome 1 and the LW system is on 19, there appears to be a close phenotypic association since rare individuals who do not express Rh antigens (called Rh_{null}) also lack LW antigens (Blancher and Socha 1997b).

For simplicity's sake, red cells are often subdivided into "Rh+" and "Rh−." This reductionism belies the complexity of the Rh system. The Rh+ designation comes from the presence of the major D antigen and Rh−, from the absence of D antigen. Two different models of the Rh system have been suggested. Wiener (1944) proposed a one-locus, eight-allele theory, while Fisher and Race 1946, (Race 1944) suggested a three-gene locus model. With either of these systems, there are eight recognizable gene complexes, or **haplotypes** (Table 4.7). We will use the Fisher-Race nomenclature, which defines the several antigens as a series: D, C/c, and E/e (The use of capital and lower-case letters does not, in this case, designate a dominant-recessive system but indicates antigens.) Genomic studies and Southern blot analysis have recently demonstrated that there are

Table 4.7 Common Rh gene complexes (haplotypes) according to the Fisher-Race and Wiener systems

Haplotypes		Antigens Produced
Fisher-Race	Wiener	
Dce	R^1	D, C, e
dce	r	c, e
DcE	R^2	D, c, E
Dce	R^O	D, c, e
dcE	r″	c, E
dCe	r′	C, e
DCE	R^Z	D, C, E
dCE	r^y	C, E

Adapted from Cartron and Agre 1995 and Blancher and Socha 1997a, b.

two structural RH genes: *RHD*, encoding the D polypeptide, and *RHCE*, coding for Cc and Ee proteins (Cartron and Agre 1995, Blancher and Socha 1997b). These genes are closely linked and inherited together as a single haplotype. The absence or presence of a functional *RHD* gene determines the Rh+/Rh− (D/d) polymorphism. In most cases, the *RHD* gene is not present; in some cases, it is partially deleted; and in other cases, it does not encode for the RHD protein (i.e., it is silent). These results suggest that rare mutations that completely or partially inactivate the *RHD* gene result in the RhD− phenotype. The order of the genes on chromosome 1 is not known, but with respect to the three-locus theory, the suggested transcription order is D → C → E.

Forty-five antigens have been identified in the Rh system, making it one of the more complex of the polymorphic systems in humans. The antigens are known by different names (Table 4.8). First assigned as Rh25 and Rh38, respectively, the LW and Duclos antigens are now considered separate blood group systems. Some of the 45 antigens of the Rhesus system are found in high frequency in many populations and are considered "public" antigens (e.g., total Rh, Hr, hr^s, Bas). Others (e.g., C^w, C^x, D^w, E^w, Hill-Hawd,

Table 4.8 Nomenclature for the RH antigens

Rosenfeld (Numerical)	Fisher-Race	Wiener
RH1	D	Rho
RH2	C	rh′
RH3	E	rh″
RH4	c	hr′
RH5	e	hr″
RH6	*cis* ce(f)	hr
RH7	*cis* Ce	rh_i
RH8[3]	C^w	rh^{w1}
RH9[3]	C^x	rh^x
RH10[3]	V (ce^s)	hr^v

Table 4.8 (*cont'd*)

Rosenfeld (Numerical)	Fisher-Race	Wiener
RH11[3]	E^w	rh^{w2}
RH12	G	rh^G
RH17[2]	(Cc)	Hr^0
RH18[2]	(Ee)	Hr, Hr^s
RH19		hr^s
RH20[3]	e^s (VS)	
RH21	C^G	
RH22	*cis* CE	rh
RH23[3]	D^w	Rh^{Wi}
RH24	E^T	rh^T
(RH25)[1]	LW	
RH26	c-like	hr^A
RH27	*cis* cE	rh$_{ii}$
RH28		hr^H
RH29[2]	total Rh	RH
RH30[3]	Go^a (D^{Cor})	
RH31	e-like	hr^B
RH32[3]	Troll-Reynols	R^N
RH33[3]	Hill-Hawd	Ro^{Har}
RH34[2]	Total Bastiaan (Bas)	Hr^b
RH35[3]	1114	R^N-like
RH36[3]	Be^a	
RH37[3]	Evans	
(RH38)[1]	Duclos	
RH39	C-like	Ht$_o$-like
RH40[3]	Tar	
RH41	Ce-like	
RH42[3]	Ce^s	hr^H-like
RH43	Crawford	
RH44[2]	Nou.	
RH45[3]	Riv.	
RH46[2]	Sec.	
RH47[2]	Dav.	
RH48[3]	JAL	
RH49[3]	STEM	
RH50[3]	FPTT	
RH51[2]	MAR	

RH13 to RH16 are now obsolete; RH13, RH14, RH15, and RH16 are Rh^A, Rh^B, Rh^C, and Rh^D, respectively. Defined as "cognates" of Rh$_o$ by Wiener (1943, 1944).

[1]LW(RH25) and Duclos(RH38) are now not part of the RH system.

[2]High-incidence antigens.

[3]Low-incidence antigens.

Adapted from Cartron and Agre 1995; Blancher and Socha 1997a,b; and Reid and Lomas-Francis 1997.

Go[a]) are found in low frequency and are called "private" antigens. This extreme complexity allows researchers to use the Rhesus system to analyze patterns of possible gene flow, genetic drift, and natural selection in many human populations.

The Rh System and Selection

HDN occurs when antibodies, derived from the mother by placental transfer, shorten the life span of the infant's red blood cells. The effect of the antibody transfer is variable. Some infants die in utero, while others appear normal. Rh-induced incompatibility between mother and fetus, known as **erythroblastosis fetalis**, is caused by antibody D, which crosses the placenta and reacts with red cells of the fetus (Fig. 4.8). This is the most common cause of severe HDN.

The possibility of producing a child with HDN occurs when the mother is Rh– and the father is Rh+, i.e., an incompatible mating because the mother lacks an antigen present in the father (Table 4.9). Because fetal red cells enter the mother's circulatory system primarily during labor, HDN is not common in the first pregnancy. In fact, during the

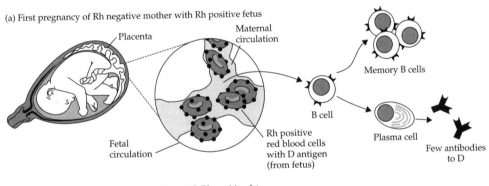

(a) First pregnancy of Rh negative mother with Rh positive fetus

Placenta

Maternal circulation

Memory B cells

B cell

Fetal circulation

Rh positive red blood cells with D antigen (from fetus)

Plasma cell

Few antibodies to D

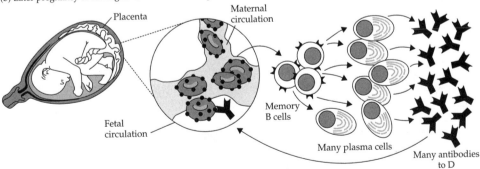

(b) Later pregnancy of Rh negative mother with Rh positive fetus

Placenta

Maternal circulation

Fetal circulation

Memory B cells

Many plasma cells

Many antibodies to D

Figure 4.8. The development of hemolytic disease of the newborn. (a) In a first pregnancy, some fetal red blood cells with the D antigen leak into the Rh-negative mother, mostly at childbirth. In response, the mother makes some anti-D and memory cells, but the baby is born before any significant damage is done. (b) In later pregnancy, leakage of fetal red blood cells from an Rh-positive fetus during pregnancy stimulates the mother's memory cells to quickly make much anti-D. Quickly entering the fetus, the anti-D destroys the Rh-positive fetal blood cells. (From Mange and Mange 1999, 405.)

Table 4.9 Mating types and possible progeny

Mother	Father		
	DD	**Dd**	**dd**
DD	DD	DD, Dd	Dd
Dd	DD, Dd	DD, Dd, dd	Dd, dd
dd	(Dd)	(Dd,) dd	dd

Those with potential Rh incompatibility (Dd) are circled. Note that these circled progeny are heterozygotes result-ing from the mating of a male who is DD with a female who is dd or a male who is Dd and a female who is dd.

first pregnancy, anti-D is found in only about 1.5% of dd mothers and the antibody con-centration is usually low (Hughes-Jones 1992). In western Europe, erythroblastosis fetalis occurs in about 5% of all Dd children with dd mothers. Because only Dd children are affected, this is selection against the heterozygote. Given this type of natural selection, we would expect the gene frequency of *d* in a population to decrease over time (unless other mechanisms counteract this trend). Presently, the gene frequency of *d* in western Europe is about 0.35. This is a high figure, given selection against the heterozygote. However, not all heterozygotes are subject to selection (see Table 4.9) since the mother is not always dd. ABO incompatibility may also provide some protection in that maternal anti-A and anti-B antibodies will destroy fetal red cells that escape into the maternal circulation before they can cause maternal immunization to the D antigens that are present in the same cells (Levine 1943, Race and Sanger 1950). These factors may explain why the Rh polymorphism (e.g. **RhE** and **RhC** varies in frequency in many populations (Fig. 4.9 A and B). Also, there may be other modes of selection that counteract the selection against the heterozygotes, or random mating may not have occurred long enough for the effects of selection against heterozygotes to be manifest. The population dynamics and the persistence of heterozy-gotes (Dd) in European and Europe-derived human populations remain to be fully explained.

THE MNSs BLOOD GROUP SYSTEM

Landsteiner and Levine (1927a,b) discovered two human red cell antigens (which they named M and N) by immunizing rabbits with human group O cells. The resultant anti-bodies defined three MN types in humans: M, MN, and N. The two codominant alleles, *M* and *N*, defined three genotypes: *MM, MN,* and *NN.* Walsh and Montgomery (1947) then discovered an antibody in the serum of a woman who gave birth to a baby with HDN. The antibody (anti-S) defined a new antigen (S). Shortly afterward, a correspond-ing antigen (s) was found. Two codominant alleles, *S* and *s,* code for these antigens. Both systems are located on the same chromosome (Fig. 4.10). Recombination occurs rarely between MN and Ss, indicating both their genetic closeness (**linkage disequilibrium**) and distinctiveness. Thus, the genes are transmitted as a unit or haplotype (*MS, Ms, NS,* or *Ns*) (Table 4.10).

The MNSs antigens are carried by major red cell glycoproteins. The protein carry-ing the M and N antigens is glycophorin A (GPA), while antigens S and s reside on glycophorin B (GPB). GPA consists of 131 amino acids. Antigens M and N differ by two

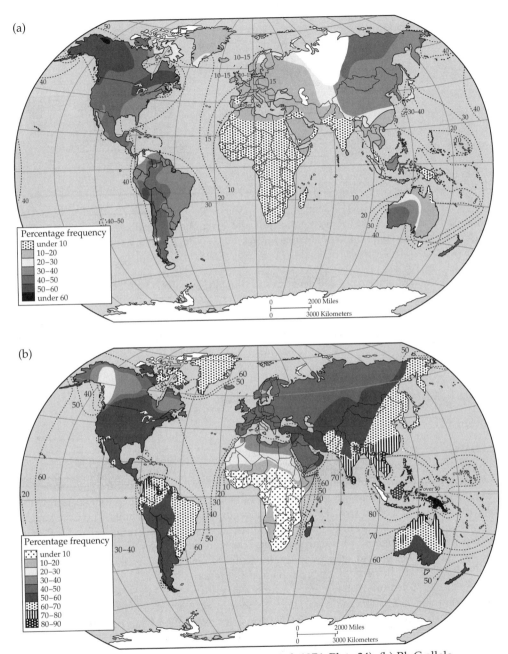

Figure 4.9. (a) Rh E distribution (From Mourant et al. 1976, Plate 24). (b) Rh C allele distribution. (From Mourant et al. 1976a. Plate 23.)

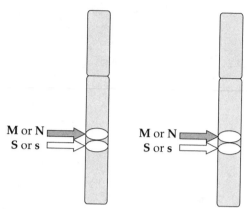

Figure 4.10. Location of MNSs (glycophorin A and glycophorin B) alleles on chromosome 4q28-q31. Possible combination of alleles *M, N, S,* and *s*: *MS, Ms, NS,* or *Ns*.

Table 4.10 Possible phenotypes and genotypes for the MNSs system

Phenotype	Genotype
MS	*MS/MS* or *MS/Ms*
Ms	*Ms/Ms*
MNS	*MS/NS, MS/Ns,* or *Ms/NS*
MNs	*Ms/Ns*
NS	*NS/NS* or *NS/Ns*
Ns	*Ns/Ns*

amino acid substitutions at positions 1 and 5 of the polypeptide chain (Fig. 4.11). Antigens S and s differ by a single substitution (methionine for threonine) at position 29 (Fig. 4.11). There is also a *GPE* gene, but it is not clear if its protein product occurs on the cell surface like the other two. The *GPA* gene occurs in anthropoid apes. The *GPB* gene is present in gorillas and chimpanzees but not in gibbons or orangutans. The *GPE* gene is found in chimpanzees but not all gorillas. Therefore, it appears that the duplication of the ancestral glycophorin gene was finally fully achieved in chimpanzees (Gahmberg 1992, Huang and Blumenfeld 1995, Socha and Blancher 1997).

At least 40 antigens have been identified within the MNSs system (Reid and Lomas–Francis 1997). The main variants of the MN system include M^c, M^r, M^z, M^a, N_2, and N^a. Most of these have limited distributions. For example, M^c has been found in a few families in Zurich and England. Antigen U is also associated with the MNSs system, and its expression appears to require the interaction of GPB and Rh-associated glycoprotein (GPRh$_{50}$) (Reid and Lomas–Francis 1997, Socha and Blancher 1997). Other rare antigens associated with the glycophorins include Wright, Vr, Ridley (Ria), Mg, M^v, St^a, SAT, Hunter (Hu), Henshaw (He), Hill, Dantu, Hop, and M^e. There are also some very rare null phenotypes, such as En(a–), S-s-U-, and M^k, which are characterized by the absence of one (either GPA or GPB) or all MNSs glycoproteins. These individuals are healthy, and their cells function normally.

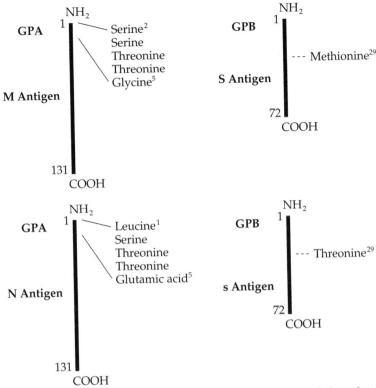

Figure 4.11. Molecular basis of the MN and Ss antigens (the structure of glycophorin A [GPA] and glycophorin B [GPB]). (Reprinted from *The Blood Group Antigenfacts Books*, Reid et al., 1997, with permission of Elsevier.)

Allele and haplotype frequencies for the MNSs system vary worldwide, suggesting a selective advantage of glycophorins during primate and human evolution. For example, allele *M* is high in Native Americans (70%–90%) and low among Australian Aborigines (10%–30%), while *S* is high in central Asia (+60%) and low in western Africa and Australia (10%–20%). Haplotype frequencies also vary, with *Ms* showing high frequencies in Southeast Asia (+70%) and low in Europe and Scandinavia (30%–40%). *NS* is high in central Asia (25%–30%), while *Ns* is high in Australia and New Guinea (70%–80%) and low (<20%) in Native Americans (Fig. 4.12) (Cavalli-Sforza et al. 1994).

In some cases, these regional frequency differences could be due to past gene flow or genetic drift. However, some microbes and parasites use glycophorin as a receptor, and bacterial adhesion to host cells is a necessary step in the establishment of many infections. Thus, it has been suggested that natural selection may be responsible for some of the geographic variation seen in the MNSs sytem. It has been suggested that glycophorin is a receptor for the parasite *Plasmodium falciparum* (Hadley et al. 1986, Hadley and Miller 1992). En(a–) individuals are relatively resistant to *P. falciparum* invasion, even though these cells can support parasite development (Pasvol et al. 1982). It has also been noted that the S–s– phenotype in central Africa correlates with the distribution and prevalence of malaria. Red blood cells deficient in GPA, En(a–), and lacking Wr[b] antigen are resistant

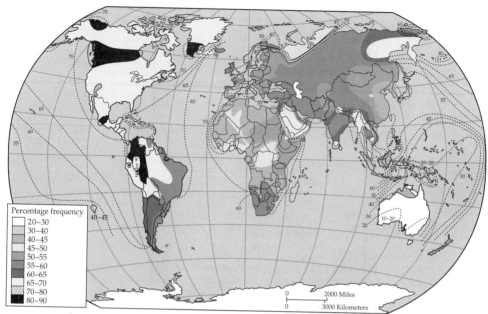

Figure 4.12. M blood group distribution. (From Mourant et al. 1976a. Plate 17.)

to *P. falciparum* invasion (Jokenin et al. 1985, Miller 1994). *E. coli* strain IH 11165 binds only to M-type molecules of GPA (Jokenin et al. 1985). GPA may also be a receptor site for influenza virus.

THE DUFFY BLOOD GROUP SYSTEM

The Duffy system consists of two principal codominant alleles, Fy^a and Fy^b, that differ by a single amino acid (Table 4.11). A third allele, Fy^o, corresponds to the Fy(a–b–) phenotype (i.e., absence of Fy antigen). Thus, antisera (anti-Fy^a and anti-Fy^b) give rise to four

Table 4.11 Duffy blood group system

Antigens	Antisera	Phenotypes
Fy^a	Anti-Fy^a	Fy(a+b–)
Fy^b	Anti-Fy^b	Fy(a–b+)
Fy^a/Fy^b	Anti-Fy^a/Anti-Fy^b	Fy(a+b+)
—	—	Fy(a–b–)
Fy3	Anti-Fy3	Fy3
Fy5	Anti-Fy5	Fy5
Fy6	Anti-Fy6	Fy6

Adapted from Pogo and Chaudhuri 1997.

major phenotypes: Fy (a+b−), Fy(a−b+), Fy(a+b+), and Fy(a−b−). In addition to the major alleles, five other antigens (Fy3, Fy4, Fy5, Fy6, and Fy^x, a variant of Fy^b) have been identified (Pogo and Chaudhuri 1997, Reid and Lomas-Francis 1997).

The frequency of the Fy^a antigen/allele is high in European and Asian populations and very low in African populations. Fy^b displays the highest frequencies in populations in Europe and of European descent. It is lower in Asian and African populations, averaging between 10% and 31%. The frequencies of Fy3, Fy5, and Fy6 are low in Africa (about 30%) and high (about 100%) in all other parts of the world (Reid and Lomas-Francis 1997).

The Duffy system provides a receptor site for the protozoal parasites *Plasmodium vivax* (causes vivax malaria) and *Plasmodium knowlesi* (monkey malaria). Individuals who do not react with either anti-Fy^a or anti-Fy^b (phenotype Fy[a−b−]) cannot be infected by *P. vivax*, and their red cells are resistant to the invasion of *P. knowlesi* parasites (Chaudhuri and Pogo 1995, Pogo and Chaudhuri 1997). In Honduras, about 59% of the population is Duffy-negative, and all vivax malaria cases are found in individuals who are Fy+. In West and East Africa, there is very little vivax malaria, and almost 100% of the population is Fy−. These findings, especially those in Africa, suggested to Livingstone (1984) that two scenarios were possible: (1) the high incidence of Fy(a−b−) indicated that over time natural selection increased the frequency of Fy− by favoring those who were Fy(a−b−) or (2) by chance, West and East African populations historically had high frequencies of Fy(a−b−) and this fact made it impossible for vivax malaria to invade the area.

Livingstone (1984) favors the second hypothesis because it is not likely that *P. vivax* originated in West Africa but in Asia and that it has been endemic to parts of Eurasia for years but the frequency of Fy(a−b−) phenotypes is low. It thus appears that the genetic makeup of a host population may be responsible for the spatial distribution of an infectious disease. This is indeed very interesting and a reversal of our typical reasoning about selective agents and the distribution of diseases. Thus, consider the following scenario. After vivax malaria (a **zoonotic** disease) arose and was established as a human disease in Asia, it probably started to expand into other parts of the Old World. As it moved into Africa, it spread until it encountered human populations that, by chance, had very high frequencies of the Duffy-negative allele. This genetic trait made it impossible for the protozoon to establish itself in these regions of Africa, thus limiting its range.

In a recent examination of DNA sequence variation around the Fy^o mutation, Hamblin and Di Rienzo (2000) explore the hypothesis that fixation of the Fy^o allele in Africa is due to directional selection (i.e., vivax malaria as the selective agent). They suggest that a "selective sweep" model may explain the patterns of variation linked to the advantageous Fy^o allele in Africa. This explanation would contradict Livingstone's hypothesis. It is also possible, but not likely, that the Fy^o allele "hitchhiked" to fixation due to selection at a neighboring gene (closely linked). Their analysis found two major haplotypes in Africa that probably originated prior to selection on the Fy^o mutation.

THE LUTHERAN BLOOD GROUP SYSTEM

Located on chromosome 19, the Lutheran system (LU) consists of at least 18 antigens. The function of this system is not known, but it may be involved in intracellular signaling and adhesion properties (Reid and Lomas-Francis 1997). There are no known disease associations. The major phenotypes are Lu(a+b–), Lu(a–b+), Lu(a+b+), and Lu(a–b–). The antigens or alleles of primary anthropological interest are Lu^a and Lu^b (the most frequent worldwide). The other antigens occur in either very high or very low frequency and are thus not that useful for describing human variation (Rao and Telen 1995). The Auberger system (Au^a, or Lu18, and Au^b, or Lu19, antigens) is also part of the Lutheran system. These genetic markers vary in frequency from area to area and have been used by some anthropologists and geneticists for population variation studies.

As a historical note, in 1951, the first autosomal linkage relationship to be suggested in humans was between the Lutheran and secretor systems (Mohr 1951a, b). In addition, in 1961, a Lutheran-negative phenotype (Lu[a–b–]) was reported (Salmon et al. 1961). This phenotype is inherited as either a dominant or a recessive trait; however, it is independent of the Lutheran genes. The gene was designated *In(Lu)* and became the first human regulatory gene to be described. The *In(Lu)* gene now appears to regulate not only Lutheran expression but also a number of other high-frequency antigens, including P1, i, AnWj, and In^b. AnWj antigen has been implicated as a *Hemophilus influenzae* receptor. In addition, an X-linked recessive Lutheran regulatory (suppressor) gene, *XS2*, was discovered, demonstrating that Lutheran antigens are involved in the expression of other proteins and antigens (Telen 1992, Rao and Telen 1995).

THE KELL SYSTEM

The Kell blood group system is located on the long arm of chromosome 7. Other blood group systems on chromosome 7 include Colton (CO, on the short arm) and Cartwright (YT). The genetic marker for phenylthiocarbamide tasting (see Chapter 7) has also been linked to the Kell system. Next to the ABO and Rh systems, the Kell system is the next most immunogenic, causing HDN and transfusion reactions if incompatibility exists. In

Table 4.12 Major antigens and phenotypes of anthropological interest of the Kell system

Antigens	Phenotypes
K and k[1]	K–k+[1]
	K+k–
	K+k+
Kp^a and Kp^b [1]	Kp(a+b–)
	Kp(a–b+)[1]
	Kp(a+b+)
Js^a and Js^b [1]	Js(a+b–)
	Js(a–b+)[1]
	Js(a+b+)

[1]Most frequent antigens and phenotypes. Some common or colloquial names for antigens in the Kell system include Kell (K), Cellano (k), Penny (Kp^a), Ratenberg (Kp^b), Peltz (Ku), Sutter (Js^a), Matthews (Js^b), Class (KL), Karhula (Ul^a), Coté, Bockman, Szro, Santini, Kx, K-like, Weeks (Wk^a), Marshall, Sublett, Km, Levay (Kp^c), and Ikar.

fact, this is how the system was discovered in 1946 because of HDN and the accompanying antibodies made by a patient named Mrs. Kell (Coombs et al. 1946). The product of the *KEL* gene is a glycoprotein whose function on the red blood cells is not known. Twenty-four antigens have been identified, and the major ones of anthropological interest are depicted in Table 4.12 (Marsh and Redman 1990, Redman and Marsh 1992). The Kell system may be associated with some microbial infections; however, details are lacking (Reid and Lomas-Francis 1997).

THE KIDD BLOOD GROUP

The Kidd (JK) blood group system, discovered in 1951, comprises three antigens (Jk^a, Jk^b, and Jk^{ab} or Jk3) and four major phenotypes, Jk(a+b−), Jk(a−b+), Jk(a+b+), and Jk(a−b−), which is very rare. The Kidd alleles are codominantly inherited in simple Mendelian fashion. As with Kell, antibodies were detected in the serum of a Mrs. Kidd, who gave birth to a child with HDN (Gunn et al. 1992).

OTHER BLOOD GROUPS

A number of other blood groups are of anthropological interest because they vary in frequency from area to area and can be used in describing the population structure, genetic affinities, and migration patterns of human populations. These blood group systems include the Diego, Cartwright, Scianna, Dombrock, Colton, Xg, and Landsteiner-Wiener. Antigens and chromosomal locations for these systems are listed in Table 4.13.

CHAPTER SUMMARY

As we have seen, the human blood group antigens are inherited as simple Mendelian traits. These traits allow anthropologists and geneticists to distinguish among individuals because of the complex molecules (antigens) residing on the surface of the red blood cells. The number of antigens and the complexity of the systems make the blood groups useful to anthropologists for a number of reasons. These genetic markers (polymorphisms) have been used for (1) describing the amount of genetic variation within a population, (2) comparing and contrasting the genetic diversity and structure between populations, (3) modeling population structure and population history (see Chapter 12 for more details on these subjects and the methods), (4) estimating admixture, and (5) studying natural selection and adaptation.

On a more practical basis, before the advent of DNA fingerprinting and other molecular genetic techniques, these "classic" genetic markers were also used for identifying paternity and eliminating suspects in criminal cases.

Why we possess these inherited antigen systems and why they are so diverse are difficult questions to answer. For example, what is the advantage of having O blood over A blood? More simply, why is there more than one blood type? We are now starting to learn that some of these variations have advantages over others in the efficiency of transport of various molecules throughout the body. Other variations may make it easier or more difficult for pathogens and parasites to establish a home within our bodies. That is,

Table 4.13 Other blood group systems of anthropological interest

System	Antigens/Alleles	Chromosome Location
Xg	Xg^a	Xp22.32
	Xg^b	
Diego	Di^a	17q12–21
	Di^b	
Cartwright	Yt^a	7p22
	Yt^b	
Scianna	Sm	1p36.2–22
	Bu^a	
	Sc3	
Dombrock	Do^a	Unknown
	Do^b	
	Gy^a	
	Hy	
	Jo^a	
Colton	Co^a	7p
	Co^b	
	Co^{ab}	
Landsteiner-Wiener	LW^a	19p13.2-cen
	LW^{ab}	
	LW^b	

some of these polymorphisms may act as receptor sites for bacteria and viruses entering the body. Given the number of associations between the blood groups and various diseases, some of the blood group antigens may give certain individuals advantages over others (differential morbidity and mortality) in certain environments.

Examples of how these blood group systems and other polymorphisms are used by anthropologists in studying diversity, population structure, genetic history, and adaptation will be presented at the end of Chapter 5 and in Chapter 12.

SUPPLEMENTAL RESOURCES

Agre P C and Cartron J-P, eds. (1992) *Protein Blood Group Antigens of the Red Cell: Structure, Function, and Clinical Significance.* Baltimore: Johns Hopkins University Press.

Blancher A, Klein J, and Socha W W, eds. (1997) *Molecular Biology and Evolution of Blood Group and MHC Antigens in Primates.* Berlin: Springer-Verlag.

Cartron J-P and Rouger P, eds. (1995) *Red Cell Biochemistry: Molecular Basis of Human Blood Group Antigens*, volume 6. New York: Plenum Press.

Cavalli-Sforza L L, Menozzi P, and Piazza A (1994) *The History and Geography of Human Genes.* Princeton: Princeton University Press.

Mourant A E, Kopec A C and Domaniewska-Sobczak K (1978) *Blood Groups and Diseases: A Study of Associations of Diseases with Blood Groups and Other Polymorphisms.* Oxford: Oxford University Press.

Reid M E and Lomas-Francis C (1997) *The Blood Group Antigen FactsBook.* San Diego: Academic Press.

Wallace M E and Gibbs F L, eds. (1986) *Blood Group Systems: ABH and Lewis.* Arlington, VA: American Association of Blood Banks.

5

Plasma Proteins and Red Cell Enzymes

When blood is placed in a test tube and spun in a centrifuge, a number of factors separate, revealing the structure of this vital substance (Fig. 5.1). Red blood cells **(erythrocytes)** settle at the bottom of the tube, white blood cells **(leukocytes)** and platelets occupy the middle, and the **plasma** rises to the top. When blood is allowed to clot, the clear liquid that separates from it is called the **serum**. Thus, *plasma* is the unclotted parent fluid of serum. Both the red cells and the plasma contain proteins and enzymes that are of interest to anthropologists in their studies of human biochemical variation. In this chapter, we examine the diversity in plasma proteins and red cell enzymes; but first, we will explore how these genetic polymorphisms are detected and how some are related to the immune system.

DETECTION OF GENETIC POLYMORPHISMS

To reveal the genetic variation present in plasma and red blood cells, researchers use a technique called **electrophoresis**. Blood is drawn from a number of individuals in a population that is being studied. A small sample of either red cells or plasma of each individual is then placed into separate slots or wells cut into a thin rectangular block of starch, agarose, or polyacrylamide along with a buffer solution (Fig. 5.2). The best analogy is to think of this block of starch or gel as a batch of potato Jell-O™. For several hours, an electrical current is passed through the gel. It must be noted that control of both the amps and the length of time are two important factors in this process. Since the amino acids making up the protein are both negatively (primarily aspartic acid and glutamic acid) and positively (mainly lysine, arginine, and histidine) charged, the protein moves, in response to the electricity, across the gel. The ratio of these positive and negative charges and their molecular sizes determine how far any protein migrates across the gel

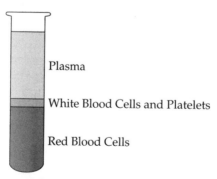

Figure 5.1. Major components of the blood. Plasma, which is about 92% water, contains such things as albumin, globulins, fats, proteins, fibrinogen, enzymes, amino acids, salt, and sugars.

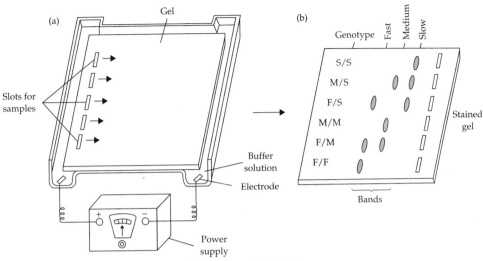

Figure 5.2. Electrophoresis. (From Strickberger 1995, 220.)

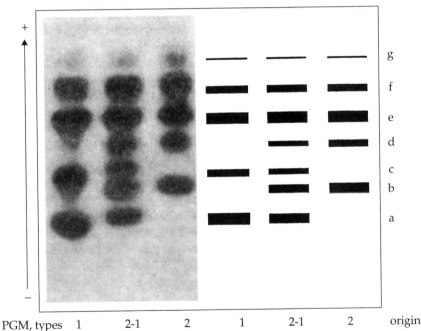

Figure 5.3. Example of starch gel electrophoresis, showing the actual bands and a diagram of the bands (resembling grocery store bar codes). Depicted are the three common types of red cell phosphoglucomutase (PGM$_1$ types 1, 2-1, and 2). (From Harris 1971, 46.)

(its *electrophoretic mobility*). After electrophoresis, the gel is stained to reveal a particular protein polymorphism. Like a grocery store bar code, the proteins appear as bands across the gel (Fig. 5.3). If one applies a stain for a specific protein, two or fewer bands appear; but if one uses a general protein stain (e.g., Coomassie blue), all proteins present in the gel can be detected. If there is any alteration in the amino acid content of the protein (e.g., a point mutation resulting in an amino acid substitution), the position on the gel will differ, representing genetic variation.

Enzyme polymorphism in natural populations is relatively well documented. Variation at these **loci** is often used as an indication of the amount of genetic variation in a population. Extensive enzyme variation has been found in almost all natural populations that have been studied using electrophoresis. Invertebrates have the highest average amount of genetic variation, followed by plants, and then vertebrates. Humans are typical of large mammals in having the least amount of variation. A number of studies (Harris 1966, Harris and Hopkins 1972, Harris et al. 1977) indicate that the average percentage of loci that are polymorphic in humans ranges from about 28% to 38%. The average heterozygosity per locus is about 0.07.

THE IMMUNE SYSTEM

To understand some of the genetic polymorphisms (e.g., immunoglobulins, human leukocyte antigen [HLA] system) discussed in this and other chapters, a very brief description of the human immune system is essential. As humans, we are constantly assaulted by a vast number of potentially infective agents. To combat these pathogens, the human genome has developed a complex defense system. This immune system is a collection of organs, cells, and molecules that collectively work to protect the human body from outside invaders such as bacterial, parasitic, fungal, and viral infections. It also protects us from our own altered internal cells (e.g., from the growth of tumor cells). As an interacting complex, it is often divided into two major components: (1) innate immunity and (2) acquired immunity (sometimes also referred to as "adaptive" or "specific" immunity).

Innate immunity is responsible for two things: (1) it prevents microorganisms from entering the body and (2) if the invaders have already gained access to the body, it eliminates them before they can cause disease. The innate system is the one we are born with, it is nonspecific (all invaders are treated in a similar fashion), and it does not become more efficient on subsequent exposure to the same invaders. Some components of this immune response include surface barriers such as the skin, mucus in the respiratory and gastrointestinal tracts, the complement system, and phagocytes that engulf and ingest foreign bodies. For our purposes, acquired immunity, or the specific immune response, is the most important because the responses are antigen-specific in nature.

Antigens are protein or carbohydrate molecules, or parts of molecules, that are attached to viruses, bacteria, or even pollen grains that can stimulate the immune system. Antigens can also be lipoproteins (protein and fat), nucleoproteins (protein and nucleic acid), and glycoproteins (protein and carbohydrate). The immune system is able to (1) recognize these types of substance as either "self" or "nonself" (i.e., foreign); (2) react in a specific manner to the substance, if it is foreign, and destroy it; and (3) remember a specific foreign substance so that it can react faster if future contacts occur.

The antibody-mediated (humoral) immune response involves the production of antibodies (proteins produced in response to specific antigens and capable of binding with those specific antigens), which are then secreted into the circulatory system. This defense system is responsible for primarily protecting the body against bacteria, fungi, and free viruses (those that have not infected cells). These antibodies bind with the

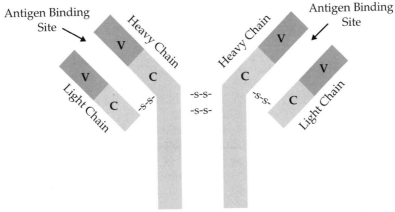

Figure 5.4. Basic structure of an immunoglobin (IgG) molecule, depicting the heavy (H) and light (L) chains with their variable (V) and constant (C) regions.

specific foreign antigens that have entered the body, allowing white blood cells to ingest and destroy this antibody–antigen complex, thus preventing any damage to the host.

Humoral defense antibodies are a class of proteins called immunoglobulins (Igs). They consist of four polypeptides, two identical heavy chains, and two identical light chains. Each chain has a constant region and a variable region (Fig. 5.4). The variable region contains the antigen-combining site and is responsible for most of the unique specificity of the antibodies. These antibodies are grouped into five classes: IgG, IgM, IgA, IgD, and IgE. Each class exhibits structural and functional differences (Table 5.1) that are based primarily on the makeup of the constant region of the heavy chain. The constant region of the heavy chain is of anthropological interest because this is the location of the Gm-Am polymorphisms. Also, the constant region of the light chain contains the Km system (see below).

Table 5.1 Classes and selected characteristics of immunoglobulins

Class	Relative Proportion	Location	Function
IgA	14%	Body secretions: saliva, tears, milk	Defense against antigens at entry sites into the body
IgD	1%	B cells, blood	May stimulate B cells to make other antibodies
IgE	<1%	Mast cells, tissues	Receptors for antigens, leading to histamine release by mast cells
IgG	80%	Blood, plasma cells, macrophages	Activates complement[1] during secondary immune response
IgM	5%	Blood, B cells	Activates complement during primary immune response

[1]The complement system consists of about 20 soluble proteins that circulate in the blood. In response to the antibody–antigen complexes, these proteins form large clumps that kill cells.
Adapted from Singer and Hilgard 1978, 329; Paul 1989, 226; Kuby 1992, 114; Leffell et al. 1997, 68–70.

The T cell–mediated (cellular) immune response is primarily responsible for protecting the body from viral infections and aberrant cells that could become cancerous. This system recognizes and kills infected host cells. Antigen specificity in the cellular defense system is provided by T-cell receptors. The T-cell receptors have a single binding site, not two as for antibodies. The T- cells do not work alone but are assisted by cell surface macromolecules called *histocompatibility antigens*. These antigens are able to distinguish foreign ("nonself") cells from "self" cells. The most important of these antigens belong to what is called the major histocompatibility complex (MHC). In humans, these are the *HLAs*, which were discovered on the surface of white blood cells (leukocytes). We now know that these antigens reside on virtually all cells of the human body. Organ transplants between unrelated individuals are usually rejected because of the actions of this part of the immune system, which recognizes the transplanted tissue as nonself and then attempts to eliminate it from the body. The HLA system is highly polymorphic and very useful to anthropologists studying disease–gene associations, population diversity, origins, migrations, and natural selection (see Chapter 7).

SOME PLASMA PROTEINS

There are a number of plasma proteins (also called "serum" proteins) that are of interest to anthropologists studying human diversity, population structure, group affinities, and migration patterns (Table 5.2). All are detected with various electrophoretic techniques (Catsimpoolas 1980). For a comprehensive listing of the plasma proteins and their global distribution, see Roychoudhury and Nei (1988). This section will discuss a number of these proteins, their properties, and population distributions.

Table 5.2 Some serum (plasma) proteins

Name	Some Common Alleles
Albumin	AL^A, AL^M
α_1-Antitrypsin (α_1-protease inhibitor)	PI_1^M, PI_2^M, PI_3^M, PI^S, PI^Z
β-Lipoprotein allotypes	AG^x, AG^y
Lp system (apoliproprotein)	LP^a, LP
Ceruloplasmin	CP^B, CP^A, CP^C
Cholinesterase (CHE2)	
Complement component-3	$C3^S$, $C3^F$
Group-specific component	GC^{1F}, GC^{1S}, GC^2 (VDBP, VDBG)
Haptoglobins	HP^{1S}, HP^{1F}, HP^2
Immunoglobulins	
Gm-Am	$G1m^3$, $G3m^5$, $G1m^1$, $G1m^{1,2}$
Km (Inv)	Km^1, Km^2
Placental alkaline phosphatase	PL^F, PL^S
Properdin factor B (glycine-rich-β-glycoprotein)	BF^S, BF^F
Pseudocholinesterase (butyrylcholinesterase)	E1, E2 (CHE1)
Transferrins	TF_1^C, TF_2^C, TF_3^C, TF^B, TF^D
Xm group (α-macroglobulin)	XM^a, XM

Adapted from Crawford 1973, Mourant et al. 1976a, Vogel and Motulsky 1997.

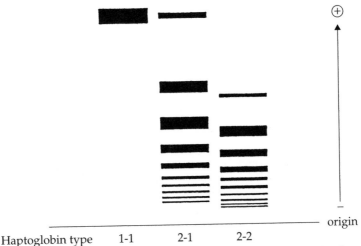

Figure 5.5. Diagram depicting the electrophoretic pattern in starch gel of the three common haptoglobin types. (From Harris 1971, 68.)

Haptoglobin (α_2-globulins)

Haptoglobins are glycoproteins (proteins with an attached carbohydrate) that are able to bind with dissolved hemoglobin. This action prevents hemoglobin from being excreted by the kidneys, accounting for a minor part of hemoglobin retention. The haptoglobin locus, on chromosome 16, is like many other proteins in that the molecule is composed of two kinds of polypeptide chain, α and β. It was initially thought that two loci were involved, one for the production of the α chain and the other for the β chain, but it now appears that there is a single locus (called Hp) for haptoglobin. Structural and functional similarities exist between the α-haptoglobin chain and the light chains of γ-globulins, suggesting a common evolutionary origin.

In 1955, Smithies, using the technique of electrophoresis, demonstrated that the haptoglobin system was polymorphic, consisting of three phenotypes: Hp 1-1, Hp 1-2, and Hp 2-2 (Fig. 5.5). Other variants have since been identified that are thought to be the result of unequal crossing over and independent mutation. There is also a modified Hp 2-1 that can be confused with Hp 2-1. In some individuals, there is no detectable haptoglobin pattern after electrophoresis. This is because the stain used relies on a hemoglobin/haptoglobin complex being recognized. If no hemoglobin binding occurs, then presumably no haptoglobin pattern is seen in the gel, even though hemoglobin will appear. This phenotype has been called Hp 0 or hypohaptoglobinemia (Kirk 1968). There are also fast (Hp^{1F}) and slow (Hp^{1S}) forms of α_1 that differ slightly in their amino acid composition. Black and Dixon (1968) report that the difference is due to a substitution of lysine (fast) for glutamic acid (slow) at position 54. However, Maeda (1991) suggests that Hp^{1F} and Hp^{1S} differ by two amino acids at positions 52 and 53 (Hp^{1F} has aspartic acid and lysine, while Hp^{1S} contains asparagine and glutamic acid, respectively).

The α_2 chain is found only in humans (it probably arose as a combination of two Hp1 alleles). The α_2 chains, because of their increased size, appear to reduce the loss of the hemoglobin–haptoglobin complex by the kidneys. This feature of the α_2 chain may

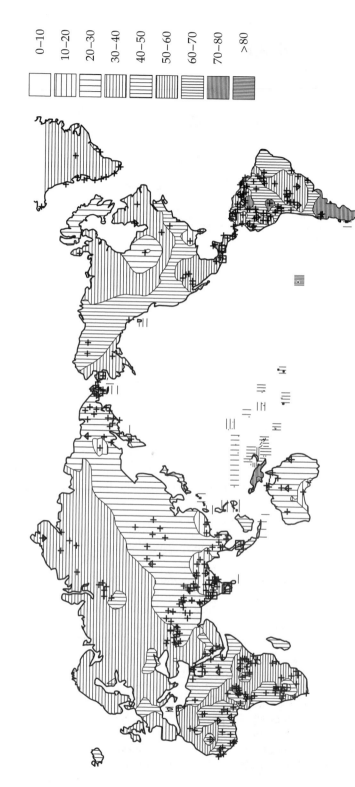

Figure 5.6. World map showing distribution of haptoglobin (Hp¹ allele). (From Cavalli-Sforza, Luca, *The History and Geography of Human Genes.* © 1994 Princeton University Press. Reprinted by permission of Princeton University Press.)

0–10
10–20
20–30
30–40
40–50
50–60
60–70
70–80
>80

explain the success of the mutation and possible selective advantage of Hp 2-2. However, Hp^1 is more efficient than Hp^2 at removing free hemoglobin (Mourant et al. 1978). Haptoglobin may play another significant selective role by limiting the usage of free, uncomplexed hemoglobin (not bound by haptoglobin) by pathogenic bacteria, thus guarding against potentially lethal infections (Eaton et al. 1982). There is a possibility that limiting the amount of free, uncomplexed hemoglobin may also deny iron to bacteria, which is essential to reproduction.

Serum haptoglobin levels rise in infection and inflammation. Specific disease associations with haptoglobin types are limited. For example, Hp^2 may give some protection against typhoid, a bacterial infection. A few studies suggest an association between Hp 1-1 and leukemia as well as breast carcinoma (Naik et al. 1979, Tsamantanis et al. 1980). These associations are complex and need more clarification. Clinical data demonstrate that individuals with Hp 2-2 phenotypes have significantly more severe myocardial infarctions than those with Hp 1-1 or Hp 2-1 phenotypes (Chapelle et al. 1982). However, no predisposition to myocardial infarction was associated with any of the phenotypes.

Allele frequencies (Fig. 5.6) for Hp^1 show a weak south to north trend in Europe, averaging about 40%. In India, the allele frequency is low, 15%, while in eastern Asia it is about 25%. Sub-Saharan Africans show much higher frequencies, averaging about 70%. There is, however, regional and cultural variation, with San (Bushmen) populations showing only 30% and Mbuti Pygmies exhibiting frequencies near 40%. In native North American populations, the frequencies vary between 30% and 50%, while in South America they are much higher, about 70%. Hp 0 averages between 0% and 5% in most parts of the world. Its frequency is, however, high in some parts of Africa, e.g., Liberia (about 20%), Gambia (40%), Uganda (24%–33%), and the Solomon Islands (27%) (Kirk 1968). High frequencies of Hp 0 seem to occur in regions with high frequencies of malaria. How these two factors interact or are associated is not known.

Transferrin

Transferrin (Tf), a glycoprotein consisting of 679 amino acid residues, combines with inorganic iron in the plasma and transfers it to all cells in the body as needed, especially the bone marrow, where hemoglobin is formed (Mourant et al. 1978, OMIM 2001). Since the time the system was first shown to be polymorphic (Smithies 1957), at least 30 genetic variants (mutational transitions, nucleotide substitutions, and insertions) have been identified (Welch and Langmead 1990).

Perusal of world Tf allele frequencies indicates that *TfC* occurs at a frequency of about 90% or more in most populations. Aside from this most common form, an electrophoretically faster migrating form (*TfB*) and a slower form (*TfD*) are often identified in low frequencies in many populations. Geographic variants (e.g., D_{Chi} and $D_{Montreal}$) and other rare types (e.g., TfC_1, TfC_2, TfD_1, TfB_2, $TfBv$, and TfB_{0-1}) have been identified by isoelectric focusing and nucleotide analysis (Mourant et al. 1976a, OMIM 2001). TfD_1 is found in sub-Saharan Africa, New Guinea, and Australia. TfD_{Chi} occurs in Finns, Lapps, southern and southeast Asia, some Pacific Islands, and Central and South America. TfB_{0-1} is rare, occurring in some native populations of North and Central America. Finally, TfB_2, also rare, is a variant found in European populations.

There is apparently little, if any, difference in iron binding capabilities among Tf variants. If the iron binding capabilities are essentially the same, why is *TfC* so frequent? In cattle, Tf is highly polymorphic, showing a number of alleles in high frequency. These

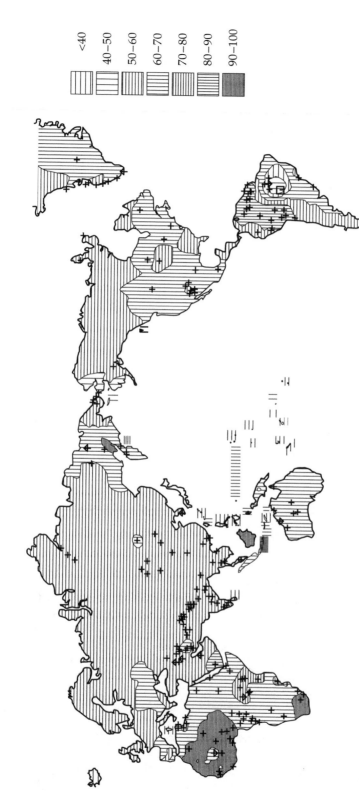

Figure 5.7. World map showing distribution of the Gc plasma protein system (Gc¹). (From Cavalli-Sforza, Luca, *The History and Geography of Human Genes.* © 1994 Princeton University Press. Reprinted by permission of Princeton University Press.)

cattle variants have been associated with a complex system of selective fertility and temperature. So, it is theoretically possible that some of the rare human Tf variants could have a selective advantage under certain circumstances (e.g., possibly in regions where humans exhibit chronic anemia) and may be transient polymorphisms. Possibly the most important feature of Tf (and Hp) is that it is what can be called an *acute-phase reactant* protein; i.e., the levels can rise very quickly. The chief role of this protein may occur during pregnancy, when levels rise and fall in different trimesters.

Group-Specific Component or Vitamin D–Binding Protein

In 1959, Hirschfeld, using electrophoresis, discovered a polymorphism of the serum α_2-globulin. Three phenotypes, Gc1-1, Gc2-2, and Gc2-1, were distinguished using starch gel or agar electrophoresis. The name *Gc*, or group-specific component, was given to this protein. During the same year, another plasma protein, vitamin D–binding protein or vitamin D–binding α-globulin, was found. These proteins were later shown to be identical (Daiger et al. 1975). Aside from the three common alleles (Gc^2, Gc^{1F}, Gc^{1S}), more than 120 variant alleles have been identified (Kofler et al. 1995).

With some exceptions, high frequencies of Gc^1 are found in populations where it is sunny and lower frequencies where it is not as sunny (does not hold in Ireland, but this is a small area geographically) (Mourant et al. 1976b). This relationship suggested that Gc types might be molded by natural selection and the availability of vitamin D. More research must be done to confirm this possibility.

Disease association studies are limited in number. Gc1-1 has been tentatively associated with psoriasis. There is a very weak association between Gc1-1 and cancer of the uterus, diabetes mellitus, and rheumatoid arthritis (Cleve 1973).

Frequencies for the vitamin D–binding alleles vary across the world. As can be seen in Figure 5.7, the frequencies for Gc^1 vary from lows of around 40% in South America to highs of above 80% in parts of Africa. Many indigenous Australian and Papua New Guinean populations have the Gc^{Ab} allele in addition to Gc^1 and Gc^2. North American populations are relatively uniform, however; e.g., the Chippewa have about 10% Gc^{Chip} and about 20% Gc^2.

Pseudocholinesterase (Butyrylcholinesterase and Cholinesterase)

Cholinesterase that is present in the serum is called butyrylcholinesterase (BCHE) or pseudocholinesterase (E1 or CHE1 locus). Four allelic forms of the gene for pseudocholinesterase are known: (1) the "normal," E_1^u ($CHE1^*U$), which is in high frequency in most populations of the world (Fig. 5.8); (2) E_1^a, which shows reduced BCHE activity; (3) E_1^f allele; and (4) the E_1^s "silent gene," responsible for the absence of cholinesterase activity. The fourth allele is the one most often responsible for suxamethonium sensitivity.

Mutations at the E1 locus (located on 3q26.1–q26.2) cause sensitivity to suxamethonium, a presurgery muscle relaxant. Individuals who are homozygous ($E_1^s E_1^s$) show prolonged apnea after being administered suxamethonium. Pseudocholinesterase activity in the serum is low. Reaction to succinylcholine and length of apnea vary, suggesting phenotypic diversity in pseudocholinesterase deficiency. This is the only known disadvantage of this phenotype.

Figure 5.8. World map depicting distribution of pseudocholinesterase. (From Cavalli-Sforza, Luca, *The History and Geography of Human Genes.* © 1994 Princeton University Press. Reprinted by permission of Princeton University Press.)

There is also an *E2* gene (*CHE2*) that produces an extra enzyme component, C5. In most populations, the *E2* gene is silent. However, some European and Europe-derived populations carry the gene that specifies the C5 component in low frequency (about 10%). This gene has been mapped to 2q33–q35 (OMIM 1998a).

Immunoglobulins (Gm–Am and Km)

Genetic markers of the Gm-Am system are located on the constant region of the heavy chains of immunoglobulin molecules (see Fig. 5.4). The Gm system involves IgG (or *IGHG*), while the Am system is associated with IgA (or *IGHA2*). These genes appear to be very closely linked on chromosome 14 (q32.33). The Gm system is a highly polymorphic system that has been successfully used to interpret human population variation and origins. The heavy chain that is unique to immunoglobulin μ (IgM or *IGHM*) is also coded for on chromosome 14q32.33 (OMIM 1998b).

Originally called the Inv locus, the Km system is associated with the constant region of the κ light chain (*IGHK*) of immunoglobulin molecules. A gene on chromosome 2 encodes this system. Three alleles, *Km1*, *Km2*, and *Km3*, have been confirmed. Family and population studies have shown numerous associations between specific Gm and Km **allotypes** or haplotypes (groups of markers that are inherited as a block) and a large number of autoimmune, malignant, and infectious diseases (see Whittingham and Propert 1986 for a summary). The selective significance and impact of these associations is still unclear.

Human populations often exhibit unique Gm allotypes, or they display markedly different frequencies of a number of different allotypes. Recombination within the Gm system is rare, and these haplotypes show considerable time depth in many populations, making them excellent markers for studying human population differentiation, origins, and migrations. For example, Zhao and Lee (1989) used Gm haplotypes to study the origins of the Chinese. They suggest that the Chinese nation is composed of two distinct subgroups, one originating in the Yellow River valley and the other in the Yangtze River valley. The northern group is characterized by haplotype Gm (1;21), while the southern is distinguished by Gm (1;3;5). They also attribute the presence of Gm (3;5) to admixture with various populations along the Silk Road and in northwest China. Other studies have used these systems to examine Basque diversity and possible origins (Calderón et al. 1998, 2000), delineate Mennonite and Amish population structures (Martin et al. 1996), reveal genetic differentiation among four Sardinian populations (Hubert et al. 1991), and estimate admixture rates among Mexican Americans for a study of gallbladder disease (Tseng et al. 1998). For a listing of Gm population haplotype frequencies, see Roychoudhury and Nei (1988).

Apolipoproteins, β-Lipoproteins, and β₂-Glycoproteins

A number of plasma proteins that are associated with lipid metabolism and components of low density (LDL), high density (HDL), or very low density (VLDL) lipoproteins are of anthropological value because of their polymorphic nature and association or susceptibility with various disorders (e.g., arteriosclerosis, coronary artery disease, and Alzheimer's disease). These genetic systems include apolipoprotein (a) or LPA (also called a β-lipoprotein, specifically the Lp system), A-I (*APOA1* gene), A-II (*APOA2*), A-IV

(*APOA4*), A-V (*APOA5*), B (*APOB,* also known as a β-lipoprotein, specifically the Ag system), C-I (*APOC1*), C-III (*APOC3*), C-IV (*APOC4*), D (*APOD*), E (*APOE*), H (*APOH* or β$_2$-glycoprotein I, specifically the Bg system), and a β-lipoprotein, specifically the Ld system. These systems have been used in some anthropological studies to explore population variation, relationships, and natural selection. We focus here on only one of these systems, apolipoprotein E (APOE), because it illustrates a number of lines of research related to human variation. This focus does not negate the usefulness of the other systems (e.g., see Kamboh et al. 1996, Ali et al. 1998, Corbo et al. 1999).

Apolipoprotein E (apoE is the protein) is important in determining plasma lipid levels, lipid transport within tissues, and cholesterol absorption for the intestine. The gene (*APOE*) has been mapped to chromosome 19q13.2, where three common alleles (*APOE*2*, *APOE*3*, and *APOE*4*) code for three isoforms of apolipoprotein E (apoE2, apoE3, and apoE4) (OMIM 2001, Corbo and Scacchi 1999). There are more variants; in fact, de Knijff et al. (1994) list 30 that have been characterized. Allele *APOE*3* (or simply *ε3*) is, however, the most frequent in most populations, ranging 48%–91% (Corbo and Scacchi 1999). Except for the Inuit, native Americans are characterized by the absence of the *ε2* allele (Corbo and Scacchi 1999). The ubiquity, frequency, and negative correlation of *ε3* with the *ε4* allele suggest that *ε3* may confer some unknown selective advantage on its carrier relative to *ε4* (Gerdes et al. 1996). The mutation (112 arginine → cystine) creating the new *ε3* allele appears to have progressively replaced the older *ε4* allele. They further suggest that the advantage may be related to the food system and climate in some manner. Aboriginal populations residing south of the 20th or north of the 40th latitude and hunting–gathering and nomadic societies have higher frequencies of *ε4* than do long-established agricultural groups (Gerdes et al. 1996). Corbo and Scacchi (1999) suggest that *ε4* could be identified as a "thrifty" allele (à la Neel 1962). That is, upon exposure to a Western diet (and greater longevity), individuals carrying the *ε4* allele would be susceptible to coronary artery disease and Alzheimer's disease. These authors state that the absence of an association between *ε4* and these diseases in sub-Saharan Africans coupled with the presence of *ε4* in African Americans confirms this "thrifty" allele hypothesis.

The relationship between apolipoprotein E and plasma cholesterol and triglyceride levels is complex. In general, regardless of the location or populations, individuals with genotype *ε2ε2* or *ε2ε3* have higher triglyceride levels and lower mean plasma cholesterol levels than homozygous *ε3ε3* individuals. Further, carriers of *ε4* (*ε4ε4* or *ε3ε4*) generally have both high cholesterol and high triglyceride levels (Gerdes et al. 1996). Apolipoprotein E appears to be associated with cardiovascular disease, abnormalities in blood lipids, Alzheimer's disease, and other neurological disorders. The relations are complex and very controversial and may involve other genetic or environmental factors either in addition to or modifying the action of the *APOE* gene. A number of studies (e.g., Corder et al. 1993, Poirier et al. 1993, Talbot et al. 1994, Myers et al. 1996) have shown a relationship between Alzheimer's disease and apolipoprotein E. The *ε2* allele may confer some advantage by protecting against Alzheimer's disease. However, patients who were homozygous or heterozygous for *ε4* showed an increased risk for Alzheimer's disease and other dementias. Compared to the *ε3* allele, *ε2* and *ε4* were associated with an increased risk of ischemic heart disease (Smit et al. 1987). Studies of *ε4* and its role in Creutzfeldt-Jakob disease show conflicting results (e.g., see Saunders et al. 1993, Amouyel et al. 1994). Many of these are late-onset diseases and, as such, play little role in natural selection and evolution since they affect individuals in the postreproductive period.

Ceruloplasmin

The ceruloplasmin, CP, locus has been mapped to chromosome 3 (q23–q24). At least three variants can be detected by starch gel electrophoresis. CP is a glycoprotein or plasma metalloprotein that is involved in plasma copper transfer and peroxidation of Fe(II) transferrin to Fe(II) transferrin. It is also essential in iron homeostasis and neuron survival in the central nervous system. Because of these functions, researchers have suggested that certain alleles may be at a selective advantage in copper-deficient environments (Mourant et al. 1976a). CP may also play an important role in the normal cellular release of iron because individuals deficient in it accumulate iron in their tissues (Mukhopadhyay et al. 1998).

The CP^B allele is the most common, occurring in European populations at frequencies hovering around 99%. African and African American populations tend to have the highest frequencies of CP^A, about 5% (Mourant et al. 1976a).

Other Plasma Proteins

There are other plasma proteins that are of general anthropological interest and which can be used to explore the genetic variation of populations. These polymorphisms can be used in population structure studies, selection studies, migration studies, and other examinations of the relatedness and origin of populations. These plasma proteins include, but are not limited to, the following. Placental alkaline phosphatase (chromosome 2q37) appears in the plasma of some women during pregnancy. It has its origin in the placenta. The Xm system is an X-linked (Xq28) serum protein that has been used in a few studies as a marker. Properdin factor B (6p21.3) is polymorphic, with four confirmed alleles. These alleles vary in frequency in different populations, but GB^F occurs at frequencies around 0.73, while GB^S shows frequencies about 0.25. Located on chromosome 19 (p13.3–p13.2), complement component-3 (C3) may have important functions in immune mechanisms. Individuals who are deficient in C3 are very susceptible to **pyogenic** (pus-producing) infections such as otitis media, pneumonia, septicemia, and meningitis. In fact, C3 increases during inflammation (Alper et al. 1972, de Bruijn and Fey 1985). The gene frequencies of these systems in world populations can be found in Roychoudhury and Nei (1988).

RED CELL ENZYMES

Red cell enzymes (Table 5.3) are often used in conjunction with the plasma proteins and other classical markers (blood groups) to study human diversity, population structure, group affinities, migration patterns, and microevolution. These polymorphisms are detected with various electrophoretic techniques, and in most cases, the full molecular sequences and differences of the variants are now known. Because of its role in malarial resistance, we highlight and discuss only one of these red cell enzymes, glucose-6-phosphate dehydrogenase (G6PD).

Glucose-6-Phosphate Dehydrogenase

During the Second World War and the Korean War, many American soldiers were given drugs to combat malaria. After administering the antimalarial drug primaquine, about

Table 5.3 Some red cell enzymes

Name	Some Common Alleles
Acetyltransferase	AC^S, AC
Acid phosphatase-1	$ACP1^A$, $ACP1^B$, $ACP1^C$
Adenosine deaminase	ADA^1, ADA^2
Adenylate kinase-1	AK^1, AK^2
Esterase D (S-formylglutathione hydrolase)	ESD^1, ESD^2 (FGH^1, FGH^1)
Glucose-6-phosphate dehydrogenase (present in other tissues also)	
G6PD (deficiency)	$GD+$, $GD-$
G6PD (structural)	GD^A, GD^B
Lactate dehydrogenase-A	LDH^A
Lactate dehydrogenase-B	LDH^B
Malate dehydrogenase	MDH^1, MDH^2, MDH^3
Nicotinamide adenine dinucleotide (NADH) diaphorase (also called methemoglobin reductase)	DIA^2, DIA^4
Peptidases	
Peptidase-A	$PEPA^1$, $PEPA^2$
Peptidase-B	$PEPB$
Peptidase-C	$PEPC$
Peptidase-D (proline dipeptidase)	$PEPD^1$, $PEPD^2$, $PEPD^3$
Peptidase-E	$PEPE$
Phosphoglucomutases	
PGM1	$PGM1^{a1}$, $PGM1^{a2}$, $PGM1^{a3}$, $PGM1^{a4}$, $PGM1^{a5}$
PGM2	$PGM2^1$, $PGM2^2$
PGM3	$PGM3^1$, $PGM3^2$
6-Phosphogluconate dehydrogenase (6PGD)	PGD^A, PGD^B
Phosphoglycerate kinase	PGK^1, PGK^2

Adapted from Crawford and Workman 1973, Mourant et al. 1976a, Vogel and Motulsky 1997.

10% of African American soldiers had a hemolytic reaction, while only about 1%–2% of those of European descent were affected (primarily only those of Mediterranean origin). A similar hemolytic condition, called **favism,** was known to occur when some individuals in Mediterranean countries consumed broad, or fava, beans (*Vicia faba*). Exposure to bean pollen may also cause a hemolytic reaction. Further research revealed that individuals who had reactions were deficient in the enzyme G6PD (Mourant et al. 1978, Vogel and Motulsky 1997). G6PD is called a *housekeeping* enzyme because of the vital functions it performs within all cells of the body, such as protecting against harmful oxidative products of cell metabolism. It is important within red blood cells, where it is involved in the hexose monophosphate pathway by catalyzing the first step in the pentose phosphate pathway. This pathway keeps the coenzymes glutathione and nicotinamide adenine dinucleotide (NADP) in their reduced form so that they can be used to repair and prevent cellular oxidative damage (for further discussion of G6PD's role within erythrocytes, see Friedman and Trager 1981, Beutler 1983, Greene et al. 1993).

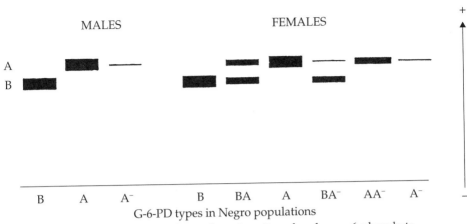

Figure 5.9. Diagram showing the electrophoretic pattern for glucose-6-phosphate dehydrogenase (G6PD) of different phenotypes. (From Harris 1971, 124.)

Family studies quickly revealed that the polymorphism was located on the X chromosome (Xq28), explaining the differing frequencies of G6PD deficiency in males and females.

Electrophoresis (Fig. 5.9) soon revealed the differences between the African and Mediterranean variants (Harris 1971). The "normal" enzymatic condition was designated B. About 20% of African males tested showed a more rapidly moving variant, which was called A. Those individuals with the enzyme deficiency exhibited a very weak band with the same mobility as the A variant and were called A–. However, individuals of Mediterranean origin who were deficient showed a weak band corresponding to the mobility of the B variant. This is sometimes referred to as *G6PD B–* or *G6PD (Med)*. Soon, many additional G6PD variants were found in populations across the world. In fact, over 400 variants were described by the early 1990s, with 77 reaching polymorphic frequencies (~1%) (Greene 1993). The numerous variants are now divided into five classes (Table 5.4) according to the level of enzyme activity (Beutler 1990). Recent DNA analysis has reduced the number of variants detected by biochemical means to about 130, with at least 34 being polymorphic (Luzzato and Mehta 2001). The majority of the differences are caused by missense mutations that cause single–amino acid substitutions that do not result in the complete loss of enzyme activity.

The G6PD enzyme found most frequently across the world in humans, as well as in chimpanzees and gorillas, is G6PD B. Outside tropical or semitropical regions, variant B is virtually the only one found. No other variants reach polymorphic frequencies in these populations. There are, however, rare private polymorphisms. The 34 polymorphic variants are all found in populations where falciparum malaria is endemic (Luzzato and Mehta 2001). This marked geographical association between the prevalence of malaria and frequencies of the various low–enzyme activity alleles of G6PD strongly suggest that G6PD-deficient individuals are at a selective advantage in a malarial environment. Ruwando et al. (1995) showed that both male hemizygotes and female heterozygotes are at a reduced risk (46%–58% reduction) of having severe malaria. The exact mechanism that makes G6PD-deficient individuals protected from *Plasmodium falciparum* infections is debated. However, Friedman and Trager (1981) suggest that G6PD-deficient red cells are more sensitive to the hydrogen peroxide generated by the parasites than "normal" red cells. The oxidative stress that is produced by the peroxides compromises the red blood

Table 5.4 Classification of glucose-6-phosphate dehydrogenase (G6PD) variants

Class	Description	Some Characteristics
I	Enzyme deficiency. Chronic nonspherocytic hemolytic anemia	Very rare. Does not reach polymorphic proportions
II	Severe enzyme deficiency. Less than 10% of the activity of G6PD B	44 of 109[1] (40.4%) that are classified as class II variants reach polymorphic frequencies. Includes G6PD A– and G6PD (Med)
III	Mild to moderate deficiency. Between 10% and 60% of the activity of G6PD B	22 variants out of 75[1] (29.3%) reach polymorphic frequencies
IV	Mild or no deficiency. 60% to 150% of the activity of G6PD B	Includes G6PD A
V	Increased enzyme activity. Greater than 150% of the activity of G6PD B	—

[1]Luzzatto and Mehta (2001) list 77 variants that have reached polymorphic frequencies. Sixty-six of these are deficient variants, and 11 have normal levels of enzyme activity.
Adapted from Greene 1993.

cells' integrity. In turn, the cells lose potassium; and because the parasites need ample potassium to thrive, the result is death of the parasite. Friedman and Trager (1981) also suggest that ingestion of fava beans causes an increase in the sensitivity to oxidants. If this is correct, eating fava beans may increase the level of protection against malaria for those individuals who are heterozygous for G6PD deficiency. Subsequent studies of fava beans identified two B-glycoside compounds, vicine and convicine, that increase the oxidant stress on the red cells. Thus, there is good evidence that G6PD-deficient red cells provide optimal antimalarial protection when they are also stressed by the consumption of fava beans (Greene et al. 1993).

Greene and colleagues (1993) hypothesize that the genetic trait for quinine taste sensitivity (see Chapter 7) may regulate the dietary intake of bitter-tasting antimalarial substances found in a variety of plants. Those individuals least sensitive would ingest more of these antimalarial compounds than those whose taste was more sensitive. They further postulate that the quinine-tasting locus is on the X chromosome and closely linked to G6PD. They speculate that through coevolution in endemic malarial environments there was an increase in haplotypes $G6PD^-/Q^+$ and $G6PD^+/Q^-$. These individuals would be less susceptible to falciparum malaria than the other haplotypes, $G6PD^-/Q^-$ and $G6PD^+/Q^+$. For example, the $G6PD^-/Q^+$ haplotype would be more fit than haplotype $G6PD^-/Q^-$ in an environment where both malaria was endemic and bitter substances were consumed because individuals would limit or avoid their intake of bitter-tasting substances. More empirical testing and the mapping of the quinine-tasting locus would enhance this hypothesis.

Cappadoro et al. (1998) suggest another possible mechanism for malaria resistance. In a study of normal and G6PD-deficient (Med) erythrocytes, they found that if the *Plasmodium* was in what is called the "ring stage" of maturation, the parasitized red blood cells were phagocytized 2.3 times more intensely than normal cells. In G6PD-deficient ring stage–parasitized erythrocytes, the level of reduced glutathione was much lower than in parasitized normal ring cells. This reduction in glutathione impaired the antioxidant defense of the parasitized cells and compromised their cell membrane. These cells were then nontoxic to the phagocytes and easily removed, thus decreasing the number of parasites that could mature. They suggest that this removal would be an efficient mechanism of malaria resistance in G6PD-deficient individuals.

Using linked restriction fragment length polymorphisms (see Chapter 8) and mathematical modeling, Tishkoff et al. (2001) suggest that the A variant arose first about 6,357 years ago (range 3,840–11,760 years ago), while the Mediterranean variant arose about 3,330 years ago (range 1,600–6,640 years ago). These dates correspond well with the timing of the rise of the most recent common ancestor of all living *P. falciparum* at about 3,200–7,700 years ago (Volkman et al. 2001, Rich and Ayala 2000). These dates are consistent with archaeological and historical records that suggest that malaria was not significant until the rise of agriculture about 10,000 years ago. The resulting G6PD polymorphisms are a nice example of the interaction of culture, environment, genes, and history shaping the human genome over the last 10,000 years.

Other Enzyme Systems of Anthropological Interest

Anthropologists often use enzymes other than G6PD in population studies. In most cases, the evolutionary mechanisms maintaining these polymorphisms are unknown. These polymorphisms include acid phosphatase-1 (ACP1), which has three common alleles, *P(a)*, *P(b)*, and *P(c)*, and a rare allele, *P(r)*. This enzyme has been mapped to chromosome 2 (p25) and may play some role in favism incidence and severity in G6PD-deficient individuals. The esterase D (ESD) polymorphism, mapped to chromosome 13 (q14.11), has been shown to be the same as the S-formylglutathione hydrolase (FGH) polymorphism. There are two common alleles, *FGH(1)* and *FGH(2)* or *ESD(1)* and *ESD(2)*. Lactate dehydrogenase-A (LDHA) is located on chromosome 11 (p15.4), while LDHB is located on chromosome 12 (p12.2-12.1). Deficiencies in either LDHA or LDHB can occur and are considered by many researchers as a "nondisease." Five distinct peptidase enzymes have been identified. All appear to be unlinked. Peptidase-A (PEPA) has been mapped to 18q23, peptidase-B to 12q21, peptidase-C to 1q42, peptidase-D to 19cen-q13.11, and peptidase-E to17q23-qter (which is monomorphic). There are three unlinked phosphoglucomutase loci, PGM1 (1p31), PGM2 (4p14-q12), and PGM3 (6q12). For PGM1, at least five alleles have been identified; for PGM2 and PGM3, there are at least two alleles. The PGM1 and PGM2 polymorphisms are in red blood cells, while the PGM3 polymorphism has been detected in white blood cells. There are two other loci, PGM4 (found in human breast milk) and PGM5 (found in many body tissues). Structural and functional similarities of PGM1, PGM2, and PGM3 indicate that they arose through gene duplication. Because of its diversity, PGM1 is often used in anthropological studies of human variation. Located on chromosome 1 (p36.2-p36.13), 6-phosphogluconate dehydrogenase (PGD) consists of two distinct red cell types, PGD^A and PGD^B. A rare variant called PGD Mediterranean has been identified. Additional red cell enzymes and gene frequencies for various populations and their allelic variants can be found in Roychoudhury and Nei (1988).

EXPLORING POPULATION STRUCTURE WITH CLASSICAL MARKERS

Anthropologists, using plasma proteins and red cell enzymes, have conducted numerous population-level studies. Sometimes these genetic systems are used alone to answer a specific question or detail the distribution of a single trait within a population or region of the world. More often, however, many systems are combined to explore the population structure of a region and detail its variation (see Chapter 12). In this section, we present two examples of the uses of some "classical genetic systems" in exploring human evolution and diversity. Please note that the second case study uses not only the classic markers that have been discussed in this chapter and Chapter 4 but also the HLA system, which is a genetic marker of white blood cells (to be discussed in Chapter 7).

Example 1: The Irish Travelers or Tinkers

The population structure of Ireland has been investigated in numerous studies (e.g., see Tills 1977, Bittles and Smith 1994, Relethford and Crawford 1995). These studies have provided valuable insight into the role of gene flow, historical events (e.g., the Irish famine), religion, population size and distribution, and economic factors in shaping the contemporary genetic diversity of Ireland. Less attention has been paid to an itinerant population, the Travelers or Tinkers, and their origin and relationship to the larger Irish population. A number of different hypotheses have been put forth to explain the origin of the Travelers. These include the idea that they are Romany Gypsies, native chieftains, displaced laborers and farmers, and even prehistoric outcasts of early Irish populations. Using a univariate analysis of gene frequencies, Crawford (1975) suggested that the Travelers were of Irish origin. Recently, a multivariate analysis was conducted to re-examine this suggestion (North et al. 2000). Using 10 blood systems (ABO, Rhesus, Duffy, Kell, MNS, P, transferrin, phosphoglucomutase, adenylate kinase, and haptoglobin), these researchers confirmed the earlier genetic findings that the Travelers are an Irish social isolate. This population was probably formed slowly, over time, rather than suddenly by in-migration (e.g., founder effect). It is suggested that the initial group was composed of artisans and craftsmen who were forced to abandon their monasteries. The Traveler population grew as additional local peoples joined the group because of political upheavals (e.g., British occupation and repression) and historical events (e.g., Irish potato famine). The study also demonstrated that population history had a major influence on shaping the population structure at the county level. That is, the Irish population is divided into distinct subpopulations while at the same time it is a genetically homogeneous population. An additional finding of this study suggests that the distinctiveness of the midland counties is due to Viking influence since this area is genetically similar to Norway. This study nicely illustrates how a number of genetic markers can be used to help elucidate the origin of a socially distinct population.

Example 2: The Origins of Indo-Europeans

In 1987, archaeologist Colin Renfrew challenged the prevailing theory on the origin of Indo-European languages. Using a **demic diffusion** model (after Ammerman and Cavalli-Sforza 1971, 1973), he suggested that, instead of arriving 6,000 years ago from the Pontic Steppes north of the Black Sea, as Gimbutas (1973, 1977, 1980, 1985, 1997, 1999) maintained, the Indo-European languages were brought to Europe with the spread of agriculture from Turkey and Asia Minor about 9,000 years ago (see also Renfrew 2000). Using linguistics, mythology, and archaeology, Gimbutas had earlier argued that the peopling of Europe occurred in three waves. Wave number one occurred between 4,400 and 4,200 B.C., wave two between 3,400 and 3,200 B.C., and wave three between 3,000 and 2,800 B.C. The first wave (called Kurgan I and II, the Khvalynsk and Svednij Stog cultures) had the advantage of domesticated horses and a militaristic social structure. These features, in addition to a rudimentary agriculture, allowed these seminomadic peoples to merge and dominate the Old Europeans (indigenous peoples) in what Gimbutas calls the "Indo-Europeanization of Europe." Renfrew's initial theory was criticized on the basis of its oversimplification, the lack of considering contact-induced linguist change, and the focus on a single proto-Indo-European dispersal. Many researchers, however, accepted the idea that proto-Indo-European language(s) arrived in Europe with the dispersal of farming (e.g., Zvelebil 1995, 1998). Renfrew (1999, 2000) has since modified his theory to take into account many of the earlier criticisms.

Archaeologist Ammerman and geneticist Cavalli-Sforza suggested that, if Renfrew is correct about the migration patterns, the spread of Neolithic farmers from south-west Asia should still be evident in the contemporary genetic structure of the region (Ammerman and Cavalli-Sforza 1971, 1973, 1984). Initially, using 38 independent alleles from 10 loci (classic markers such as those discussed in Chapters 4–7) and then increasing to 34 loci and 95 alleles, Cavalli-Sforza and colleagues began to explore the genetic origins of Europeans (Menozzi et al. 1978, Piazza 1993, Piazza et al. 1995, Cavalli-Sforza 1997). They produced what they called "synthetic genetic maps." The first map generated was interpreted as reflecting the spread of agriculture from Anatolia to Europe (Fig. 5.10). The second might reflect an expansion from the Iberian Peninsula. The third synthetic map portrays the movement of peoples (or genes) from the east, adding to Gimbutas's argument of an early Kurgan expansion into Europe.

Were these maps incorrectly interpreted simply using the prevailing theories of the time? Did they really reflect the movement of peoples into Europe with the Neolithic expansion, or were they vestiges of earlier Pleistocene movements? Was there an expansion of peoples or just the ideas (e.g., new farming techniques)? Was the model really one

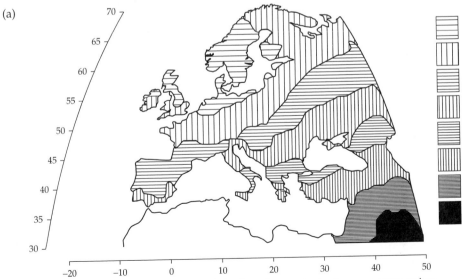

Figure 5.10. Synthetic maps of Europe of the first three principal components using 95 gene frequencies. (From Cavalli-Sforza, Luca, *The History and Geography of Human Genes*. © 1994 Princeton University Press. Reprinted by permission of Princeton University Press.)

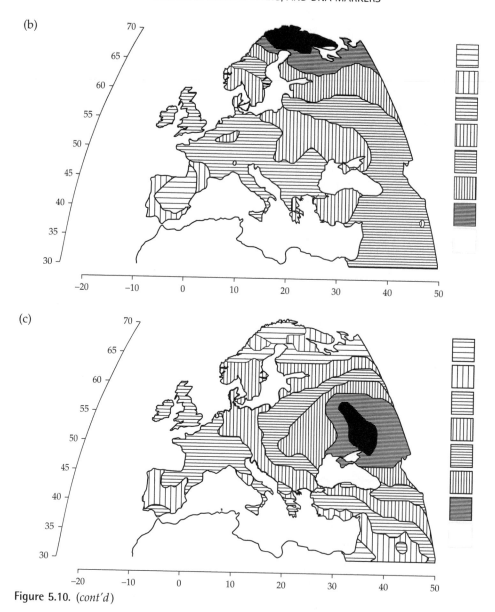

Figure 5.10. (*cont'd*)

of acculturation—contact-induced change? How stable is the genetic structure (variation) of a region given the dynamics of microevolution and historical processes (e.g., bottle-necks in populations) that have occurred over the millennia? Can there be language replacement without gene flow? Is the origin of Indo-Europeans one of much more complexity—could both Renfrew's and Gimbutas's models be operating, or were there other processes besides migration (e.g., see Fix 1996)? These and other questions (see

Brown and Pluciennik 2001) were asked and new data, analyses, and theories were expanded.

Using 120 alleles from the classical genetic systems (non-DNA polymorphisms), Cavalli-Sforza et al. (1988) suggested that the correlation between genetic data and linguistic data was evidence of a "coevolution" between genes and language in Europe. Strengthening this view was the finding of Sokal and colleagues (Barbujani and Sokal 1990, Sokal et al. 1992) of a significant partial correlation between linguistic distance and genetic distance when geographic distance was controlled. These geneticists also used classical genetic markers as the basis of their arguments. Initially, using 19 loci (63 allele frequencies) from 3,119 European localities, they found that the genetic structure of Europe was determined primarily by gene flow and admixture. A later study (Sokal et al. 1992) used 25 genetic systems (plasma proteins, serum enzymes, immunoglobins, histocompatibility alleles, and red cell antigens) from 2,111 Indo-European-speaking samples. Their genetics–language correlations suggested that neither Renfrew's nor Gimbutas's model appears to explain the origin of Indo-Europeans.

With the advent of molecular techniques, the use of classical markers to examine questions of population origins, expansion, regional diversity, and replacement such as has been detailed here has largely taken a back seat. DNA (nuclear, mitochondrial, and ancient) and Y-chromosome analyses are now being used to explore these and similar questions not only for Europe but also for the rest of the world (e.g., see Torroni et al. 1998, Santos et al. 1999, Stone and Stoneking 1998, Sykes 1999). Studies of the peopling of Europe using the Y chromosome, mitochondrial DNA (mtDNA), and nuclear DNA are now adding to and in, some cases, both corroborating and challenging the interpretations based on classic markers. Studies of mtDNA and Y-chromosome data have suggested that population expansions into Europe occurred in several waves during the late Paleolithic, with a Neolithic component accounting for only 25% of the mtDNA pool of modern Europeans, and that there was considerable back-migration into the Near East (Richards et al. 2000). However, Chikhi et al. (1998) argue that clines of nuclear DNA markers largely reflect a westward and northward expansion of Neolithic peoples out the Levant and that the lineages attributed to earlier Paleolithic expansions actually arrived in Europe at a much later date. Richards et al. (2002), using larger sample sizes, report that clades of mtDNA show clinal gradients similar to those of other genetic markers, but they also suggest a much more complex picture involving post-Neolitic gene flow from the Near East and Africa. Barbujani and Bertorelle (2001) argue that the main event in the peopling of Europe was the spread of early farmers from the Levant; however, older and more recent processes have also left their marks on the genetic structure and gene pool of Europe. Using Y-chromosome data, Chikhi et al. (2002) argue that the demic diffusion model introduced by Ammerman and Cavalli-Sforza in 1984 fits the data nicely.

Clearly, there are many processes and events that have molded the genetic diversity that we see in Europe today; and our goal, as anthropologists, is to unravel and explain these complex and intriguing patterns. This example shows how an understanding of genetic variation and evolution can contribute to answering questions relevant to a variety of disciplines, in this case, genetics, linguistics, historical demography, and archaeology. By combining the theories, data, methods, and interpretations of these various fields, questions of culture change, diversity, contact, and dispersal can be explored and clarified. The question of the peopling of Europe clearly exemplifies how an interdisciplinary and holistic perspective can enhance our understanding of human diversity, evolution, and behavior (see, e.g., Bellwood 2001).

CHAPTER SUMMARY

In this chapter, we saw how the red cell and serum enzymes and proteins are detected by electrophoresis, revealing their diversity. Because more than 30 of these genetic systems are polymorphic, anthropologists have often used these proteins and enzymes to characterize the genetic variation of human populations and to explore questions of population affinity or uniqueness. Even though some of these systems were only briefly mentioned while others were accorded more detail, they are all useful in delineating genetic similarity and diversity as was seen in the two examples detailed at the end of the chapter. Some of these systems may also be important in terms of natural selection and human adaptation. G6PD is a very good candidate as a buffer against malaria. However, because of its high number of unique haplotypes, the Gm system has been used extensively to explore population origins and relationships.

These polymorphisms have often been used in conjunction with the blood groups (Chapter 4) to examine the population structure (see Chapter 12) of regions and continents. They are, however, not being used as much now as they were in the past. This is because molecular markers have now taken a "front seat" in anthropological studies of genetic diversity.

SUPPLEMENTAL RESOURCES

Bellwood P (2001) Early agriculturalist population diasporas? Farming, languages, and genes. *Annual Reviews in Anthropology* 30:181–207.

Friedman M J and Trager W (1981) The biochemistry of resistance to malaria. *Scientific American* 244:154–164.

Greene L S (1993) G6PD deficiency as protection against falciparum malaria: An epidemiologic critique of population and experimental studies. *Yearbook of Physical Anthropology* 36:153–178.

Mourant A E, Kopeć A C and Domaniewska-Sobczak K (1976) *The Distribution of the Human Blood Groups and Other Polymorphisms*, London: Oxford University Press.

Mourant A E, Kopeć A C and Domaniewska-Sobczak K (1978) *Blood Groups and Diseases: A Study of Associations of Diseases with Blood Groups and Other Polymorphisms*, London: Oxford University Press.

Neel J V (1962) Diabetes mellitus: A "thrifty" genotype rendered detrimental by "progress"? *American Journal of Human Genetics* 14:353–362.

6

Human Hemoglobin Variants

BEGINNINGS OF HEMOGLOBIN RESEARCH

During the course of a medical examination in 1904 of an anemic 20-year-old from Grenada, West Indies, an intern, Earnest E. Irons, noticed that the patient's red cells were misshapen (Savitt and Goldberg 1989). This observation was eventually reported by Herrick in 1910. It quickly became apparent that the condition, which we call **sickle cell anemia**, was fairly common among African Americans. Aside from having hemolytic anemia, sickle cell patients suffer from recurrent bouts of muscular and skeletal pain, are physically weak, and may experience heart, lung, spleen, and kidney damage. Without medical treatment, most die before reaching adulthood. By 1923, Taliaferro and Huck demonstrated that the condition was hereditary; and in 1956, Ingram, using electrophoretic and chromatographic techniques, showed that sickle cell hemoglobin differed from normal hemoglobin by only one amino acid. Shortly afterward, the full amino acid sequence and the three-dimensional structure of hemoglobin was established.

Over the years, research on hemoglobin has furthered our understanding of a number of genetic processes and relationships. Hemoglobin serves as a model for our understanding of gene action at the molecular level. It has aided in our understanding of gene structure–function relationships and in the detection of various mutations (e.g., deletions and frameshifts). Hemoglobin research has aided our understanding of regulatory mutations and developmental gene switching. It also serves as the classic model for natural selection in humans, particularly heterozygote advantage (balanced polymorphism).

THE HEMOGLOBIN MOLECULE AND HEMOGLOBIN GENES

About 85%–90% of the protein content in human red blood cells is hemoglobin. Hemoglobin transports oxygen from the lungs to the tissues and is thus crucial to the normal functioning of the human body. At the cellular level, hemoglobin also picks up and binds to (at a different site from the one that binds with oxygen) about half of the carbon dioxide that is generated in the tissues and transports it to the lungs for release from the body. Adult human hemoglobin is a complex molecule consisting of four chains of amino acids (polypeptide chains) and four **heme** groups (Fig. 6.1). Two of these polypeptide chains are identical and are designated α (α_2). Two others are also identical and are called β (β_2). Each chain carries a heme group, a large nonprotein molecule containing an atom of iron that can carry a molecule of oxygen. The α chains are composed of 141 amino acids, while the β chains have 146 amino acids. The major hemoglobin of adults and children is designated HbA, or adult hemoglobin ($\alpha_2\beta_2$). During growth and development, all humans also possess what is called "embryonic" and "fetal" hemoglobin.

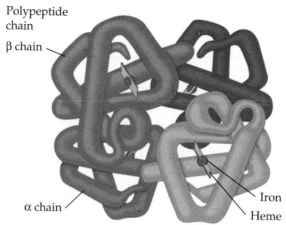

Polypeptide
chain

β chain

α chain

Iron

Heme

Figure 6.1. The hemoglobin molecule.

Embryonic hemogloblin is also composed of two pairs of identical chains (designated ζ and ε). The ζ chain, which appears very early during embryonic development, is similar to the α chain in amino acid composition, and the ε chain is similar to the β chain. These chains disappear after about 8–10 weeks of embryonic life and are "replaced" by fetal hemoglobin (Fig. 6.2). Fetal hemoglobin (HbF, $\alpha_2\gamma_2$) is replaced by adult hemoglobin shortly after birth. All adults also carry a small amount (2%–3%) of HbA$_2$ ($\alpha_2\delta_2$). Most "normal" hemoglobins (fetal and adult) have identical α chains. What makes a difference in these hemoglobins is the amino acid structure of the non-α chains. All of these chains probably have a common evolutionary origin and occurred by genetic duplication.

Each human carries at least one α, β, γ, δ, ε, and ζ gene in the haploid state. These genes are found in clusters on two separate chromosomes. The α gene cluster is located on the short arm of chromosome 16, and the β gene cluster is found on the short arm of

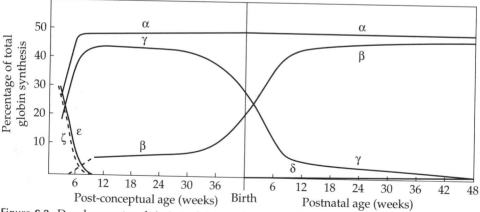

Figure 6.2. Development and timing of the production of the various globin chains from conception to 48 weeks of life. (From Serjeant and Serjeant 2001, 3.)

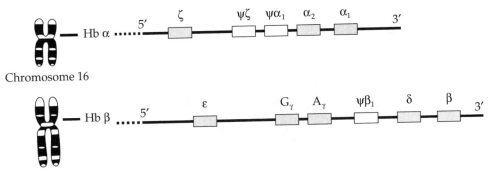

Figure 6.3. Location and organization of human α- and β-globin gene clusters on chromosomes 16 and 11. White boxes with ψ indicate pseudogenes. Introns and exons not shown. (Adapted from Vogel and Motulsky 1997, 304.)

chromosome 11 (Fig. 6.3). Gene expression is complex, and the genetic mechanisms that regulate the timing and production of these various chains are now just starting to become clear. In most instances, there are two *α-globin* genes for every *β-globin* gene. Because these genes are on two different chromosomes, their expression is controlled independently. During protein synthesis, there are slightly more α chains than β chains produced. These chains then incorporate a heme group and join together to form functioning hemoglobin. Interestingly, β chains are also capable, on their own, of associating with a heme (and no α chain) to produce HbH. This resulting product (tetramer) is functionally useless because it cannot release the oxygen it binds. However, heme-bound α chains that do not combine with a heme-bound β chain form precipitates (called α-inclusion bodies) that harm the red blood cells. Kihm et al. (2002) suggest that a protein they call α-hemoglobin-stabilizing protein (AHSP) stabilizes and aids in delivering the newly synthesized α chains to their β-chain counterparts. The slight excess of α chains that is normally produced can be neutralized by AHSP. However, if too many α chains are produced, AHSP is overwhelmed and the excess α chains precipitate. In this case, a condition called β-thalassemia occurs. β-Thalassemia is caused when there is a reduction in β-chain synthesis. We will discuss the thalassemias in a later section in this chapter.

On both chromosomes, there are **pseudogenes** that have DNA sequences similar to those functioning loci. These pseudogenes have experienced various mutational alterations that have inactivated their transcription; hence, there is no functional expression of these loci. These genes are probably products of duplication that arose during evolution.

HEMOGLOBIN VARIANTS

The "normal" structure of the β chain of hemoglobin is coded for by the allele *HbA*. The frequency of the *A* allele in many populations reaches almost 100%; as a result, most individuals are genotypically *AA* and have what we call "normal" adult hemoglobin. Hemoglobin variants arise when mutations affect the various hemoglobin genes. The most common variants are caused by simple amino acid substitutions that change the

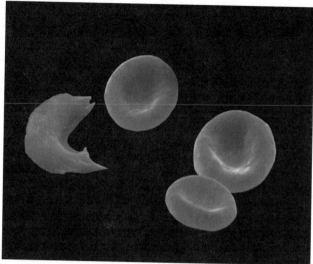

Figure 6.4. Sickle cell anemia. Comparison of normal red blood cells with a sickle cell. (© Dr. Stanley Flegler/Gettyimages.)

globin chain. Over 600 of these substitutions have been identified. Most are caused by point mutations (single nucleotide substitutions). There are also some deletions, insertions, and frameshifts; however, these are relatively rare. Most of the mutations are rare and have little or no effect on hemoglobin function. There are a few (e.g., HbS, HbC, HbD, and HbE) that reach relatively high frequencies in various populations, and these are discussed in this chapter.

The S allele, also known as the sickle cell allele, causes sickle cell anemia if the individual's genotype is SS. In this condition, the red blood cells are altered under conditions of severe **hypoxic** stress (low levels of oxygen in the blood) (Fig. 6.4), increasing viscosity and severely impairing oxygen transport and normal circulation in small blood vessels. Without medical attention, most individuals with this genotype will not survive to adulthood. The S allele is the result of a point mutation where the amino acid valine (GTG) replaces glutamic acid (GAG) at the sixth position of the β chain. Individuals who are heterozygous AS (carriers of the sickle trait) have about 25%–40% HbS and are clinically normal. The red cells of these individuals have both HbA and HbS, and sickling occurs only under severe hypoxic stress.

Because the S allele is harmful in the homozygous state, causing premature death and other complications, we would expect it to be eliminated by the action of natural selection, making its frequency very low in all populations. This is indeed the case in many populations where the balance between mutation and selection is maintained. However, in numerous populations, we find that the frequency of the S allele is high, ranging 5%–20% in some regions. The frequency of heterozygotes (AS) ranges 15%–40% in these populations. These frequencies are too high to be accounted for by recurrent mutation, especially if the average mutation rate is about 10^{-5}–10^{-6}. Also, since genetic drift is random and affected by population size, not environmental factors, we would expect to see very high frequencies in some small, isolated populations and very low frequencies in others.

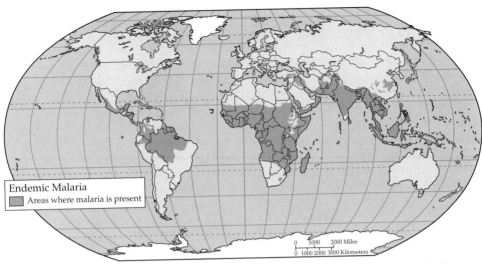

Figure 6.5. Map showing distribution of falciparum malaria. (From Allen and Shalinsky (2004), *Student Atlas of Anthropology*. New York: McGraw-Hill Companies. Used by permission.)

THE GEOGRAPHIC DISTRIBUTION OF THE SICKLE CELL ALLELE

The geographic distribution of *HbS* is related to the prevalence of malaria. Malaria is a vector-borne (*Anopheles* mosquito) infectious disease, caused by the protozoal parasite *Plasmodium*. As discussed in Chapters 4 and 5, there are a number of different species of *Plasmodium* that cause malaria in humans and other primates. The World Health Organization (1997) estimates that between 300 and 500 million people are affected by some form of malaria and between 1.5 and 2.7 million people die yearly from the disease. As can been in Figures 6.5 and 6.6, the correspondence between HbS and the prevalence of falciparum malaria (*P. falciparum*) is striking. This geographic correspondence suggests that high frequencies of sickle cell anemia and malaria are related to high frequencies of the *S* allele.

The *S* allele became frequent in these tropical and subtropical areas because heterozygotes (*AS*) have a selective advantage in an environment where malaria is prevalent. This is a case of balancing selection (balanced polymorphism) in which the heterozygotes (*AS*) have the highest fitness in a malarial environment. The fitness of both homozygotes is lower because individuals who are *AA* die more frequently from malaria and those who have sickle cell anemia (*SS*) usually die before reproductive age.

If the mortality from malaria and sickle cell anemia were equal, we would expect that the frequencies of the *S* allele and the *A* allele in a population would, over time, become equal. However, because sickle cell anemia is a much more severe disease than malaria, the selection pressures are not equal and those individuals who are genotypically *SS* are selected out of the population at a greater rate than those who are genotypically *AA* and die from malaria. Thus, the balance between these two diseases is not equal, and the maximum fitness of the population occurs when the frequency of the *S* allele is between 0.10 and 0.20 (Fig. 6.7).

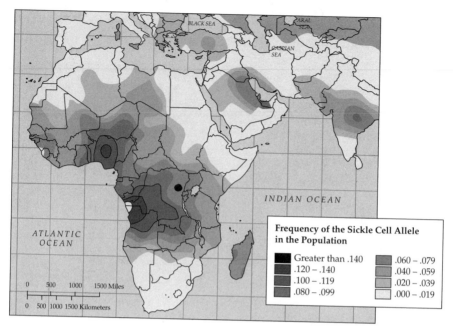

Figure 6.6. Distribution of sickle cell in the Old World. (From Allen and Shalinsky (2004), *Student Atlas of Anthropology*. New York: McGraw-Hill Companies. Used by permission.)

In an analysis using data from a survey (Edington 1959) of the Yoruba of Ibadan, Nigeria, Bodmer and Cavalli-Sforza (1976) estimate that for every 100 people with genotype *AS* who survive to adulthood, 88 who are *AA* and 14 who are *SS* survive. These figures clearly show the differential effect of mortality from malaria and sickle cell anemia. That is, the mortality (selection) from malaria and sickle cell anemia is not equal. Differences such as these are capable of rapidly changing the genetic structure of a population. Figure 6.8 illustrates the pattern and allele frequency changes over time using the above data. The initial starting frequency of the *S* allele is set at 10^{-5} (or 0.00001), a very reasonable value given known mutation rates. Note that during the first 40 generations there is very little change. This is expected because of the low starting frequency. As the frequency of *S* increases, so does the rate. This is due to the fact that there are more heterozygotes, *AS*, in the population. By about 100 generations (or about 2,500 years), the *S* allele reaches equilibrium. This equilibrium is based on the balance between the selective effects of malaria and sickle cell anemia acting on the two homozygotes, while favoring the heterozygote. The term *balanced polymorphism* describes this equilibrium. Eventually, the *S* allele will reach an equilibrium frequency of 0.122 (frequency of allele *A* is then 0.878). Please note that this example does not include the action of the other mechanisms of evolution.

We also present an additional illustration of how the sickle cell balanced polymorphism works. Using adult genotype frequency data and comparing it to Hardy-Weinberg expectations, Allison (1956) estimated the relative fitness (*w*) of the three genotypes:

$$w_1 \text{ (or } w_{AA}) = 0.7961 \quad w_2 \text{ (or } w_{AS}) = 1.0000 \quad w_3 \text{ (or } w_{SS}) = 0.1698$$

Using the simple formula, $s = 1 - w$, the corresponding selection coefficients, reflecting the probability of being selected against, are:

$$s_1 = 0.2039 \quad s_2 = 0.0000 \quad s_3 = 0.8302$$

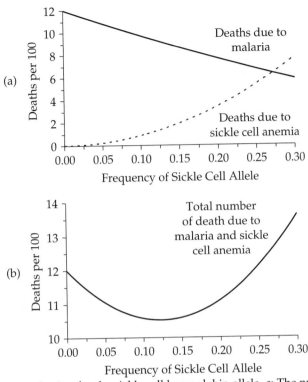

Figure 6.7. Balancing selection for the sickle cell hemoglobin allele. *a*: The number of deaths per 100 people in a hypothetical population due to malaria and sickle cell anemia as a function of the frequency of the sickle cell allele. These data were derived using the relative fitness values from Bodmer and Cavalli-Sforza (1976): *AA* = 0.88, *AS* = 1.00, and *SS* = 0.14. Thus, 1 − 0.88 = 0.12 of individuals with genotype *AA* will die from malaria and 1 − 0.14 = 0.86 of individuals with genotype *SS* will die from sickle cell anemia. As the frequency of the sickle cell allele increases, the number of deaths due to malaria decreases because the number of heterozygotes (*AS*) resistant to malaria increases but, at the same time, the number of deaths due to sickle cell anemia increases because of the increased number of sickle cell homozygotes (*SS*). *b*: What happens when deaths due to malaria *and* sickle cell anemia are added together. Here, as the frequency of the sickle cell allele increases, the total number of deaths decreases for a while and then increases afterward. Balancing selection will ultimately lead to an equilibrium sickle cell allele frequency associated with the *minimum* total number of deaths, which in this case is associated with a frequency of 0.122. See Figure 6.8 for an example of how these changes occur over time.

Equilibrium values under heterozygote advantage can then be estimated by the following formula:

$$\hat{q} = \frac{s_1}{s_1 + s_3}$$

$$\hat{q} = \frac{0.2039}{0.2039 + 0.8302} = 0.1972$$

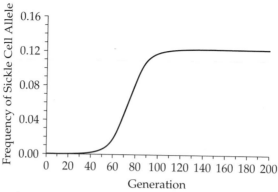

Figure 6.8. Evolution of the sickle cell allele over time starting with an initial frequency of 0.00001 due to mutation and assuming relative fitness values over time of $AA = 0.88$, $AS = 1.00$, $SS = 0.14$.

Remember that $p^2 + 2pq + q^2 = 1$, so we can estimate the frequency of heterozygotes in the populations as $2pq = 0.317$, or almost 32%, which is a frequency found in some populations where malaria is prevalent. Note that these estimates are consistent with those presented above.

When examining the effects of natural selection on a population, sickle cell clearly shows the importance of the specific environment. If there is no malaria prevalent in the environment, the selective pressures on the phenotypes would be very different. The fitness of the heterozygous AS individuals would be less than that of the homozygous AA individuals. In this case, the S allele would remain very low in the population. This example also shows that the S allele is not inherently "bad" or "good." It clearly depends upon the circumstances.

How Resistance to Falciparum Malaria Works

Individuals with one S allele do not have anemia, but since 25%–40% of their hemoglobin is capable of sickling (lysing), this apparently creates an environment which is not hospitable to malarial parasites. In 1981, Friedman and Trager were able to elucidate the mechanisms that possibly provide the heterozygote with this selective advantage (see also Friedman 1978). The malarial parasite invades red blood cells (at this stage it is called a *merozoite*), and as it develops into a trophozoite, it causes the red cells to develop knobs on their surfaces. These knobs allow the red cells to attach to the walls of capillaries, where they remain while the parasite matures and releases more merozoites into the bloodstream. A large accumulation of these red cells in a vital organ can result in death (e.g., cerebral malaria is very deadly). Normal red blood cells are able to withstand short periods of low oxygenation as they move through the capillaries. However, an infected cell, with surface knobs, can become lodged in the capillaries. These cells spend a longer period of time in low oxygenation. If the sequestered red cell contains hemoglobin S, it sickles because of the reduced oxygen tension on its surface. This sickling causes its cell membrane to become permeable and to release its potassium. Without potassium, an essential metabolic factor, the parasites perish. Removal of these parasites apparently allows the body's immune system ample time to mount an effective defense and rid the body of the remaining parasites.

Other possible explanations of resistance involve (1) enhanced removal by the host's immune system and (2) impaired growth of the malarial parasite (Evans and Wellems 2002, Weatherall et al. 2002). However, many of the studies documenting possible mechanisms of resistance are fraught with problems (Akide-Ndunge 2003).

THE POSSIBILITY OF DIFFERENTIAL FERTILITY

Theoretically, natural selection molds and shapes populations through both differential mortality and differential fertility. Sickle cell provides a persuasive argument for differential mortality affecting populations in malarial environments. However, the evidence for differential fertility is far from unequivocal.

Early on, a number of researchers suggested that women who are heterozygotes (*AS*) have greater fertility than homozygotes (*AA*) (Foy et al. 1954, Allison 1956). Livingstone (1957) suggested that heterozygotes were in some manner buffered from the ill effects of malaria during pregnancy. In a study among Central American Black Caribs, Firschein (1961, 1984) found significantly higher fertility among *AS* women than among *AA* women. He suggested that parasitemia of the placenta resulted in more abortions among *AA* women than among *AS* women. Another study failed to find any fertility differences among Black Caribs (Custodio and Huntsman 1984). Madrigal (1989) studied the reproductive histories of women in Limón, Costa Rica. She was unable to find any difference between *AS* and *AA* women and a number of measures of differential fertility, including number of pregnancies, number of spontaneous abortions, number of live births, and family sizes. However, a recent study by Hoff et al. (2001) of about 10,000 African American women in Mobile, Alabama, revealed that *AS* women had more live births than *AA* women. Since Alabama is a nonmalarial environment, these results are intriguing and challenging. Also intriguing is the discovery that parasitized red blood cells collected from placentas apparently have unique adhesion and antigenic properties different from those found in nonpregnant women (Beeson 1999). How this finding may relate to and affect fertility of *AA* and *AS* individuals is not known.

These contradictory results suggest that differential fertility may be influential in shaping the *S* allele in some populations. Remember that natural selection is environment-specific and that selection pressures vary greatly from region to region. In addition, *S* is not the only allele that confers a selective advantage on individuals living in malarial environments.

ORIGIN OF THE SICKLE CELL ALLELE

Single-Mutation Theory

For many years, a single-mutation theory for the origin of the *S* allele was favored by researchers (e.g., Lehmann 1954; Livingstone 1958, 1989; Gelpi 1973; Lehmann and Huntsman 1974). The classic study by Livingstone (1958), which combined archaeological evidence, vector ecology, cultural behavior, genetics, and parasite evolution, provides an excellent example of the interaction of culture and biology.

During much of human prehistory, the African landscape was neither hospitable nor conducive to the spread of malaria. Much of the continent consisted of dense tropical and subtropical forests. The canopy of the rain forest prevented sunlight from reaching the ground, and water did not readily accumulate on the surface in pools. The vector for malaria, the *Anopheles* mosquito, thrives best where it can reproduce in sunny, stagnant pools of water. Hence, the malarial parasite and its vector did not flourish under these conditions.

With changing climatic conditions and the introduction of iron implements during the Neolithic, prehistoric horticulturalists began to migrate south into Africa. As **slash and burn (swidden) horticulture** expanded south and as populations increased in size, the ecology of the region shifted, favoring the growth and spread of mosquito populations. As a consequence, the malarial parasite also expanded its territory. Prior to this southern expansion of peoples into Africa, the frequency of the *S* allele was probably low (as it is today in many nonmalarial regions of the world). As malaria increased in frequency, it became an evolutionary advantage to have the *AS* genotype. Selection pressures increased, and as a consequence, the frequency of the *S* allele increased as well. This sequence of events is depicted in Figure 6.9. Gene flow also spread the allele from population to population. A number of simulations suggest that the change could have taken place in about 100 generations. Some researchers (e.g., Lehmann 1954) favor a single mutation in the Arabian Peninsula and subsequent spread south and east, while others favor an equatorial African origin and then diffusion to India, Arabia, and the Mediterranean as a result of the east African slave trade (e.g., Kamel and Awny 1965, Gelpi 1973).

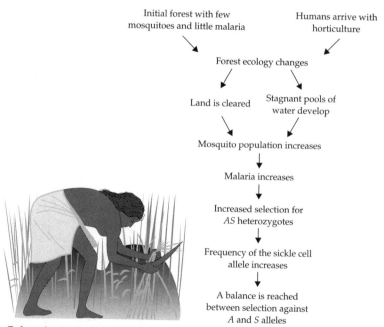

Figure 6.9. Cultural and environmental factors resulting in changes in the frequency of the sickle cell allele in Africa. (From Relethford 2003b, 159.)

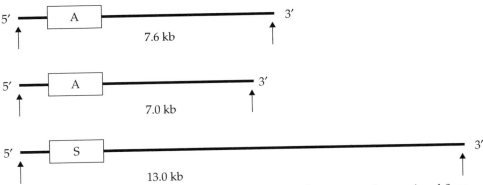

Figure 6.10. The three *HpaI* fragments containing the α-globin structural gene. *A* and *S* are "normal" and sickle cell genes, respectively. The *arrows* indicate *HpaI* recognition sites. (From Kan and Dozy 1978.)

Multiple-Mutation Theory

Studies using DNA polymorphisms indicate that the *S* allele arose multiple times in various regions of the Old World by independent mutational events. As will be discussed in Chapter 8, restriction endonucleases identify and cut DNA at specific recognition sites. If the *S* allele arose only once in prehistory and geographically spread through gene flow, we would expect the genes and DNA sequences surrounding the mutated *β-globin* gene to be identical. When DNA is digested by the restriction endonuclease HpaI, the *β-globin* structural gene is usually contained in a DNA fragment that is 7.6 kb in length. In 1978 Kan and Dozy reported finding two variants, one 7.0 kb long and the other 13.0 kb long (see also Labie et al. 1986). Of the 46 individuals tested, only those of African descent displayed these two polymorphisms. The 13.0 kb fragment was predominantly found in persons with *HbS*, while the 7.6 kb and 7.0 kb fragments were present in individuals with *HbA* (Fig. 6.10). It was also discovered that the 13.0 kb fragment was linked to the *β^C* gene (Feldenzer et al. 1979). If both the *β^C* and *β^S* genes are associated with the 13.0 kb fragment, the HpaI mutation that produced the 13.0 kb fragment must have occurred earlier in time. Also, since the distribution and frequency of *HbC* suggest that it originated in Upper Volta and Ghana, it is also assumed that the HpaI mutation arose in the same region.

The early method of using only one restriction enzyme has now been replaced with the use of a series of different enzymes to identify multiple recognition sites. Using an array of enzymes (e.g., HincII, HindIII, XmnI, AvaII, HapI, TaqI, PvuII, HinfI, RsaI, and BamHI), four African haplotypes and one Asian haplotype have been identified (Fig. 6.11) (Antonarakis et al. 1984, Nagel and Labie 1985, Chebloune et al. 1988, Lapouméroulie et al. 1992, Zeng et al. 1994). These β-globin haplotypes have been named after the areas where they were initially found: Senegal, Benin, Bantu, Cameroon, and Asian. These results suggest that the *S* allele arose independently at least five times. Then, because of its selective advantage in malarial environments, it increased in frequency and spread through gene flow (Fig. 6.12). The Benin haplotype spread to north Africa and the Mediterranean (Mears et al. 1981, Pagnier et al. 1984, Ragusa et al. 1988). One analysis even suggests that the Benin haplotype initially spread to Portugal between the eighth

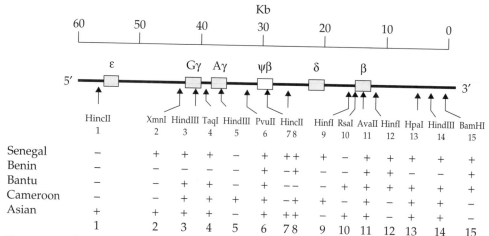

	HincII		XmnI	HindIII	TaqI	HindIII	PvuII	HincII		Hinfl	RsaI	AvaII	Hinfl	HpaI	HindIII	BamHI
	1		2	3	4	5	6	7 8		9	10	11	12	13	14	15
Senegal	–		+	+	+	–	+	+ +		+	–	+	+	+	+	+
Benin	–		–	–	–	–	+	– +		–	–	+	+	–	–	+
Bantu	–		–	+	+	–	+	– –		–	+	+	+	+	+	+
Cameroon	–		–	+	+	+	+	– +		+	–	+	–	+	+	–
Asian	+		+	+	+	–	+	+ +		–	+	+	–	+	+	–
	1		2	3	4	5	6	7 8		9	10	11	12	13	14	15

Figure 6.11. The α-like gene cluster on chromosome 11 is shown in the top part of the diagram. The polymorphic sites detected by different restriction enzymes are shown by the *arrows*. The bottom of the diagram defines the five different restriction fragment length polymorphism haplotypes according to recognition sites. (After Nagel and Labie 1985, Labie et al. 1986, Lapouméroulie et al. 1992.)

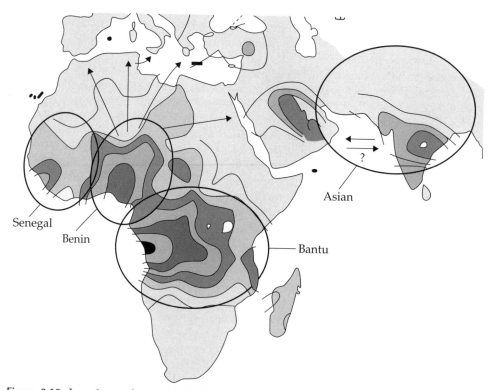

Figure 6.12. Location and variation in frequency of the four major β-globin haplotypes. Shading indicates changes in haplotype frequencies over space, probably reflecting gene flow from the center of origin of each haplotype. Arrows suggest movement patterns of haplotypes to other areas of the old world (From *The American Journal of Hematology*, vol. 27, page 140. © John Wiley and Sons, Inc., 1998. Reprinted with the permission of Wiley-Liss, Inc., a subsidiary of John Wiley and Sons, Inc.)

and thirteenth centuries. Then, sometime after the fifteenth century, both the Bantu and Senegal haplotypes arrived in Portugal (Lavinha et al. 1992). The Bantu, Senegal, and Benin haplotypes have been found in the Americas, suggesting that they were spread by the slave trade (Pante-De-Sousa et al. 1999).

These haplotypes are tightly linked to HbS. If they represent relatively recent independent mutational events, then they were probably subjected to malarial selection within the last 2,000–3,000 years. The chance of three or more independent mutations arising at about the same time (or at least being present in neighboring African populations) as malaria was becoming endemic is slight, according to some researchers. These scientists suggest that rare recombination (e.g., double crossing over) and **gene conversion** are much more likely explanations for the occurrence of multiple HbS haplotypes (Antonarakis et al. 1984, Flint et al. 1993). Gene conversion results from mismatched repairs to the DNA. That is, a direct copy of part of a gene can be inserted into the middle of another gene that resides on the other DNA strand. In the case of the S allele, the mutational event could have occurred once. Due to malarial selection, this HbS mutation could reach high frequencies within the population where it originated. Neighboring populations could easily have diverse haplotype composition due to genetic drift acting on small, geographically separated populations. With gene flow among these populations, it is possible for gene conversion and recombination to spread HbS onto these preexisting divergent haplotypes. This gene conversion could have occurred in numerous haplotypes in a number of populations. Through the processes of genetic drift and natural selection, one or more of these haplotypes could dominate in a population. If this process occurred in a number of populations, we would not expect the same haplotype to predominate in each area. Thus, we would expect to find different haplotypes bearing the HbS allele in different regions. It should also be noted that the mutation causing valine to replace glutamic acid involves the transversion of A to T. This is a relatively rare type of base replacement compared to others. This fact lends additional support to the argument that HbS arose as a single mutational event.

Anthropologists are interested not only in identifying the location of specific mutational events but also in the timing of these events. Early estimates suggested that the S allele arose between 3,000 and 6,000 generations ago, corresponding to between 70,000 and 150,000 years ago (Kurnit 1979, Solomon and Bodmer 1979). Nagel and Labie (1985) argued that these estimates were incorrect and suggested that 150 generations was ample time, corresponding to the expansion of malaria between 2,000 and 3,000 years ago. Recent simulations by Currat et al. (2002) set the origin of the Senegal haplotype not any earlier than 3,000 years ago (fewer than 100 generations ago). The majority of their simulations suggested a time frame of between 45 and 70 generations or between 1,350 and 2,100 years ago. These dates correspond reasonably well with the timing of the rise of the most recent common ancestor of all living P. falciparum at about 3,200–7,700 years ago (Volkman et al. 2001, Rich and Ayala 2000).

Understanding the interaction of parasite, vector, and host is vital to explaining the origin and spread of the S allele across Africa and into semitropical regions of the world. Clearly, the timing and complexity of the spread of the various haplotypes depends on numerous factors operating over the centuries since the origin of these mutations. Using molecular analyses such as these, migration patterns, historical connections, and the spread of specific alleles among dispersed populations will undoubtedly become much clearer.

OTHER HEMOGLOBINOPATHIES

There are several other structural mutations of the β-globin chain that are of interest to anthropologists and geneticists studying adaptation, evolution, and population structure. These are the result of specific single nucleotide substitutions (mutations) that change the amino acids in the β chain. Like *HbS*, the distribution of these hemoglobinopathies corresponds to areas where malaria is present, suggesting that these conditions provide some protection against falciparum malaria.

Hemoglobin E

HbE is caused by a structural mutation at position 26 in the β-globin chain ($\alpha_2\beta_2^{26\,Glu \rightarrow Lys}$). In contrast to the *HbβS* mutation which is a rare **transversion**, the mutation causing *HbβE* is a relatively common **transition** (GAG to AAG). *HbE*, the second most common abnormal hemoglobin in the world, is relatively frequent in parts of Southeast Asia (Fig. 6.13), with gene frequencies reaching 0.5 in some isolated groups (Flatz 1967). It is also found in low frequency in Turkey, where malaria was historically endemic. Migrants from Southeast Asia have also introduced *HbE* to Madagascar and the Philippines.

Haplotype data suggest that gene flow has probably been responsible for much of the distribution of *HbE*. Whether *HbE* arose multiple times in prehistory or only once is

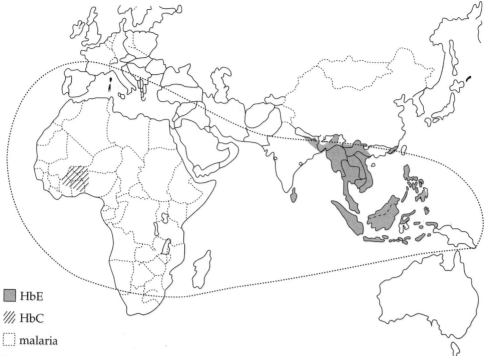

■ HbE

▨ HbC

▢ malaria

Figure 6.13. Distribution of HbC, HbE, and malaria. (Honing and Adams 1986, 238.)

debated. By analyzing restriction enzyme polymorphisms, Antonarakis et al. (1982) found evidence that the HbE mutation occurred at least twice in Southeast Asia. Other scenarios, such as gene conversion, double crossing over, or two single crossing-over events were rejected in favor of recurrent mutation. As with *HbS* and multiple origins, we must ask why the *HbE* mutation occurred at least twice in Southeast Asia and not once in Africa and why the *HbS* mutation occurred multiple times in Africa and not once in Southeast Asia.

The geographic overlap of the hemoglobinopathies and malaria provide circumstantial evidence that these genetic conditions provide protection against malaria. Clinical research suggests that the severity of malaria is reduced in individuals with *HbE* (Hutagalung et al. 1999). The protective role of *HbE* may be some unidentified membrane abnormality that makes most of the erythrocytes in *HbAE* heterozygotes resistant to invasion by *P. falciparum* (Chotivanich et al. 2002). It should also be noted that homozygous *HbEE* individuals are not as severely affected as are *HbSS* or *HbT/T* (thalassemia) homozygotes, suggesting that this mutation provides antimalarial protection at less cost compared to sickle cell anemia or thalassemia protection.

Hemoglobin C

HbC ($\alpha_2\beta_2^{6\ Glu \rightarrow Lys}$) is common in west Africa, reaching high frequencies in Burkina Faso and Mali (Fig. 6.13). Haplotype data suggest a unicentric origin of the HbC mutation in central west Africa, with later mutational modification in a few instances (Boehm et al. 1985, Trabuchet et al. 1991). The data also support the idea that gene flow was responsible for spreading β^C to north Africa.

As with *HbS*, individuals with *HbC* were also assumed to be protected against malaria. Early studies, however, failed to show reduced infection rates or reduced parasite densities in heterozygous *HbAC* individuals (e.g., Thompson 1962, Ringelhann et al. 1976). Two recent studies have, however, demonstrated that *HbC* protects against severe cases of falciparum malaria. Agarwal et al. (2000) found, in vitro, that *P. falciparum* parasites were unable to replicate within *HbCC* red blood cells and that *HbC* protects against severe malaria (an 80% reduction) in the Dogon population in Mali. Modiano et al. (2001) showed that the risk of clinical malaria in *HbAC* heterozygotes is reduced by 29%, while *HbCC* homozygotes show a 93% reduction. Unlike *HbS* and *HbE*, these data suggest that *HbC* protects individuals from malaria in the homozygous state (*HbCC*) better than in the heterozygous state (*HbAC*). The fact that *HbC* protects better in the homozygous state also helps explain its rather limited distribution in west Africa. That is, since the selective advantage of *HbC* would be proportional to its gene frequency, the homozygous *HbCC* would decrease with increasing distance from the center of origin.

Given the selective disadvantage *HbSS* and *HbSC* and the low frequency of *HbS* in the geographic epicenter of *HbC*, Modiano et al. (2001) suggest that, in the absence of malaria control, *HbC* would eventually replace *HbS* in west Africa. This replacement is contrary to the deterministic and stochastic models of Livingstone (1967) that show *HbC* being replaced by *HbS*. However, they support the modeling done by Cavalli-Sforza and Bodmer (1971) that show that *HbC* will eventually replace *HbS* in west Africa. Modeling the evolutionary outcome of triallelic systems with different fitness values for the various genotypes is difficult. Equilibrium values are often elusive, and the inaccuracy of fitness estimates for all genotypes creates problems in analysis. *HbC* may, in fact, be a transient polymorphism.

Other Hemoglobinopathies of Anthropological Interest

A number of other hemoglobin variants have been used to study human population structure, adaptation, and dynamics. These include *HbD* (also called hemoglobin Los Angeles or hemoglobin D-Punjab [$\alpha_2\beta_2^{121\ Glu \rightarrow Gln}$]), which reaches its highest frequencies of about 1% among the Sikhs of Punjab, India. It also occurs in England and in African American populations in North America and the Caribbean. This mutation does not carry any severe clinical consequences like the other hemoglobinopathies discussed above.

HbD and other variants have been used to explore such things as the historical role of population movements in shaping the genetic structure of Asia (Li et al. 1990). For example, the most frequent of 24 identifiable abnormal hemoglobins found along the Silk Road included *HbD*, *HbG-Taipei* ($\alpha_2\beta_2^{22\ Glu \rightarrow Gly}$), and *Hb G-Coushatta* ($\alpha_2\beta_2^{22\ Glu \rightarrow Ala}$). These all probably arose in central Asia and spread out as various ethnic groups interacted and traded along the Silk Road. *HbE*, however, probably arose in Southeast Asia and spread to central Asia with traders and travelers on the Silk Road. Similarly, it is argued that variants of *HbS*, *HbJ-Lome* ($\alpha_2\beta_2^{59\ Lys \rightarrow Asn}$), and others came from Africa as peoples moved and mixed in this trade network.

THALASSEMIAS

As we have seen, hemoglobinopathies, such as *HbS* and *HbE*, are caused by single nucleotide mutations that alter the β chain of hemoglobin. These are structural mutations. However, a group of genetically determined conditions, the thalassemias, are caused by a variety of mutations that either diminish or eliminate the synthesis of the hemoglobin chains. If the α chain is reduced or absent, the condition is called α-thalassemia. If the β chain is impaired, the disorder is known as β-thalassemia. If synthesis is reduced but not eliminated, the condition is designated as α^+-thalassemia or β^+-thalassemia. If the synthesis of a chain is eliminated, the designations are α^0-thalassemia and β^0-thalassemia. The term *thalassemia* comes from the Greek *thalassa*, meaning "Mediterranean Sea." The term was originally picked because many carriers of the gene were of Mediterranean origin. As can be seen, however, the actual distribution of β-thalassemia is much greater than just the Mediterranean region (Fig. 6.14).

α-Thalassemias

Most α-thalassemias are caused by deletions of one or more of the four α-globin genes. The most common form of α-thalassemia is called α^+-thalassemia, involving loss of a single α-globin gene. This form is probably due to unequal **crossover** and recombination (Fig. 6.15A). The crossover results in the formation of one chromosome with a single α gene (α–) and the other with three α genes ($\alpha\alpha\alpha$). The triple α chromosome does not cause any problems, nor does it appear to confer any advantages to the carriers. Since hematological measurements are often normal in heterozygotes carrying a single α deletion, the only reliable diagnosis is with DNA analysis. The α-thalassemias caused by deletions are listed in Table 6.1.

If one Hb α gene is deleted ($-\alpha/\alpha\alpha$), there is little, if any, effect on the synthesis of hemogloblin due to the fact that three *Hb* α genes remain active. The deletion of two *Hb*

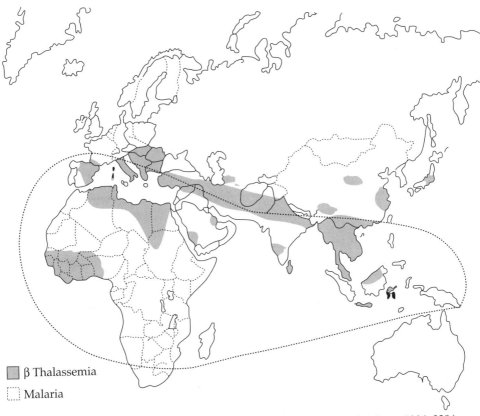

Figure 6.14. Distribution of β Thalassemia and malaria. (Honing and Adams 1986, 238.)

α genes (either –α/–α or ––/αα) produces severe anemia, while the deletion of three Hb α genes (–α/––) results in severe anemia and is characterized by the production of *Hb H* (α Hb β⁴ tetramer). Deletion of all four α genes is fatal, resulting in stillbirth.

A variety of nondeletion α-thalassemia mutations (including single nucleotide substitutions, frameshift, terminator codon, initiator codon, nonsense, RNA processing, and RNA cleavage) have been identified (Honig and Adams 1986). These are not as common as the deletion-type mutations and are found in different parts of the Old World (Vogel and Motulsky 1997).

The most common type of α-thalassemia in Africa and the Mediterranean is the rightward crossover (Fig. 6.15B). In Asia and Melanesia, both leftward and rightward crossovers have been documented. However, the rightward crossover is the most frequent. There are, in fact, four common deletions. These are differentiated by the amount of DNA lost (3.7 or 4.2 kb) and the position of the crossover (for $-\alpha^{3.7}$ only). Hence, the four deletions are designated $-\alpha^{4.2}$, $-\alpha^{3.7I}$, $-\alpha^{3.7II}$, and $-\alpha^{3.7III}$ (Embury et al. 1980). The single Hb α gene (–α), in its various forms, has apparently been amplified by malarial selection. Extensive research, especially in the South Pacific, has shown that α⁺-thalassemias show a clinal distribution, with the gene frequencies being proportional to the prevalence of malaria (Flint et al. 1986, 1998).

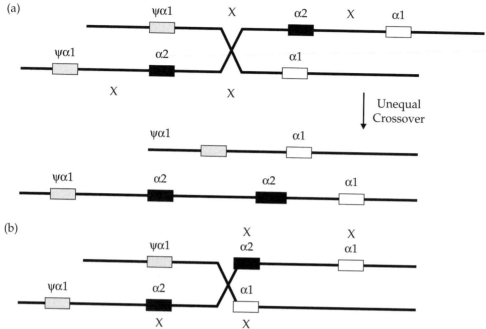

Figure 6.15. (a) Misalignment followed by crossing over, resulting in one chromosome carrying a single *α1* gene and the other having two copies of the *α2* gene in addition to a single *α1* gene. The crossing over depicted here is a leftward crossover, resulting in the creation of a single Hb *α* gene with a 4.2 kb deletion due to recombination. *X* indicates sequence homology. (b) A rightward crossover inside the *α* gene resulting in the creation of an *α2–α1* fusion gene with a 3.7 kb deletion due to recombination. *X* indicates sequence homology. Note that the area of recombination is further upstream in the leftward crossover depicted in *A*, hence the terms *leftward* and *rightward*. (Adapted from Vogel and Motulsky 1997, Honig and Adams 1986.)

Table 6.1 Deletion α-thalassemias

Clinical Condition	Hemoglobin (Hb) Symbols
Normal	αα/αα
Mild thalassemia (α-thalassemia-2 haplotype)	−α/αα
Severe thalassemia (α-thalassemia-1 haplotype)	−−/αα, −α/−α
HbH disease	−α/−−
Hydrops fetalis	−−/−−

Individuals with haplotype (−α/αα) show little or no hematological problems or impairment because three α genes remain active. Deletion of all four α genes is fatal (the name *hydrops fetalis* indicates the extreme edema of the stillborn infant). Deletion of two or more α genes produces more severe anemia and the production of hemoglobin H (Hb β⁴ tetramer).

After Vogel and Motulsky 1997, 321.

Table 6.2 Some frequent β-thalassemias in various populations

Group/Region	Mutation	Type	Frequency
African Americans	TAT box (−29)	β^+	39%
	Poly A site	β^+	26%
Mediterranean area	Intron 1 (position 110)	β^+	35%
	β^{39} terminator	β^+	27%
East Indians	Intron 1 (position 5)	β^+	36%
	Deletion (619 bp)	β^0	36%
Chinese	Frameshift (position 71/72)	β^0	49%
	Intron 2 (position 654)	β^0	38%

Adapted from Vogel and Motulsky 1997.

β-Thalassemias

Unlike α-thalassemias, which are caused primarily by deletions, β-thalassemias are caused by various mutational events affecting the synthesis of the β chain. About 150 mutations have been identified that cause β-thalassemia. These include transcription or promoter mutations, RNA cleavage and polyadenylation mutations, terminator (nonsense) mutations, frameshift mutations, and deletion mutations (rare).

Mild thalassemia results from regulatory mutations that affect the noncoding 5' upstream regions of the Hb β gene. These mutations affect gene transcription and diminish hemoglobin synthesis. Other β+-thalassemia mutations, such as the one that occurs in the downstream flanking sequences of the Hb β gene (AATAAA → AACAAA), affect transcription efficiency and are found in relatively high frequency among African Americans. Deletions are relatively rare, but interestingly a 619 bp deletion within intron 2 past the end of the Hb β gene accounts for more than one-third of the β-thalassemias found among East Indians (Table 6.2).

Most of the common β-thalassemias probably arose as unique haplotypes followed by expansion of the chromosomes due to malarial selection. These thalassemias are widespread throughout tropical and subtropical areas of the world (Table 6.3). The heterozygotes have mild anemia, elevated amounts of HbF (fetal hemoglobin) and HbA$_2$

Table 6.3 Some clinically important β-thalassemias

Disease	Genetics
β^0-Thalassemia major (Cooley's anemia[1])	Homozygote
β^+-Thalassemia major (Cooley's anemia)	Homozygote
β^0/β^+-Thalassemia	Compound heterozygote
Hb Lepore heterozygote	δ–β fusion
β^0, β^+, and δ β^0-thalassemia trait	Heterozygote
HbE-β-thalassemia	Compound heterozygote

[1]Named after its discoverer, a pediatrician in Detroit.
Adapted from Vogel and Motulsky 1997.

$(\alpha_2\delta_2)$, and smaller red cells. These heterozygotes usually do not need medical treatment. However, homozygotes often require blood transfusions because HbA is completely absent (β^0-thalassemia) or greatly reduced (β^+-thalassemia). Most of the hemoglobin in homozygotes is HbF. Growth failure is often followed by death during or in early adolescence.

Researchers have identified about 90 different β-thalassemia point mutations. This number of mutations results in frequent compound heterozygotes (individuals who inherit different thalassemias from each parent) being produced. For example, the β^{39} nonsense mutation accounts for about 27% of all thalassemia mutations in most Mediterranean populations (Table 6.2). The frequency of these heterozygotes is less in population isolates where single mutations account for the majority of the thalassemias. For example, the β^{39} nonsense mutation accounts for most of the β-thalassemia found in Sardinia (Vogel and Motulsky 1997).

THALASSEMIA HAPLOTYPES

Patterns of DNA polymorphisms (linked genes) that are on a single chromosome are called **haplotypes**. Because these DNA polymorphisms are close together, recombination is very infrequent. As a result, the polymorphisms are inherited as single alleles. Using restriction endonucleases, geneticists have detected seven restriction sites that define the α-globin haplotypes (Fig. 6.16). In addition, there are some differences in DNA sequence lengths between the ζ genes and the presence or absence of the ψ ζ gene. Seventeen polymorphic sites have been identified on the β-globin gene cluster. In most studies, however, only seven of the haplotypes are commonly used (Fig. 6.17). As we saw with *HbS*, haplotypes are very useful in reconstructing the origin, timing, and dispersal history of mutations. Because the mutations are passed from generation to generation on the haplotype where they arose, we can trace the migration of specific mutations over time and space. Also, we would expect that genetically separated populations should have unique haplotypes because of genetic drift and the slow movement of haplotypes across the

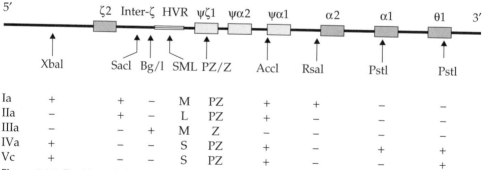

Figure 6.16. Position of the seven commonly used polymorphic restriction enzyme sites and the hypervariable region (HVR) on the α-globin gene cluster. Allele sizes for the inter-ζ HVR are part of the haplotype (designated as S small; M, medium; and L, large). PZ indicates the presence of $\psi\zeta1$ and Z, the presence of $\zeta1$ variants. Pseudogenes indicated by ψ (Adapted from Flint et al. 1998.)

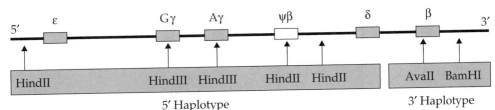

Figure 6.17. Position of the seven commonly used polymorphic restriction enzyme sites used to define the α-globin haploptypes. ψ indicates a pseudogene. (Adapted from Flint et al. 1998.)

Table 6.4 Geographic distribution of seven 5′ β-globin haplotypes

	Haplotypes						
Area	+----	-++-+	-+-++	----+	-+--+	-++--	-+---
Italy	61.8	10.9	25.3			1.0	0.3
Germany	43.8	6.3	31.3	6.3		12.5	
Greece	59.3	18.5	8.6	1.2	3.7	6.2	1.2
Britain	43.2	13.5	40.5		2.7		
Sardinia	58.7	21.5	13.2		3.3	3.3	
Cyprus	70.0	7.5	20.0			2.5	
North Africa	56.5	8.7	23.3	4.3		2.9	
West Africa	4.8	2.9	12.5	55.8	17.3		1.9
South Africa	8.0	12.6	14.9	34.5	25.3		2.3
U.S. blacks	21.4	7.1	21.4	26.2	14.3	2.4	2.4
China	79.7	10.0	6.9				
Thailand	75.7	3.3	17.3	1.2			
Cambodia	72.3	4.3	19.1		4.3		
Japan	66.7	11.1	11.1				
India	53.7	10.3	25.6		4.7	1.2	0.2
Indonesia	90.0	1.7	3.3				1.7
Melanesia	66.1	2.2	15.6	1.1	3.2		0.5
Micronesia	93.6					2.0	
Polynesia	90.7	1.1	4.5		1.5		0.7

Note the uniqueness of specific areas. Numbers are percentages; however, since there are actually more haplotypes than reported here, the figures will not sum to 100%.
Adapted from Flint et al. 1998.

geographic space. The distribution of the α-globin and β-globin haplotypes in selected world populations is shown in Tables 6.4 and 6.5. Note that in many populations there are a few common haplotypes and a number of unique haplotypes. These haplotype patterns can aid in distinguishing population from population since what is common in one population may not be common in another. The haplotypes are also of use in population structure studies such as those discussed in Chapter 12.

Table 6.5 Geographic distribution of five α-globin haplotypes

Area	Ia	IIa	IIIa	IVa	Vc
Iceland	41.7	8.3	25.0		
Britain	53.3	16.7	6.7		
Mediterranean	56.0	20.0	6.0		
Asian Indian	54.2	16.7	6.3		
Sri Lanka	25.0	35.7	21.4		
Saudi Arabia	35.7	21.4	14.3		
Nigeria					4.2
Gambia			7.1	14.3	
Zambia	8.3				
Senegal	11.4		1.0		
Thailand	19.4	30.6	16.7		
Burma	13.3	50.0	3.1		
South China	33.3	28.1	7.0		
Taiwan	34.5	16.7	4.8		
Brunei	11.4	20.0	14.3		
Philippines	31.8	9.1	9.1		
Tarawa[1]	24.2	3.0	6.1	21.2	3.0
Ponape[1]	31.0	21.4	19.0	4.8	7.1
Guam[1]	57.7		7.7		5.3
Palau[1]	36.8	26.3	10.5		2.6
Majuro[1]	13.9	9.7	15.3	5.6	
PNG[2] (highlands)[3]			57.9	5.3	5.3
PNG[2] (coast)[3]			60.0	20.0	20.0
Vanuatu[3]	3.7		4.3	5.3	
Fiji[3]	22.0		22.0	16.9	13.6
Tonga[4]	37.9	17.2	27.6		
Somoa[4]	26.9	4.2	19.5	14.9	0.3
Niue[5]	75.9	1.9	3.7	11.1	
Cook Islands[5]	36.4	3.6	12.7	5.5	9.1
Tahiti[5]	28.8	6.4	2.6	3.2	4.5
Maori[5]	23.2	10.4	11.6	40.0	

Note the uniqueness of specific areas. Numbers are percentages; however, since there are actually more haplotypes than reported here, the figures will not sum to 100%.
[1]Micronesia, [2]Papua New Guinea, [3]Melanesia, [4]West Polynesia, [5]East Polynesia.
Adapted from Flint et al. 1998.

THE THALASSEMIAS AND NATURAL SELECTION

The thalassemias are common in malarial regions of the world. This association gives us the first line of evidence suggesting that these hemoglobinopathies may provide protection against malaria. In fact, the frequency of α^+-thalassemias ranges 10%–80% in many of these populations. The highest frequencies of $-\alpha^{4.2}$ are found in Southeast Asia and the Pacific; the $-\alpha^{3.7III}$ variant is found exclusively in Oceania; and the $-\alpha^{3.71}$ variant is common in Africa and the Mediterranean.

The nondeletion and α^0-thalassemias, though relatively rare, are also found almost exclusively in areas with malaria. Similarly, the β-thalassemias are found in malarial regions of the world. The various β-thalassemia mutations have population-specific distributions, but all are found predominantly in areas with malaria.

Some of the most convincing evidence for an association between thalassemia and the endemicity of *P. falciparum* comes from the Pacific. In malarial regions, the gene frequency of thalassemias is clinal, mirroring the prevalence rates of malaria (Flint et al. 1986, 1993, 1998). In Papua New Guinea, for example, as malaria becomes less frequent in the interior in comparison to the coast, the frequency of α^+-thalassemia gradually declines. In the Vanuatu archipelago, the decline in α^+-thalassemia rates follows a similar decline in malaria rates. Other regions in the Pacific show a similar clinal distribution (both on latitude and altitude), following the prevalence of malaria.

The frequency of thalassemia in nonmalarial regions in the Pacific helps support the idea that these mutations are advantageous in a malarial environment. In areas where there is no malaria, thalassemia frequencies range 0%–12% and vary from island to island, in a random pattern, showing no clinal variation. It also appears that the $-\alpha^{3.7III}$ deletion that is found in Polynesia arose as a single mutation (on one haplotype) and spread over a very large area by migration. Genetic drift then molded the frequencies over time, producing the random pattern observed today in the central Pacific (Flint et al. 1993, 1998).

At least 10 specific mutations that are regionally specific have high frequencies in malarial regions. For example, the $-\alpha^{4.2}$ deletion is high in the northern parts of Papua New Guinea and the $-\alpha^{3.71}$ deletion is high in the south (Hill et al. 1985). At least six mutations produce very similar, if not identical, phenotypes. We know that natural selection acts at the phenotypic level. So, natural selection has elevated the frequencies of these independent thalassemia mutations in malarial regions of the Pacific. This pattern provides additional support for the malaria hypothesis.

Further evidence supporting the idea that the various thalassemias provide protection against *P. falciparum* comes from morbidity studies. A study in Papua New Guinea by Allen et al. (1997) revealed that, compared to "normal" children, α^+-thalassemia homozygous children ($-\alpha/-\alpha$) had a 0.40 risk of having severe malaria while heterozygotes ($-\alpha/\alpha\alpha$) had a 0.66 risk. In addition, it was found that the risk of hospitalization for infections other than malaria was also less for those with α^+-thalassemia (homozygote risk, 0.36; heterozygote risk, 0.63). The mechanism for this additional protection remains elusive but may be associated with an enhanced cellular immune response. Research also lends support for the selective advantage of α-globin deletions in Nepal (Modiano et al. 1991), where malaria morbidity was markedly reduced.

The mechanisms responsible for the distribution of the β-thalassemias have not been systematically studied using haplotypes as has been done with the α-thalassemias. As with the α-thalassemias, however, the various β-thalassemia mutations show clear associations with particular haplotypes. These various mutations are regionally specific, are in many cases phenotypically similar, and have elevated frequencies in malarial regions. All of these features strongly suggest an association with malaria. The reasons for the selective advantage of the various haplotypes are not clear but may have something to do with impairing the growth of the parasite or enhancing removal by the immune system (Evans and Wellems 2002, Akide-Ndunge et al. 2003). More research is needed to clarify the relationships among the various mutations and their role in protecting against malaria.

CHAPTER SUMMARY

The study of human hemoglobin has aided in our understanding of many genetic processes and relationships. It has helped us in detecting and identifying various types of mutation and their geographic spread, it has aided in our understanding of the relationship between gene structure and gene function, it has provided a model for comprehending regulatory mutations and developmental gene switching, and it has served as the classic model for natural selection in humans.

The hemoglobin molecule is composed of four polypeptide chains of amino acids and an associated heme group. Two of these chains are designated α (141 amino acids long) and two β (146 amino acids long). Normal hemoglobin is responsible for transporting oxygen throughout the body. Mutations affecting this structure resulted in the hemoglobinopathies. These hemoglobin variants include sickle cell (HbS), HbC, HbE, and HbD. These variants arose as single-point mutations and have historically increased in frequency because of the action of natural selection in malarial environments coupled with gene flow. The S allele became frequent in tropical and subtropical areas because heterozygotes (HbA HbS) have a selective advantage in environments where falciparum malaria is endemic. This is a case of balancing selection (balanced polymorphism) in which the heterozygotes have the highest fitness.

Two theories have been put forth to explain the origin of sickle cell anemia. The single-mutation theory posits that a random mutation occurred in sub-Saharan Africa sometime during the Neolithic. As populations moved south, opening the rainforest for horticulture, mosquitoes and malarial parasites expanded their ranges. Prior to this southern expansion of peoples, the frequency of the S allele was probably very low. As malaria increased, it became an evolutionary advantage to have the AS genotype. As selection pressures increased, so did the frequency of sickle cell anemia. Gene flow spread the mutation even further.

The multiple-mutation theory suggests that the sickle cell mutation arose independently at least five times in various regions of the Old World. The selective advantage of these mutations caused the allele to increase in frequency in malarial regions. Gene flow was also responsible for spreading these various mutations from population to population.

Other structural mutations on the β chain are responsible for other hemoglobinopathies, such as HbE, which is prevalent in Southeast Asia; HbC, which is common in west Africa; and HbD, which reaches its highest frequency in Punjab, India. These mutations also provide protection against falciparum malaria.

Aside from these structural changes of the β chain, a group of genetically determined conditions, the thalassemias, are due to a variety of mutations that either diminish or eliminate the synthesis of the hemoglobin chains. If the α chain is reduced or absent, the condition is called α-thalassemia. If the β chain is reduced or absent, the disorder is known as β-thalassemia. As with sickle cell anemia, these various mutations provide varied degrees of protection against malaria.

SUPPLEMENTAL RESOURCES

Cavalli-Sforza L L and Bodmer W F (1971) *The Genetics of Human Populations*. San Francisco: W H Freeman.

Flint J, Harding R M, Boyce A J, and Clegg J B (1998) The population genetics of the haemoglobinopathies. *Baillière's Clinical Haematology* 11:1–51.

Flint J, Harding R M, Clegg J B, and Boyce A J (1993) Why are some genetic disorders common? Distinguishing selection from other processes by molecular analysis of globin gene variants. *Human Genetics* 91:91–117.

Honig G R and Adams J G III (1986) *Human Hemoglobin Genetics*. Vienna: Springer-Verlag.

Huisman T H J, Carver M-F, and Efremov G P (1996) *A Syllabus of Human Hemoglobin Variants*. Augusta, GA: Sickle Cell Anemia Foundation, 1996.

Lehmann H and Huntsman R G (1974) *Man's Haemoglobins*, 2nd ed. Amsterdam: North-Holland.

Livingstone F B (1958) Anthropological implications of sickle cell gene distribution in West Africa. *American Anthropologist* 60:533–562.

Rich S M and Ayala F J (2000) Population structure and recent evolution of *Plasmodium falciparum*. *Proceedings of the National Academy of Sciences USA* 97:6994–7001.

7

Human Leukocyte Antigen and Some Polymorphisms of Anthropological Interest

In this chapter, we examine four traits of simple inheritance in humans that are of considerable anthropological interest. Firstly, we look at the major histocompatibility complex (MHC) or the human leukocyte antigen (HLA) system, the most polymorphic of all human genetic systems yet found. It is intimately associated with the immune system and provides humans with great flexibility in dealing with a variety of disease conditions. Its extensive polymorphic nature also makes it an excellent genetic system to use to study population diversity, evolution, origins, and relationships. Secondly, we explore lactase restriction and persistence. This interesting characteristic has dietary implications that require researchers to examine numerous lines of evidence (genetic, environmental, archaeological, and sociocultural) in order to understand its distribution and evolution. Thirdly, we explore the genetics of taste variation by looking at the phenylthiocarbamide system. Again, dietary features come into play with this genetic trait. Fourthly, we look at the diversity that is displayed in human ear wax.

THE MAJOR HISTOCOMPATIBILITY COMPLEX

As we saw in Chapter 5, the immune system is extremely complex, consisting of four major components. For anthropological purposes, the humoral and the cellular defense systems are of the most interest because of the genetic variation they contain and their association with diseases. The immune system is responsible for recognizing substances as either "self" or "nonself" and reacting in a specific manner if the substances are foreign to the body. As we have already seen, the humoral immune response is involved in the removal of antigens by antibodies in the body fluids. However, pathogens that have invaded cells cannot be removed by this system. Hence, the cellular immune response is responsible for recognizing body cells that have been invaded or altered by foreign pathogens. This recognition is achieved by various antigens that are part of the MHC. The antigens in humans are the HLAs. These antigens were initially discovered on the surface of white blood cells, or leukocytes. We now know that these antigens reside on virtually all cells of the body.

The history of the discovery of the MHC dates back to the recognition that skin grafts from one person to another were usually rejected. In addition, while studying tumor transplants between mice, Little and Tyzzer (1916) suggested that the graft rejection was due to a number of dominant genes. Then, in 1927, Bauer discovered that skin grafts between identical twins were not rejected, supporting the earlier suggestion that genetic mechanisms were responsible for the rejections. Research continued with the eventual discovery in the 1930s and 1940s of the MHC (called H-2) in mice. In the 1950s, Dausset

(1958) and von Rood et al. (1958) outlined the MHC in humans by identifying the poly-morphic gene loci (*HLA-A*, *HLA-B*, and *HLA-C*) responsible for the antigens that caused transplantation rejections. The discovery of the *HLA-D* locus followed (Jones et al. 1975, Park et al. 1978).

LOCATION AND ORGANIZATION OF THE MHC

The MHC is located on the short arm of chromosome 6, spanning a region of about 4 mil-lion base pairs of DNA. The MHC genes are divided into three major classes: I, II, and III. The class II genes are located nearest the centromere. In between these genes and the class I genes are the class III genes (Fig. 7.1). The entire region includes about 200 genes;

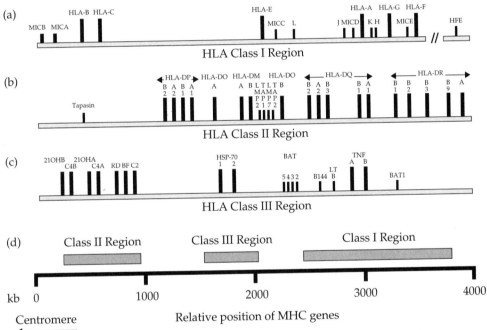

Figure 7.1. Genetic map of the human leukocyte antigen (HLA) region on the short arm of chromosome 6. (a) Map of the class I region. *HLA-H*, *HLA-J*, *HLA-K*, and *HLA-L* are pseudogenes that are not expressed. The *MIC* gene family is composed of five genes. *MICA* and *MICB* are expressed genes, while *MICC*, *MICD*, and *MICE* are pseudogenes. *HFE* on the far right is a functional class I-like gene. In addition, there are over 50 other genes interspersed within the class I region. (b) Map of the class II region genes. As with the class I region, there are other genes interspersed in this region. The *LMP2* and *LMP7* genes code subunits of protease proteasome. The *TAP1* and *TAP2* genes code for a peptide transporter, while tapasin aids in peptide delivery. (c) Map of the class III region. This region consists of a heterogenous group of about 75 genes. Some of these genes are important in the immune defense system, e.g., complement components C4, C2, and factor B[4], while others are not. (d) Simplified map of the HLA complex showing the relative location of the three regions. (Adapted after Marsh et al. 2000, Vogel and Motulsky 1997, Ahnini et al. 1997.)

Table 7.1 Human leukocyte antigen (HLA) class I and class II genes

Class I Genes	Number of Alleles[1]	Class II Genes	Number of Alleles[1]
HLA-A	290	*HLA-DMA*	4
HLA-B	553	*HLA-DMB*	6
HLA-C	140	*HLA-DOA*	8
HLA-E	6	*HLA-DOB*	8
HLA-F	2	*HLA-DPA1*	20
HLA-G	15	*HLA-DPB1*	106
HLA-H	—	HLA-DPA2	—
HLA-J	—	HLA-DPA3	—
HLA-K	—	HLA-DPB2	—
HLA-L	—	*HLA-DQA1*	25
HLA-N[2]	—	*HLA-DQB1*	56
HLA-S[2]	—	HLA-DQA2	—
HLA-X[2]	—	HLA-DQB2	—
		HLA-DQB3	—
Total number	1,006	*HLA-DRA*	3
		HLA-DRB1	354
		HLA-DRB2	1
		HLA-DRB3	39
		HLA-DRB4	12
		HLA-DRB5	17
		HLA-DRB6	3
		HLA-DRB7	2
		HLA-DRB8	1
		HLA-DRB9	1
		HLA-Z[3]	—
		Total number	666

Bold indicates expressed genes and non-bold, pseudogenes.
[1]As research continues, more alleles are found and verified, so these numbers often change.
[2]Gene fragments.
[3]Class I gene fragment located within the class II region.
From Marsh et al. 2000 and European Bioinformatics Institute (EMBL-EBL) 2003.

however, only about 10%–20% are involved with the defense and immune system. The antigens (called **alloantigens**) that are encoded are the HLA. Two major classes of antigens are defined.

Class I includes the *HLA-A, HLA-B, HLA-C, HLA-E, HLA-F,* and *HLA-G* loci (Table 7.1). In addition, there are four nonfunctional pseudogenes (*HLA-H, HLA-J, HLA-K,* and *HLA-L*) located in the same general region on the chromosome. Just recently, three gene fragments have been identified and named *HLA-N, HLA-S,* and *HLA-X* (Marsh et al. 2002).

HLA class II molecules consist of α and β chains of about equal size. In most cases, the α- and β-chain loci are located next to each other on the chromosome (Fig. 7.1). There are five types of HLA in this region: *HLA-DM, HLA-DO, HLA-DP, HLA-DQ,* and *HLA-DR.* As with the class I region, there are a number of pseudogenes (e.g., *HLA-DQA2* and *HLA-DPA2*) in this area as well (Table 7.1).

The class III region consists of a heterogenous group of about 75 genes that encode a variety of proteins, including the complement components (C4, C2, and factor B[4]). These genes are important in the immune defense system by aiding such processes as **lysis**, chemotaxis, and histamine release. In addition, a number of genes within this region have no relationship with the immune system. As a group, we will not discuss this region but will focus only on the HLAs of class I and class II.

Each individual has two alleles (antigens) for each of the HLA loci. The presence of one antigen (allele) does not inhibit the expression of the other antigen. Therefore, HLAs are codominant. If an individual has two different alleles at a locus, he or she is *heterozygous*; and if an individual has two identical alleles, he or she is *homozygous*. The complete set of alleles found on one chromosome is called the HLA haplotype. In most cases, individuals inherit the complete set of alleles located on a single chromosome (Fig. 7.2). Recombination does occur, but it is rare. Note that Figure 7.2 shows only two possible offspring out of the four possible combinations of haplotypes (*AC, AD, BC, BD*).

The number of unique haplotypes, given the number of alleles at each locus of the HLA system, is simply amazing (Table 7.1). For example, using only three loci and fewer alleles than we now know exist, geneticists have calculated that over 10^9 distinct haplotypes could be generated (Bodmer 1996). The extremely high degree of polymorphism

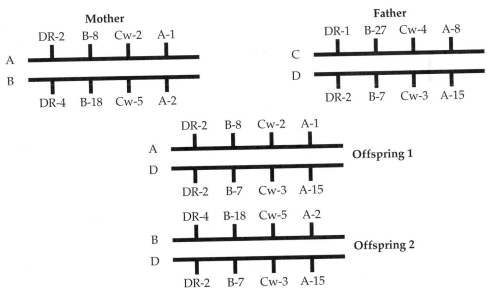

Figure 7.2. Inheritance of human leukocyte antigen (HLA) haplotypes. An individual inherits all of his or her HLA alleles together, one set from each parent. Only two possible offspring are shown here. Note that offspring 1 is homozygous for the DR locus and heterozygous for the other loci. Recombination is rare but can occur and change the haplotypes. (Adapted from Tiwari and Terasaki 1985.)

found in the MHC loci could, theoretically, be due to a number of factors: (1) very high mutation rates, (2) gene and allele conversion, (2) selection in the form of heterozygote advantage, or (4) frequency-dependent natural selection. This extremely high level of polymorphism is one of the most valuable (and often daunting) features of the MHC for anthropologists studying population history, evolution, and diversity.

LINKAGE DISEQUILIBRIUM

Aside from the high degree of polymorphism, another striking property of the MHC complex is the fact that some HLA alleles occur together more frequently than expected by chance alone. For example, *HLA-A1* and *HLA-B8* are found together about five times more often than expected by chance. This nonrandom combination or pairing is referred to as linkage disequilibrium. Remember that *linkage* refers to the fact that two genes are located within a measurable distance between each other on the same chromosome. Two main reasons are responsible for linkage disequilibrium:

1. If two populations that were homozygous for different HLA haplotypes merged recently, there may not have been sufficient time for recombination to distribute the alleles throughout the population in a random fashion.
2. Unique combinations of closely linked loci can be maintained because of the selective advantage they confer on the bearers.

Geneticists have developed techniques to test these possibilities and have found that natural selection, rather than recombination, appears to be primarily responsible for linkage disequilibrium in the HLA system (e.g., see Bodmer 1972). It is suggested that infectious diseases (viral) and autoimmune disorders are probably the most important selective forces maintaining HLA linkage disequilibrium and the extensive genetic polymorphism (Bodmer 1996, Black and Hedrick 1997, Dean et al. 2002).

HLA AND DISEASE ASSOCIATIONS

The facts that the HLA system is so polymorphic and that it plays an integral role in controlling the immune response suggest that this genetic system should have a major influence on disease susceptibility. The first disease association in humans involved *HLA-A1* and Hodgkin's disease (Amiel 1967). Other associations were soon found (Table 7.2).

The concept of disease association is an important one and needs some clarification. For the HLA system, the term *association* is used to indicate that the specific gene is found in higher frequency in a certain condition or disease. One of the most striking associations found is between *HLA-B27* and ankylosing spondylitis. Studies have shown that this disease is 87 times more likely to occur in individuals who carry the *HLA-B27* allele. Most of the strong associations are now found with the class II polymorphisms rather than the class I antigens. The list of disease associations is now relatively long (see Tiwari and Terasaki 1985, Lechler 1994, Browning and McMichael 1996).

The diseases that are associated with the HLA system fall into four broad categories: autoimmune diseases, infectious diseases, malignant disorders, and diseases with no known association with autoimmunity (Bell et al. 1989). Many HLA-associated diseases also probably have multifactorial etiologies, including genetic, environmental, and other

Table 7.2 Association between specific human leukocyte antigen (HLA) genes and some selected diseases

Disease	HLA Gene	Relative Risk[1]
Hodgkin's disease	*A1*	1.4
Idiopathic hemochromatosis	*A3*	8.2
Idiopathic hemochromatosis	*B14*	4.7
Ankylosing spondylitis	*B27*	87.4
Reiter's disease	*B27*	37.0
Subacute thryroiditis	*B35*	13.7
Congenital adrenal hyperplasia	*B47*	15.4
Psoriasis vulgaris	*Cw6*	13.3
Multiple sclerosis	*DR2*	4.1
Goodpasture's syndrome	*DR2*	15.9
Celiac disease	*DR3*	10.8
Rheumatoid arthritis	*DR4*	4.2
Pernicious anemia	*DR5*	5.4

[1]Relative risk (a measure of the strength of the association) is equal to

$$\frac{(\text{Freq. of patients with a specific antigen}) \times (\text{Freq. of healthy controls without the antigen})}{(\text{Freq. of healthy controls with the antigen}) \times (\text{Freq. of patients without the antigen})}$$

Adapted from Vogel and Motulsky 1997, 224.

factors. Most of the associations found to date involve the immune system. Examples include ankylosing spondylitis, rheumatoid arthritis, insulin-dependent diabetes, Goodpasture's syndrome, and multiple sclerosis. Both HLA susceptibility to and protection from Type 1 diabetes is one of the more thoroughly studied of the disease associations (Heard 1994, Cucca and Todd 1996). The number of associations with infectious diseases is not large but includes mononucleosis, severity of dengue, progression to tuberculoid leprosy, predisposition to lepromatous leprosy, risk and expansion of herpes simplex virus-2, progression to chronic Lyme disease, severity of malaria, and human immunodeficiency virus-1 (HIV-1, the rapid progression to acquired immunodeficiency syndrome [AIDS] of homozygotes). Malaria and HLA present a rather interesting picture since we have already discussed the relationship between the various hemoglobinopathies and malaria (Chapter 6). Allele *HLA-Bw53* and haplotype *DRB1*1302/DQB1*0501* are common in west Africa but rare elsewhere. Hill et al. (1991) demonstrated that these are independently associated with protecting their bearers from severe malaria. The elevated frequencies of this allele and haplotype in west Africa are thus attributed to the action of natural selection against malaria. Malignant disorders include Hodgkin's disease and nasopharyngeal carcinoma, which may, however, be related to an infectious agent. It is also very interesting that the association (*A2/B46/DR9*) with nasopharyngeal cancer is found only among Chinese populations and that patients with this cancer in other populations do not have this haplotype. Examples of disease conditions that are not linked to autoimmunity include narcolepsy, 21-hydroxylase deficiency (deletion of the *21-OH B* gene, which is in linkage disequilibrium with *Bw47*) and hemachromatosis. The association between B8 and hemachromatosis is most likely due to linkage with another gene located at 6p21.3 (Yaounq et al. 1994).

The HLA complex is responsible for recognizing specific antigens (antigenic peptides) and presenting them to the T cells (lymphocytes), thus starting the immune response and the eventual removal of the invading agents. Hence, the association between certain HLA loci and specific diseases may be the result of differences in the presentation or recognition of certain HLA antigens. Scientists are still not sure of the exact reasons for the disease associations that have been found. At least eight explanations have been suggested (see Hall and Bowness 1996 for a detailed discussion). While there are multiple possibilities, two reasons are most often suggested. (1) The biological function of the HLA allele itself (or possibly even the haplotype) may be the cause of the association and of the disease. (2) The disease may not be caused by the HLA locus but by another gene(s) that is tightly linked with the specific HLA locus. To complicate matters, the presence of a specific HLA antigen may be necessary for a specific disease condition to develop or progess; however, its presence is not sufficient to cause the disease. For example, only about 2% of individuals who bear the HLA-B27 antigen actually develop ankylosing spondylitis. Furthermore, this frequency increases to only about 20% if there is also a history of the disease in the family. These results suggest that some of these disease conditions are the result of multiple factors, both genetic and environmental.

The association of particular alleles with specific diseases is found in most populations around the world (nasopharyngeal cancer is an exception, as noted above). This finding suggests that the susceptibility is actually the HLA allele itself or that the gene is in tight linkage disequilibrium with the allele that is present in all populations. Most associations are, however, probably caused by the HLA specificities themselves or their subtypes.

These disease associations have also aided researchers in elucidating the pathogenesis of a variety of diseases. For example, mature-onset diabetes does not show an association with any HLA antigens, while insulin-dependent diabetes does (*HLA-DR3, DR4, DR2*). This difference suggests that these two types of diabetes probably have different etiologies. A virus or some type of autoimmunity may be involved with insulin-dependent diabetes.

HLA AND POPULATION DIVERSITY

Many HLA alleles and haplotypes are unique to specific geographic locations (Tables 7.3–7.5). Hence, there are alleles and haplotypes that are characteristically African, Asian, American, Australian, and European in origin (e.g., Klitz et al. 2003). For example, the *A1/B8/DR3* haplotype is a European marker, while *A30/B18/DR3* characterizes Sardinians, Gypsies, and Basques. The *A2/B46/DR9* haplotype is found only in populations of Chinese origin. Some haplotypes have a unique distribution and can be used as specific population markers because of either their presence or absence (Tables 7.4 and 7.5). As an example, the *DRB1*12/DQA1*0101/DQB*0501* haplotype is found in northern and southern Africa and in South America. It is, however, not found in the San or the Khoikhoi in Africa, while *DRB1*03/DQA1*03/DQB1*0201* is found in the Khoi, San, and southern African blacks (Bodmer 1996). There are fewer alleles that are geographically widely distributed (e.g., *HLA-A2* and *HLA-DRB1*). There are, however, at least 17 DNA variants of *HLA-A2*.

A number of HLA alleles show clinal distributions, suggesting the action of gene flow and/or natural selection. For example, the frequency of *HLA-B5* decreases as one

Table 7.3 Frequency distributions of some selected human leukocyte antigens (HLAs) in 87 world populations

HLA	\% Positive									AF[1]	Range
	0	1–5	6–10	11–20	21–30	31–40	41–50	51–75	76–100		
A1	4	31	23	27	2	0	0	0	0	8.2	0–29
A2	1	0	4	25	40	15	1	1	0	24.6	0–51
A3	4	33	25	25	0	0	0	0	0	7.5	0–20
A11	8	29	29	12	6	2	1	0	0	8.6	0–42
A23	21	51	8	7	0	0	0	0	0	2.7	0–17
A24	0	14	27	32	3	4	4	3	0	16.1	1–64
A26	5	60	20	2	0	0	0	0	0	4.4	0–13
A29	19	57	10	1	0	0	0	0	0	2.3	0–11
A30	10	60	6	9	2	0	0	0	0	4.0	0–25
A31	9	69	5	2	2	1	1	0	0	4.0	0–47
A32	23	56	8	0	0	0	0	0	0	2.4	0–11
A33	11	53	17	6	0	0	0	0	0	4.4	0–19
A68	16	52	17	2	0	0	0	0	0	3.0	0–13
B7	4	40	31	12	0	0	0	0	0	6.4	0–20
B8	13	43	23	8	0	0	0	0	0	4.4	0–16
B13	12	52	18	5	0	0	0	0	0	3.7	0–19
B18	18	50	16	2	1	0	0	0	0	3.7	0–27
B27	11	62	13	1	0	0	0	0	0	3.0	0–11
B35	3	31	27	23	1	2	0	0	0	8.4	0–38
B39	16	63	3	5	0	0	0	0	0	2.9	0–17
B44	6	33	30	17	1	0	0	0	0	7.6	0–24
B51	5	45	41	6	0	0	0	0	0	6.4	0–18
B52	16	58	11	2	0	0	0	0	0	3.0	0–11
B55	21	60	5	0	1	0	0	0	0	2.0	0–22
B57	17	63	7	0	0	0	0	0	0	2.2	0–8
B58	11	57	14	4	0	1	0	0	0	3.2	0–36
B60	15	48	11	11	2	0	0	0	0	4.3	0–29
B61	16	56	9	4	2	0	0	0	0	4.0	0–26
B62	4	58	14	8	2	1	0	0	0	5.4	0–34
B63	21	63	3	0	0	0	0	0	0	1.1	0–8
B70	30	47	6	3	1	0	0	0	0	1.7	0–22

The figures are the number of populations, out of the 87, that show differing frequencies of each antigen. For example, four populations do not possess HLA-A1, while the frequency of HLA-A1 in 31 populations ranges 1%–5%. Note also that gene frequencies vary greatly among populations and that HLA-A2 is present in all 87 populations except one (New Guinea), B70 is not present in 30 of the populations examined, and A24 is present in all 87 populations.

[1]Estimated population average gene frequency.

Adapted from Bodmer 1996.

Table 7.4 Presence (P) or absence of some selected three-locus human leukocyte antigen (HLA) haplotypes (HLA-A, HLA-B, and HLA-DR) in various populations

Population	Haplotype						
	A1/B8/DR3	A3/B7/DR15	A2/B44/DR4	A30/B18/DR3	A33/B44/DR7	A2/B46/DR9	A30/B42/DR3
Hungary	P						
Yugoslavia	P	P					
Sweden	P		P				
Norway	P	P					
Germany	P						
Denmark	P	P					
Canada	P	P					
Sardinia				P			
Spanish Gypsy				P			
Basque	P		P	P			
Nepal					P		
Thai					P	P	
Korea					P		
Thai Chinese					P	P	
Singapore Chinese						P	
Taiwan						P	
Miao						P	
Vietnam						P	
China						P	
Caribbean							P
South African blacks							P

In all cases, these haplotypes are also in linkage disequilibrium.
Data taken from Bodmer 1996.

goes north from the Middle East to Iceland (Fig. 7.3). Similarly, *HLA-B8* decreases in frequency from a high in Scotland to a low in southern Italy and Greece (Fig. 7.3). Most of these clinal distributions have been attributed to population movements and gene flow rather than selection and have been used to aid in elucidating the population history of Europe and other regions of the world.

The MHC is the most polymorphic of the genetic systems in humans. In addition, the majority of people tested for HLAs are heterozygous; thus, populations show a deficiency in homozygosity. High levels of heterozygosity and polymorphism can be maintained in a population by four major factors: *(1)* a very high mutation rate, *(2)* gene and allele conversion, *(3)* balancing selection (heterozygote advantage), and *(4)* frequency-dependent selection. Allele conversion occurs when small segments of an amino acid sequence are replaced with the homologous region of another allele. This recombination process is similar to gene conversion, which occurs between alleles of different but homologous loci.

Most research has concluded that natural selection (primarily balancing selection) is probably the most important mechanism in maintaining the high levels of observed polymorphism found at the various HLA loci (e.g., Hedrick and Thomson 1983, Hughes and Nei 1989, Hedrick, et al. 1991, Black and Hedrick 1997, Salamon et al. 1999, Dean

Table 7.5 Presence (P) or absence of some selected three-locus human leukocyte antigen (HLA) haplotypes (*HLA-DR*, *HLA-DQA*, and *HLA-DQB*) in various populations

Population	Haplotypes						
	DRB1*04 DQA1*03 DQB1*0302	DRB1*03 DQA1*0501 DQB1*0201	DRB1*09 DQA1*03 DQB1*0303	DRB1*08 DQA1*0103 DQB1*0601	DRB1*04 DQA1*03 DQB1*0401	DRB1*12 DQA1*0101 DQB1*0501	DRB1*12 DQA1*0601 DQB1*0301
San (Bushmen)	P						
Denmark	P	P					
Khoikhoi	P	P					
Germany	P	P					
Singapore	P	P	P		P		P
Korea	P	P		P	P		
Canada	P	P		P			
Tlingit	P		P				
Spain	P	P					
France	P	P					
Sardinia	P	P					
Romania	P	P				P	
South American black	P	P				P	
Japan (Wajin)	P		P	P	P		
United States	P	P					
South African black	P	P				P	
Buyi	P	P	P	P			P
Italy	P	P					
North American black	P	P				P	
Algeria	P	P					
Greece	P	P					
Native American	P	P	P				
Inuit		P	P				
Gypsy		P					
Papua New Guinea (highlands)		P		P	P		
Zuni						P	
Thai Chinese							P

In all cases, these haplotypes are also in linkage disequilibrium. For all serotypic alleles (e.g., *DQB1*0302*), the first two digits (03) indicate the serotype and the remaining digits (02) indicate the specific allelic sequence. By international convention, alleles are now identified by six digits.

Data taken from Bodmer 1996.

et al. 2002). Further, many of these studies have also shown that there are higher rates of nonsynonymous (changing the amino acid by base substitutions in the first and second positions of the codon) than synonymous amino acid substitutions in the antigen recognition sites for class I and class II antigens. If balancing selection is for novel patterns of antigen recognition and presentation, the increased polymorphism gives individuals a better immune system and an enhanced ability to protect against a wide variety of pathogens.

As we have seen (e.g., Chapter 5), the clarification of population origins and history is of great interest to anthropologists and human geneticists. As with blood groups and serum proteins, HLAs have been used to study the genetic relationships and origins of

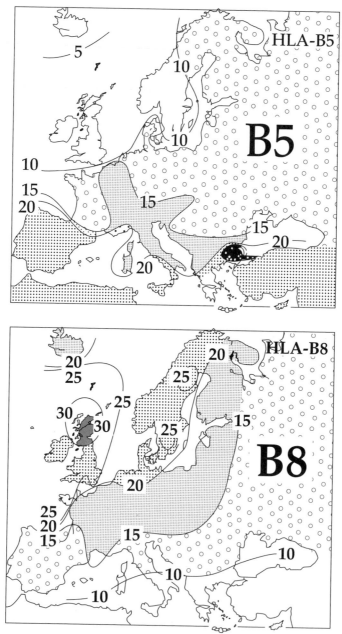

Figure 7.3. Geographic distribution of *HLA-B5* and *HLA-B8* in Europe. Numbers indicate allele frequencies. (From Tiwari and Terasaki, 1985.)

many populations. As an example of this type of research, we very briefly look at two issues to illustrate the use of the HLA system in population studies.

A rather unique set of HLA alleles (and variants) characterizes South American Indian groups (Parham and Ohta 1996, Gómez-Casado et al. 2003, Tsuneto et al. 2003). The New World was probably colonized sometime between 10,000 and 30,000 years ago by groups migrating from Asia. This colonization coupled with a drastic reduction in population numbers that occurred during European contact in the 1500s was most likely responsible for shaping the HLA landscape. It is hypothesized that marked differences between and among South American and North American groups is the result of changes in the set of founding alleles by a combination of point substitutions, allele conversions, gene conversions, recombination, and genetic drift coupled with differential selection (Bodmer 1991, Parham and Ohta 1996). Black and Hedrick (1997) examined 23 South American tribes and found strong evidence for balancing selection. As additional support for this explanation, it is interesting to note that over 25% of the total number of HLA-B alleles known to exist are unique to South and Central American populations.

The diversity and origins of native Siberian populations and their relationship to European and Native American groups continue to be of great interest to anthropologists. Analysis of alleles and haplotypes in Siberian populations has shown that no HLA allele is found in all groups or is a specific allele restricted to a single Siberian group. Population-specific HLA haplotypes are, however, found in various ethnic groups and can be used to study gene flow and group relationships. Uinuk-Ool et al. (2003), using HLA frequency data, suggested that two centers of expansion were established in Asia sometime after the initial expansion of humans out of Africa. One center in southwest Asia fueled population expansions to Europe and the Near East, while expansions from central Asia went north, east, and southeast, establishing populations in Siberia, Asia, and Southeast Asia. The separation of Siberian populations from the other Asian groups then took place sometime between 21,000 and 24,000 years ago. HLA allele frequency data (DRB1, DQA1, and DQB1) further revealed that all Asian groups clustered together, separate from European, African, and American groups. Eliminating non-Asian groups (except Native American populations) from the HLA tree supports a clear relationship between Siberian populations and Native Americans. Further, the HLA data lend support to the idea that at least three separate Siberian founder populations established themselves in the Americas at different times. Thus, HLA data, especially when coupled with other types of data (e.g., mitochondrial DNA, Y chromosome, archaeological, linguistic), can help in our understanding of population origins, relationships, evolution, and migration.

LACTASE RESTRICTION AND PERSISTENCE

Nutritionally, lactose, or milk sugar, is the most important carbohydrate a newborn mammal needs for growth and development. In order to be utilized by the body, lactose (a disaccharide) must be cleaved into glucose and galactose. The enzyme lactase accomplishes this feat by hydrolyzing lactose in the small intestine, where it is then absorbed. Lactase activity is present in the fetus during the first part of the second trimester, peaking right after birth. As a general mammalian characteristic, lactase activity remains high in the newborn, persisting until after weaning, when it drops to 5%–20% of the initial newborn level. The digestive capacity of lactose varies among individuals and with age, and some

Table 7.6 Genotypes and phenotypes for lactase

Genotype	Enzymatic Phenotype	Digestive Phenotype
LCT*P/LCT*P	Lactase persistence	High lactose digestion capacity
LCT*P/LCT*R	Lactase persistence	High lactose digestion capacity
LCT*R/LCT*R	Lactase restriction	Low lactose digestion capacity

The rare congenital lactase deficiency allele (LCT*C) is not included in table.
Adapted from Flatz 1987, 29.

undigested lactose may reach the colon, causing irritation by acidification and the formation of carbon dioxide and hydrogen gases (called *lactose intolerance*).

As a general rule, lactase activity in humans follows the mammalian pattern, starting high during infancy and declining during early childhood. This pattern, called *lactase restriction*, has gone by other names, such as *lactose intolerance, lactose malabsorption*, and *lactase deficiency*, in the older literature. It has been suggested that the switchover from high to low lactase activity is achieved in a similar manner to that which occurs with fetal to adult hemoglobin (Flatz 1987). The decline usually occurs between 2 and 5 years of age, corresponding to weaning times in many human societies. However, studies in Finland have shown late onset of lactase restriction, suggesting that environmental or other genetic factors may also be involved (Sahi et al. 1983). Adults with persistence of lactase activity and the ability to digest large amounts of lactose are common in only a limited number of populations.

The locus that determines lactase activity, called LCT (or LAC), has been mapped to chromosome 2q21 (Harvey et al. 1993). Three alleles determine lactase activity: *LCT*R*, *LCT*P*, and *LCT*C* (Table 7.6). The *LCT*R* allele causes postweaning lactase restriction (the common mammalian pattern), while the *LCT*P* allele determines lactase persistence into adulthood (Flatz 1987). The *LCT*C* variant is a rare allele (or possibly multiple alleles) that is responsible for congenital lactase deficiency and will not be discussed here.

The worldwide distribution of *LCT* alleles demonstrates that the majority of humans follow the ubiquitous mammalian pattern of lactase activity decline after weaning (Fig. 7.4). There are, however, two general categories of populations in which lactase persistence prevails: *(1)* populations in northern Europe and the British Isles and *(2)* a number of nomadic, pastoral populations in Africa and Asia. Europe shows considerable variation, with the lowest *LCT*R* frequencies occurring in Sweden, Denmark, England, and Ireland. There is a gradual increase in frequency of the *LCT*R* allele as one moves south and east across Europe. Many, but not all, nomadic, pastoral populations have high frequencies of *LCT*P* (>0.5).

Native peoples of the New World, Australia, and Oceania show a decline in lactase activity after weaning (lactase restriction). Where there is no admixture, frequencies of *LCT*R* are 1.0. New World populations that migrated during historic times reflect the frequencies of the parental populations.

Because the general mammalian pattern is lactase restriction, it is assumed that the *LCP*P* allele, which is dominant, was introduced by a mutation(s) and then increased in frequency in a number of populations. The mechanisms responsible for the present-day pattern of variation are unclear; however, most researchers argue that natural selection played a major role. In some cases, gene flow may have also played a significant role, introducing the allele to some early European populations.

Figure 7.4. Milk consumption and dairying. Dark areas in Europe, the Middle East, and northern Africa have milk consumption and frequency of *LCT*R* of <0.5. Southern and central parts of Africa, Australia, and Asia have low levels of milk consumption and frequency of *LAC*R* of >0.85 (lighter shading). Open areas (without shading) have milk-consuming populations and variable lactase phenotypes. Hence, congruence is not exact or complete. (From Flatz 1987.)

Because of many cultural and environmental differences between European populations and nomadic, pastoral populations where lactose digestion capacity is high, three different hypotheses have been proposed. In general, these hypotheses suggest that in populations that had ample supplies of milk, individuals who were capable of using the nutrients in milk would be at an adaptive advantage over those who could not.

One hypothesis suggests that as pastoralism and the consumption of milk coevolved over the millennia, those adults who could digest lactose were at a selective advantage over those who could not (Simoons 1969, 1970; McCracken 1971). This idea is sometimes referred to as the "culture–history hypothesis" or the "milk dependence hypothesis." This is a coevolutionary theory since the cultural environment affects the selection of a genetic trait. In this case, the raising and milking of livestock is the culture-selective environment. In these pastoral populations, adults who could digest lactose and use the nutrients in unprocessed milk would be at a selective advantage over those who could not. Over time, these populations would see an increase in the frequency of adults who could digest lactose. A number of pastoral populations in Africa and the Middle East fit this hypothesis. However, pastoral populations that do not milk their livestock would not be expected to have high frequencies of lactase persistence. In fact, livestock-keeping societies in China, southeast Asia, and parts of sub-Saharan Africa that do not use unprocessed milk do not have high frequencies of adult lactase persistence.

A recent study compared the frequency distribution of the lactase persistence allele and six milk protein genes in cattle in Europe and found strong geographic concordance, suggesting that there has been gene–culture coevolution between humans and cattle (Beja-Pereira et al. 2003). The advantages provided by milk consumption shaped the genetic structure of both domestic cattle and humans. Since the Neolithic, selection for increased milk yield and changes in the composition of milk protein shaped the milk protein genes in cattle at the same time they influenced the gene encoding lactase in human beings. Whether this explanation will apply to other parts of the world where milk consumption is culturally important is not known.

Another hypothesis relates specifically to high-latitude populations in Europe, where sunshine is limited and there is an increased risk of vitamin D deficiency (Flatz and Rotthauwe 1973, Durham 1991). For humans, the major source of vitamin D is obtained from exposing the bare skin to sunshine (see Chapter 11 for a more detailed discussion of vitamin D and its relation to skin pigmentation). The sunshine then converts a pro-hormone in the body into vitamin D that the body can use. During childhood, ample amounts of vitamin D are required for normal bone growth and development. If there is a deficiency in vitamin D during this time period, the risk of **rickets** (bones become deformed, e.g., compressed pelves, bow legs, twisted spine) increases dramatically. Adults who do not get ample amounts of vitamin D can develop osteomalacia (can cause deformity of the pelvis, and this can be detrimental in childbirth). The hypothesis then suggests that, like vitamin D, lactose in fresh milk promotes the uptake of calcium and thus aids in normal bone growth. If, after weaning, children can continue to digest and use milk, they would not run the risk of getting rickets. This hypothesis appears to explain the high frequency of lactase persistence in northern European populations, where large amounts of milk are consumed. In southern Europe, where more processing of milk occurs (e.g., cheese) and there is less risk of vitamin D deficiency, the frequency of lactase persistence should be much lower. The clinal distribution of allele frequencies for lactase persistence in Europe follows this hypothesis, with high frequencies in the north, intermediate ones in central Europe, and lower frequencies in the Mediterranean area. As noted above, however, this clinal distribution may have been caused by coevolution between milk proteins in cattle and the lactase persistence allele.

An additional hypothesis suggests that in arid environments those individuals with lactase persistence would be able to make better use of the water content in milk, thus increasing their chances of survival in a desert environment (Cook and Al-Torki 1975). In addition, if lactose-intolerant individuals consumed milk, they would run a greater risk of water depletion and diarrhea. Thus, there would be selection for lactase persistence and selection against those who were unable to digest lactose (lactose restriction). Pastoral populations that have high frequencies of lactase persistence and live in arid regions of Africa and the Middle East (e.g., Tuareg, Bedouin, and Fulani) fit this hypothesis. This hypothesis is probably the weakest of all because it is difficult to surmise how milk could effectively replace water and have such a selective effect on humans.

A number of researchers have modeled the evolution of lactase persistence in adults. One model suggests that with a selective advantage of lactase persistence of 4% and a starting allele frequency of 0.001%, it would take about 9,000 years (290 generations) to reach present-day levels of lactase persistence in Europe. These results seem realistic for the Middle East, where livestock were domesticated (and probably milked) about 10,000 years ago. However, farming and the use of livestock did not arrive in northern Europe until about 3,000 years ago. Thus, a higher selective advantage (0.05–0.07) would be necessary for the frequency to rise to contemporary European levels. These models are not precise and rely on a number of different assumptions about population size, initial allele frequencies, and selection coefficients. However, they do support the idea that relatively rapid natural selection is responsible for the present-day distribution of the uniquely human trait of lactase persistence in adults.

Whether one or all of these hypotheses applies to the present-day diversity in lactase persistence and lactose restriction is a matter of debate. As Holden and Mace (1997) point out, these hypotheses are also confounded by shared ancestry of many of the populations. Their analysis, correcting for this problem, supports the "culture–history" hypothesis but not the high-latitude or arid regions hypotheses. Their study also suggests that the evolution of milking herd animals came before the evolution of lactase persistence in these populations.

The variation in lactase persistence in the world and the related hypotheses remind us that a biocultural perspective in research is often necessary in order to understand the distribution of a genetic trait. Culture, the natural environment, and biology often work in concert over the millennia to shape the structure of populations. This biocultural perspective is also important in formulating policy or implementing health-care programs in other countries. In the United States, we regard milk as an essential part of daily nutrition for all ages. In the past, policy makers in the United States indiscriminately sent milk to many countries with the idea that "what is good for children in the USA, is good for everyone in the world." It soon became apparent that many of these people were lactose-intolerant and that what was good for U.S. children was clearly not good for them.

TASTE: PHENYLTHIOCARBAMIDE (6-*N*-PROPYLTHIOURACIL)

Orally, humans perceive a variety of chemical compounds found in food as bitter, sweet, sour, salty, umami (the taste of monosodium glutamate), and other. The old concept of four basic taste qualities does not appear to be valid as research on taste continues (Hladik and Simmen 1996). Some of this variation in taste sensitivity is undoubtedly genetic in origin and probably related to adaptive responses for determining nutritional content or dealing with potential toxicity of foods. Over the years, many anthropologists have focused their attention on one substance called phenylthiocarbamide (PTC). For some individuals, this substance tastes bitter and for others it has no taste at all.

As Du Pont chemist A. L. Fox was transferring some PTC into a bottle, some of the compound became airborne. Fox (1931a,b; 1932) was surprised to find that his coworker in the laboratory complained that the compound tasted bitter. Apparently, a lively discussion occurred since Fox was closer to the compound yet did not perceive it at all. Fox even placed the crystals in his mouth and found them to be tasteless. Soon, other laboratory workers were called into the room—some perceived the compound as bitter while

Table 7.7 Frequencies of nontasters of phenylthiocarbamide in a selected world sample

Area	Population	Percent Nontasters
Africa	Urban Bantu	2.3
Africa	Libyans	19.8
Africa	Arabs, Syria	36.5
Australia	Aborigines	49.3
China	Bao'an Gansu	5.1
China	Han Lan Zhou City	11.0
China	YuGru Gansu	23.0
Southeast Asia	Malay, Senoi	4.0
Southeast Asia	Thailand	9.7
Southeast Asia	Tamil, Singapore	26.8
Southwest Asia	Turkey	4.1
Southwest Asia	Turkey	20.0
Southwest Asia	Kurdish, Iran	27.5
India	Relli	1.7
India	Manne Dora	34.5
India	Wad Balgel	66.7
Russia	Siberians	5.8
Russia	Tiflis	23.0
Russia	Zagorsk	41.0
Europe	Lapps	6.9
Europe	Åland Islands	26.2
Europe	Derbyshire, England	36.8
North America	Papago, Arizona	1.4
North America	Mennonite	25.0
North America	"Caucasian"	30.9
Central America	Rama, Nicaragua	1.3
Central America	Mexican	10.4
Central America	Miskito, Nicaragua	20.8
South America	Carajas, Brazil	0.0
South America	Manaus, Amazonia	15.6
South America	Salvador (Bahia)	38.0

Information extracted from Guo and Reed (2001). For sample sizes, references, and more populations, see Table 1 in Guo and Reed (2001).

others found it tasteless. Blakeslee and Salmon (1931, 1935), Blakeslee (1932), Salmon and Blakeslee (1935), and Synder (1931, 1932) suggested that the ability to taste PTC was inherited as a simple Mendelian recessive trait. Since this initial chance discovery, thousands of people have been tested for PTC tasting. Table 7.7 lists a selected number of world populations and the percent of non-tasters found in these groups. Note that frequencies vary considerably within continents and among populations. Testing has also revealed that many nonhuman primates exhibit taste sensitivity variation similar to that of humans (Fisher et al. 1939, Eaton and Gavan 1965).

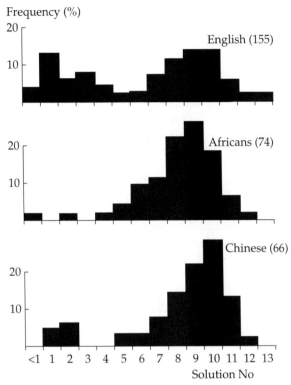

Figure 7.5. Phenylthiocarbamide taste thresholds in three populations. Note the bimodal distribution, especially in the English sample. Nontasters occupy the distribution from solution number 5 and lower, while tasters are 6 and above. (From Harrison et al. 1988)

Individuals perceive PTC at varying *thresholds* (levels of concentration in solution). For almost all populations tested, this variable sensitivity creates a population distribution that is bimodal (Fig. 7.5). The one exception is the Carajas in Brazil, which have 100% tasters (Kalmus 1957). During the early years of taste testing, researchers used PTC crystals placed directly on the tongue. Other researchers used filter paper impregnated with PTC. By 1949, a more rigorous testing method, using 13 serially diluted solutions of PTC, was developed by Harris and Kalmus. Then, during the 1960s and 1970s, most researchers switched to 6-*n*-propylthiouracil (PROP) because it lacks the sulfurous odor of PTC and safety limits on dosage could be set (Guo and Reed 2001). PROP and PTC are chemically related compounds, and the ability to taste PROP is highly correlated with the ability to taste PTC and reflects the same polymorphism. The bitter taste derives from the thiocarbamide group (S=C-N-H).

The ability to taste PROP (and PTC) is mapped to chromosome 5p15 (Reed et al. 1999). There is some evidence that PTC and the Kell blood group are linked, suggesting that a locus or region on chromosome 7q may also affect the taster phenotype (Chautard-Freire-Maia 1974, Conneally et al. 1976, Reed et al. 1999, Drayna et al. 2003, Kim et al. 2003). Twin and family studies suggest that the ability to taste PTC is inherited as a simple Mendelian characteristic. According to this monogenic hypothesis, those individuals who taste PTC are either homozygous (TT) or heterozygous (Tt) for the trait. Nontasters are homozygous (tt) recessive. There is some evidence that the tasting threshold is different for homozyous (TT) and heterozygous (Tt) tasters. That is, TT individuals can detect bitterness in more dilute concentrations than Tt individuals. Other researchers (Ibraimov and Mirrakhimov 1979) suggest that there are two codominant alleles, T_1 and T_2, both dominant to t, while some think the variability reflects the existence of a major locus with incomplete dominance coupled with a multifactorial component (Reddy and Rao 1989, Reed et al. 1995). Still others argue for a two-locus model, one for general tasting ability and the other for detecting PTC (Olson et al. 1989, see also Koertvelyessy 2000, Guo and Reed 2001). Other factors (e.g., other loci, sex, age, smoking habits) have been suggested to modify the ability to taste PTC. There is also some evidence that saccharin tastes less bitter to nontasters than to tasters (Bartoshuk 1979).

As with many genetic polymorphisms, researchers have reported associations between PTC tasting ability and a variety of diseases and conditions, all non-taste-related. The meanings and implications of these associations are complex and elusive, and some associations are likely due to chance. Reported associations include dental caries, alcoholism, diabetes, depression, schizophrenia, mental function, personality characteristics, glaucoma, gastrointestinal ulcers, malignant tumors, leprosy, cystic fibrosis, diabetes mellitus, tuberculosis, and variation in growth and development (see Mourant et al. 1978, Koertvelyessy 2000, Guo and Reed 2001). From an anthropological vantage, the most interesting association is that between PTC tasting ability and dietary intake of potentially harmful compounds.

Many cruciferous vegetables (e.g., cabbage, brussels sprouts, turnips, kale, broccoli, mustard greens) and manioc (cassava) contain isothiocyanates and goitrin compounds (have the S = C-N-H radical). Thus, researchers have suggested that tasters of PTC (PROP) would find these vegetables bitter in taste and not consume them. However, nontasters would not find these foods objectionable. In large quantities, these compounds interfere with iodine metabolism, causing enlargement of the thyroid gland (goiter). The consumption of these naturally occurring goitrogens would have a greater effect on metabolism, physiology, growth, and reproduction in regions where iodine intake is low. Thus, it is argued that tasters would be at a selective advantage over homozygous nontasters. Reduced thyroid function has also been implicated in lower fertility rates (Jackson 1993). Further research must be conducted to elucidate how the PTC genotypes influence food consumption and avoidance.

In addition to examining the bitter taste of PTC, researchers have explored the variability in taste detection of other substances in humans and other animals. Some of these tastes have a clear genetic basis, such as sucrose octaacetate and quinine sulfate sensitivity in mice (Whitney and Harder 1994) and probably quinine taste in humans (Smith and Davies 1973). Taste perception research is often interpreted in an adaptive or evolutionary frame. Evidence suggests that plants containing sugars coevolved with the taste perception of nonhuman primate fruit eaters, varying among species and the size of animals (Hladik and Simmen 1996). However, plants also evolved defense mechanisms to reduce the damage caused by plant eaters. The bitter taste of some plants indicates that they contain toxic substances such as alkaloids. Plant eaters that could perceive the bitterness would then avoid consuming these plants (and the toxins they contain). The detection of a bitter taste varies among individuals and among primate species, often varying with availability of food souces (Simmen 1994), giving weight to a coevolutionary model.

However, not all compounds that taste bitter are toxic. Thus, it may not be advantageous to evolve sensitivity to bitterness in general, limiting one's potential food sources (Glendinning 1994, Hladik and Simmen 1996). Taste thresholds for various substances (e.g., quinine hydrochloride) vary among different species. Carnivores appear to have the lowest thresholds, with omnivores being in the middle and browsers having the highest thresholds (Glendinning 1994).

The lack of salt in many environments and its importance in physiology has also led some researchers to suggest that the perception of sodium coevolved with plants. However, the nutritional needs of salt are easily met by most natural diets, and the sensitivity of primate taste does not appear to allow them to taste even the low concentrations of salt found in many plants. Hladik et al. (2002) argue that salt perception is primarily a culturally learned response that does not have a long evolutionary history.

VARIATION IN EAR WAX OR CERUMEN

Research on the variability of cerumen, or ear wax, started in the early 1930s in Japan. It was not until 1962 that Matsunaga reported to the English-speaking world that ear wax occurred in two forms, one wet and sticky (yellow to brown) and the other dry and flaky or brittle (gray to light tan). Japanese family studies revealed that the trait is controlled by two alleles (W and w), located on chromosome 16p (Tomita et al. 2002), with the wet form being dominant to the recessive dry form. The homozygous dominant form (WW) is not phenotypically distinguishable from the heterozygous form (Ww) because both result in wet ear wax.

There are no known qualitative differences in the chemical composition of dry and wet cerumen. Uncertainty surrounds quantitative differences. The proportion of lipids is reportedly less in dry cerumen than wet (Matsunaga 1962). However, Kataura and Kataura (1967b) found no qualitative or quantitative differences in lipid content between ear wax types, but the relative proportion of polyunsaturated to saturated fats was slightly higher in dry ear wax. They also found differences in the content of free amino acids but no qualitative differences (Kataura and Kataura 1967a).

Wet cerumen is associated with axillary body odor (apocrine glands), suggesting that the gene for cerumen is possibly **pleiotropic** (a gene affecting more than one trait) in its effects. Matsunaga (1962) even suggested that the frequency of wet ear wax in Japan, which is low, may be influenced by sexual selection (mating practices) because individuals with axillary body odor are considered by some Japanese as pathologic and in need of medical attention.

A number of populations have been assessed for wet and dry cerumen (Table 7.8). In general, high frequencies of dry ear wax prevail in Asia, while low frequencies occur in Europe and Africa. There is a north–south clinal change in frequency in Asia and the Americas, suggesting that ear wax type may be adaptive in varying climates. McCullough and Giles (1970) found a correlation between wet ear wax and hot–moist environments. They suggest that cerumen types may be related to disease defense mechanisms involving external ear canal infections (possibly leading to otitis media, mastoiditis, and general cranial infections). Also, the finding that immunoglobulin G and lysozyme occur more frequently in dry cerumen may indicate that ear wax type plays a role (anti-bacterial potential) in host defense mechanisms (Petrakis et al. 1971, Hyslop 1971).

Table 7.8 Frequencies of dry cerumen in various populations

Group/Population	Dry Cerumen (%)	Reference[1]
Northern Chinese	95.8	1
Southern Chinese	74.0	1
Chinese (San Francisco, USA)	58.5	4
Koreans	92.4	1
Tuvinians, Tuva, Russia	78.6	6
Altains, Altai Mountains, Russia	72.6	6
Khakass, Khakassia, Russia	63.5	6
Turkmen, Turkmenistan	54.2	6
Japanese	83.7	1
Ryukyu Islanders	62.5	1
Ainu	13.3	1
Melanesians	27.8	1
Micronesians	37.1	1
Africans	0.0	4
African Americans	0.0	4
Mozambiquans	53.0	6
Ethiopians	49.0	6
Angolans	41.5	6
Aleuts	48.6	4
Nootka	36.4	4
Papago	59.7	4
Navajo	63.3	3
Sioux	36.7	3
Mississippi Choctaw	21.0	5
Cuna	4.4	4
Quechua	76.7	4
Tzotzil Maya	6.7	2
Zinancantec Indians	4.7	2
Europeans	1.3	4
Germans (Westphalia)	8.4	1
European Americans	4.4	6
European Americans	1.3	4

[1], Matsunaga 1962; 2, Kalmus et al. 1964; 3, Petrakis et al. 1967; 4, Petrakis 1969; 5, Martin and Jackson 1969; 6, Ibraimov 1991.

Petrakis (1971) noted an association between wet ear wax and carcinoma of the breast. Even though Ing et al. (1973) could not confirm this association, it appears reasonable because both the ceruminous glands of the ear and the mammary glands are derived from the apocrine-type sweat glands. If the locus is pleiotropic, affecting the axillary apocrine sweat glands, ceruminous glands, mammary glands, it may also affect the susceptibility to breast cancer. More research needs to be done on the possible adaptive significance of the varieties of ear wax and their associations with other genetic traits (loci) and diseases. It would also be interesting to study more nonhuman primates given the report of Adachi (1937) that two of eight chimpanzees examined had dry ear wax while the others had wet.

CHAPTER SUMMARY

The MHC in humans is the most polymorphic of all systems yet described. Located in three regions on the short arm of chromosome 6, this complex system is responsible for encoding HLAs that are integral to our immune system. In addition to the high degree of polymorphism, another striking property of the MHC is that many HLA alleles occur in linkage disequilibrium. The MHC has been associated with a variety of diseases, suggesting that natural selection plays a very important role in shaping the unique distributions of many HLA alleles and haplotypes. The extensive polymorphism of this system makes it a tool to explore the origins, diversity, and relationships among various populations across the world.

The enzyme lactase breaks down milk sugar, or lactose, so that humans can take advantage of this important nutrient. Following weaning, many humans lose the ability to hydrolyze lactose and become intolerant to milk. This pattern characterizes mammals as a group. However, some human populations show a very high frequency of the persistence of lactase activity, allowing individuals with this trait to make use of the nutritional value of milk as adults. The ability to digest lactose probably arose as a random mutation sometime in prehistory as populations began to domesticate cattle. A number of hypotheses have been put forth to explain the frequencies of the *LCT*P* allele found in various populations. Most hypotheses suggest that the frequency of the persistence of lactase activity allele increased as a result of coevolution between pastoralism and milk consumption. This trait provides an excellent example of the interrelationships between culture and biology.

Some of the variation in taste sensitivity is undoubtedly genetic in origin and probably related to adaptive responses for determining nutritional content or dealing with the potential toxicity of various foods. Anthropologists have focused most of their attention on the perception of the bitter substance PTC. Whether this trait is a simple Mendelian one or quantitative is a debated issue. For anthropologists, one of the more interesting dimensions of this trait is its possible relationship to food avoidance. It has been suggested that tasters would find that certain plants tasted bitter and would not consume them; nontasters would not find these food sources objectionable and would consume them. In large quantities, these plants have compounds that can interfere with normal iodine metabolism, causing a goiter. It is thus argued that tasters would be at a selective advantage over nontasters because they would avoid ingesting these potentially harmful compounds. The genetics of taste perception is an open field of inquiry that is sure to expand.

We normally do not think about the composition of our ear wax. However, researchers have shown that there are two basic types, wet and dry. The frequency distributions of wet and dry cerumen have been interpreted to reflect both climate and disease patterns. It is also interesting to note that the locus for cerumen is probably pleiotropic, affecting the axillary apocrine sweat glands and, hence, the diversity of body odors found in human populations.

SUPPLEMENTAL RESOURCES

Blancher A, Klein J, and Socha W W, eds. (1997) *Molecular Biology and Evolution of Blood Group and MHC Antigens in Primates.* Berlin: Springer-Verlag.

Browning M and McMichael A, eds. (1996) *HLA and MHC: Genes, Molecules and Function.* Oxford: BIOS Scientific Publishers.

Durham W (1991) *Coevolution: Genes, Culture, and Human Diversity.* Stanford: Stanford University Press.

Guo S-W and Reed D R (2001) The genetics of phenylthiocarbamide perception. *Annals of Human Biology* 28: 111–142.

Lechler R, ed. (1994) *HLA and Disease.* San Diego: Academic Press.

Marsh S G E, Parham P, and Barber L D (2000) *The HLA FactsBook.* San Diego: Academic Press.

Tiwari J L and Terasaki P I (1985) *HLA and Disease Associations.* New York: Springer-Verlag.

8

DNA Markers

The development of DNA markers in human population genetic studies has been phenomenally fast-paced. As a result, anything that is written is almost sure to be dated within a year or two. To the student, this provides the uncomfortable challenge of dealing with information that is by definition incomplete and by extension subject to revision and reversals. Those who have dealt with the human paleontological record will be familiar with this challenge as each new fossil find adds complexity and ambiguity to previous interpretations. Because DNA marker analysis in human populations is a relatively new phenomenon, the field has yet to gain the maturity to admit to its own shortcomings (though see Forster's [2003] comments regarding the alarmingly high rate of error in published mitochondrial DNA [mtDNA] sequence data). Consequently, those students who branch out to the suggested supplemental readings or follow up on some of the cited references should be forewarned to always bring their own ounce of salt along, lest they uncritically accept the author's reconstructions of human history and prehistory. Indeed, Goldstein and Chikhi (2002, 130) open their review "Human Migrations and Population Structure" with the statement "Given the growing cachet associated with genetic inference in human evolution, it is perhaps wise to begin with the limitations of genetic approaches," and they close their discussion with a section titled "A note on storytelling in human evolution and the example of Europe" (2002, 143).

MOLECULAR GENETIC TECHNIQUES AND DNA VARIATION

The tremendous gains in human genetic studies stem from a handful of specific molecular genetic techniques and the identification of various polymorphic DNA markers. In this section, we discuss the polymerase chain reaction, **restriction fragment length polymorphisms**, insertions and deletions (**indels**), interspersed nuclear elements, DNA sequence analysis, and tandem repeats. We close the chapter with some examples of analyses from large geographic areas and a section on how DNA markers are being used in forensic and historical analyses.

Polymerase Chain Reaction

Polymerase chain reaction (PCR) is a procedure used to make many copies of a particular DNA segment. In some respects, it can be thought of as "replication in a test tube" (or, more properly, in an **Eppendorf tube**). The power of PCR lies in the fact that it not only makes many, many copies of DNA but does so with incredible specificity. Despite the fact that there are approximately 3.2 gigabases (International Human Genome Sequencing Consortium 2001) in the human genome, PCR can be used to pick out a

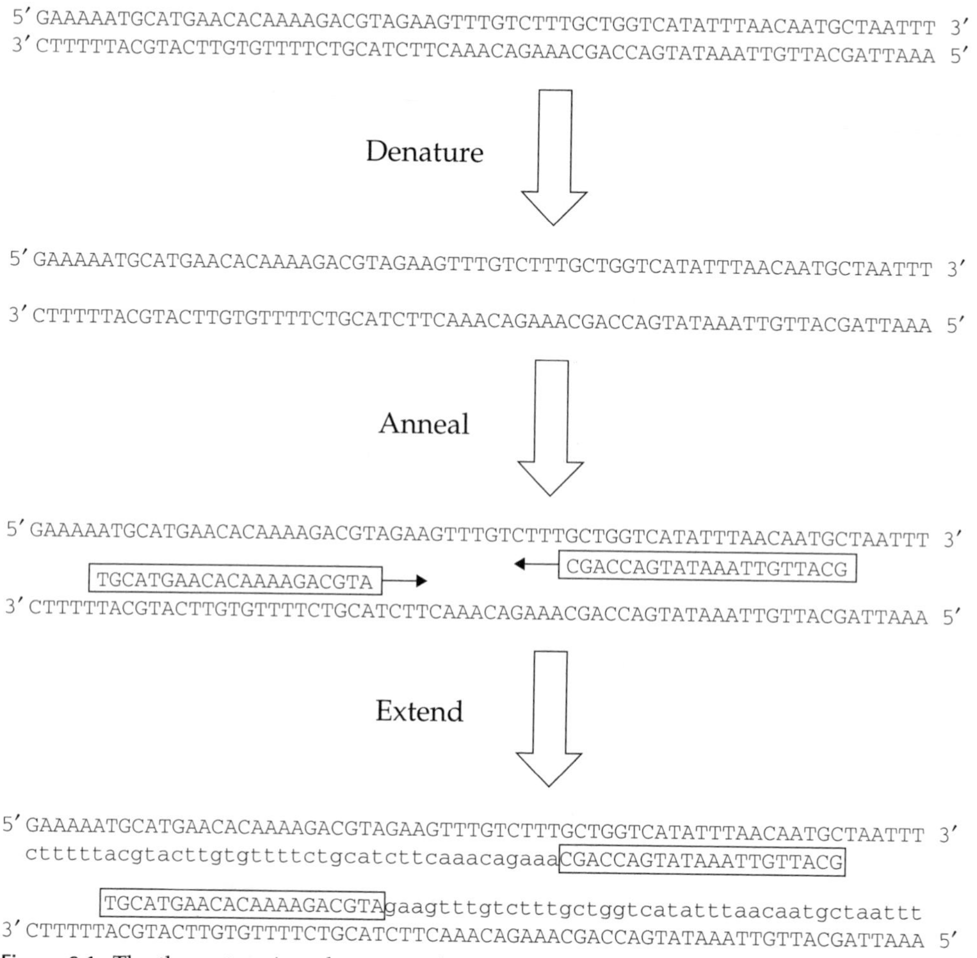

5′ GAAAAATGCATGAACACAAAAGACGTAGAAGTTTGTCTTTGCTGGTCATATTTAACAATGCTAATTT 3′
3′ CTTTTTACGTACTTGTGTTTTCTGCATCTTCAAACAGAAACGACCAGTATAAATTGTTACGATTAAA 5′

Denature

5′ GAAAAATGCATGAACACAAAAGACGTAGAAGTTTGTCTTTGCTGGTCATATTTAACAATGCTAATTT 3′

3′ CTTTTTACGTACTTGTGTTTTCTGCATCTTCAAACAGAAACGACCAGTATAAATTGTTACGATTAAA 5′

Anneal

5′ GAAAAATGCATGAACACAAAAGACGTAGAAGTTTGTCTTTGCTGGTCATATTTAACAATGCTAATTT 3′

TGCATGAACACAAAAGACGTA ➝   ⬅ CGACCAGTATAAATTGTTACG

3′ CTTTTTACGTACTTGTGTTTTCTGCATCTTCAAACAGAAACGACCAGTATAAATTGTTACGATTAAA 5′

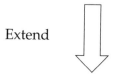

Extend

5′ GAAAAATGCATGAACACAAAAGACGTAGAAGTTTGTCTTTGCTGGTCATATTTAACAATGCTAATTT 3′
    cttttttacgtacttgtgttttctgcatcttcaaacagaaa CGACCAGTATAAATTGTTACG

TGCATGAACACAAAAGACGTA gaagtttgtctttgctggtcatatttaacaatgctaattt
3′ CTTTTTACGTACTTGTGTTTTCTGCATCTTCAAACAGAAACGACCAGTATAAATTGTTACGATTAAA 5′

Figure 8.1. The three steps in polymerase chain reaction (denature, anneal, extend). The particular example shown here is for a segment of the Y chromosome known as 92R7, and the boxed letters represent the two primers.

specific sequence in the genome and make millions of copies from one double-stranded DNA.

The steps in PCR follow the broad outline of DNA replication that occurs within cells (see Fig. 8.1 for a schematic diagram of PCR). As we discussed in Chapter 2, in order for DNA to replicate it must first be present in single-stranded, rather than double-stranded, form. In the cell, the enzyme DNA helicase ("unwindase") opens up the double-stranded DNA at replication forks. In PCR, the double-stranded DNA is heated to about 94°C (201°F), at which point it **denatures** (separates) into single strands. The addition of heat makes this denaturation phase of PCR quite harsh in comparison to the denaturation that takes place within cells during replication. As we also discussed in Chapter 2, new DNA strands built by complementation to an existent strand must first be *primed* (i.e., built on to the end of a double-stranded section). Priming is a complicated process in the cell, but in PCR all that is required is the presence of two synthetic **primers**, which are short sections of single-stranded DNA. These primers are synthesized so that they match (complement) a short segment on each of the single-stranded DNAs from the denaturation phase. In Figure 8.1, the primers are each 21 bases long and were designed (see Kwok et al. 1996, Forster et al. 2000, or http://www.icb.ufmg.br/~prodap/projetos/variabilitY/pol/92r7.html) to amplify a 55-base sequence from the Y chromosome.

The second phase of PCR is called the **annealing** phase because in this part of the reaction the PCR mixture is cooled (usually to about 55°C [131°F]), at which point the primers attach to their complementary target sites on the denatured DNA. In the third phase of the reaction, DNA polymerase adds bases by complementation, "growing" the DNA in the 5′ to 3′ direction off of the primers. Because the DNA polymerase is extending DNA off of the primers, this third phase is called **extension**. Extension is usually accomplished at a temperature of 72°C (162°F). Within the PCR mix are free deoxyadenosine triphosphate (dATP), deoxycytidine triphosphate (dCTP), deoxyguanosine triphosphate (dGTP), and deoxythymidine triphosphate (dTTP) bases (generically referred to as deoxynucleotide triphosphates, [dNTPs]), which the polymerase will "tack-in" to the growing strands by complementation. These are the adenine, cytosine, guanine, and thymine bases plus the sugar and phosphate groups that will be connected (polymerized) to form DNA strands. In sum, the three phases for one cycle of PCR are denaturation, annealing, and extension.

An interesting detail in the extension phase of PCR demonstrates the power of evolution by natural selection, one of the four evolutionary forces we discussed in Chapter 3. The extension phase of PCR does not use human DNA polymerase but instead uses DNA polymerase from the bacterium *Thermus aquaticus*. This is usually referred to as **TAQ polymerase**. The temperatures required for denaturation in PCR far exceed what human DNA polymerase would be exposed to in cells, and human polymerase quickly degrades if used in PCR. In contrast, TAQ polymerase is very heat-stable because *T. aquaticus* is a bacterium that lives in hot springs. The interesting evolutionary side-light here is that TAQ polymerase functions properly in PCR of not only bacterial DNA but also human DNA—indeed, any DNA. This shows the unity of life on earth, while at the same time demonstrating evolution by natural selection in the heat adaptations required for life in hot springs.

All told, one cycle of PCR lasts on the order of a couple of minutes. There is a great deal of variation in how much time is allotted for each of the three phases in various PCR protocols, but generally one whole cycle of denaturation, annealing, and extension will take no more than 2 minutes. A cycle of PCR can be accomplished using a machine called a **thermocycler**. The thermocycler is programmed to spend a certain amount of time at (usually) 94°C to denature the double-stranded DNA, to drop its temperature for annealing, and then to raise it to an intermediate level for extension. The thermocycler can be programmed to repeat these steps a number of times (this is the "chain" part of PCR), generally 20–30 times in most PCR protocols. Figure 8.2 shows the extension phase at the end of the first, second, and third cycles. At the end of the first cycle, the initial double-stranded DNA has been "replicated" into two double-stranded DNA molecules; at the end of the second cycle, these two molecules have been replicated into four double-stranded DNA molecules; and at the end of the third cycle, there are eight double-stranded DNA molecules. Just like computer memory, the number of PCR-manufactured DNA molecules goes up by powers of 2, so the progression is 2, 4, 8, 16, 32, 64, 128, 256, 512, on up to 1,048,576 (or 2^{20}) molecules after 20 PCR cycles and 1,073,741,824 after 30 cycles.

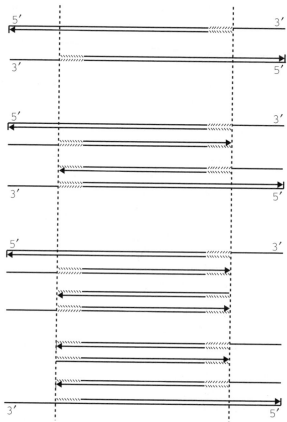

Figure 8.2. Schematic diagram of three polymerase chain reaction cycles.

Restriction Fragment Length Polymorphisms

Restriction fragment length polymorphisms (RFLPs) are a class of DNA variants that have been widely applied in human genetics research. The technology underlying RFLP analysis is based on bacterial enzymes referred to as **restriction endonucleases**. Restriction endonucleases are proteins that recognize a specific sequence in target DNA and will cleave the molecule across the sugar–phosphate backbone, leaving two fragments. The intimidating name for these enzymes refers to both their function and how they operate. They are "restriction" enzymes in that they first recognize viral DNA that could otherwise run rampant in the bacterial cell and then cleave it to deactivate it. Thus, they restrict the growth of viral DNA. A *nuclease* is an enzyme that cleaves nucleotides, while the *endo* prefix refers to the fact that these particular nucleases cleave the DNA internally (away from the ends). Endonucleases function by cutting across the sugar–phosphate backbones of DNA molecules whenever they "read" a particular sequence of bases. Table 8.1 lists a few restriction endonucleases and shows where they cut within particular DNA sequences. The enzymes are named after the bacteria in which they are produced (e.g., *AluI* is made by *Arthrobacter luteus*).

Table 8.1 Examples of restriction endonuclease recognition sites

Enzyme	Bacterial Source	Sequence
*Alu*I	*Arthrobacter luteus*	A G|C T T C|G A
*Eco*RI	*Escherichia coli*	G|A A T T C C T T A A|G
*Hind*III	*Haemophilus influenzae*	A|A G C T T T T C G A|A

The vertical bars in the sequence indicate where the sugar–phosphate backbone is cleaved for the given DNA sequence.

At first blush, it might appear that the production of these enzymes within bacterial cells would be highly counterproductive as presumably the enzymes would cut not only viral DNA but the host bacterial DNA as well. Fortunately, for the life of the bacteria, their own DNA is chemically protected from these enzymes, and the bacteria consequently do not "digest themselves." Humans and other multicellular organisms have evolved more elaborate immunological defenses (see Chapter 5), so they do not produce restriction endonucleases. Consequently, when human DNA is confronted with these bacterial enzymes, it will be cleaved wherever the particular recognition sequence occurs. These enzymes therefore provide a powerful tool for manipulating human DNA. In particular, the enzymes can be used to cut human DNA into pieces and then insert these pieces into host chromosomes or other "cloning vectors." This is the basis for recombinant DNA ("cloning") technology, which has been a cornerstone for the Human Genome Project.

In addition to their role in recombinant DNA technology, restriction endonucleases are essential for identification of RFLPs in human population genetics research. Let us presume that we have used PCR to make many copies of a particular stretch of DNA from an autosomal chromosome. If this sequence contains one *restriction site* (i.e., the recognition sequence for a particular restriction endonuclease), then when we digest the DNA using the enzyme, there will be two fragments formed from each DNA molecule. However, if there has been a mutation within the restriction site, then the enzyme will not digest the DNA and we will have one long piece of DNA between the two primer sites. When some individuals have a particular restriction site and others lack it, this is an RFLP. RFLPs act as codominant systems in that the heterozygote, who has the restriction site on one chromosome but not on the homolog, is distinguishable from both homozygotes. This raises the issue of how RFLPs are assayed (visualized).

Another molecular technique, **gel electrophoresis**, is used to visualize RFLPs. DNA is negatively charged, so if it is placed in an electrical field, it will migrate away from the negative pole and toward the positive pole. Gel electrophoresis makes use of this fact by placing sample DNA in a gel that is hooked up to a powerful electrical battery. The gel is loaded with the sample at the negative pole, and the DNA will then be pulled through the gel toward the positive pole. For those who have used jumper cables to start a car, black is negative and red is positive. The power supply for electrophoresis is similarly marked, so the DNA sample will "run to red" from black. Shorter DNA fragments will move more quickly through the gel, while longer fragments will move more slowly.

Consequently, the gel will size-sort the fragments so that the shorter fragments will be closer to the positive end of the gel and longer fragments will be closer to the negative end. Prior to running the DNA through the gel, the sample is mixed with a dye that will fluoresce under ultraviolet light, and as a consequence, bands of the size-sorted sample will be visible in each lane of the gel.

As an example of a possible restriction analysis, we return to the PCR shown heuristically in Figure 8.1. The amplified fragment is 55 bases long, with 42 bases comprised by the two primers. This leaves 13 bases in the middle:

```
G - A - A - G - T - T - T - G - T - C - T - T - T
C - T - T - C - A - A - A - C - A - G - A - A - A
```

If we look at just the second through seventh bases of this sequence and presume that the fourth base within this short six-base sequence has mutated from T to C (and the complementary strand from A to G), then we have

```
A - A - G - C - T - T
T - T - C - G - A - A
```

which is a *Hind*III restriction site. If we run the PCR shown in Figure 8.1, expose the PCR product to the *Hind*III restriction endonuclease, and then run the sample on a gel, we will get one band for individuals who lack the T-to-C mutation and two bands for individuals with the mutation. Because the mutation that causes the restriction site occurs at a single nucleotide and because this particular base is polymorphic in human populations (some individuals have the mutation, while others do not), it is referred to as a single nucleotide polymorphism (SNP, pronounced "snip"). The probability of this mutation, which is called 92R7, arising more than once in human history is thought to be so remote that the SNP is referred to as a **unique event polymorphism** (UEP). The 92R7 UEP is one of two polymorphisms that are used to define the Cohen modal haplotype, which we will discuss below.

Insertions and Deletions

The gel electrophoresis technique described above can also be used to study insertions and deletions in DNA. If there has been a mutation causing a DNA insertion or deletion, it can be demonstrated by electrophoresis, though in the absence of analyses from other species (e.g., Weber et al. 2002), we cannot tell the original state prior to the mutation. Because it is usually impossible to tell the ancestral state for insertions and deletions, these variants are often referred to as "indels." Indels can be visualized by picking primers whose sequences lie to either side of an insertion/deletion site and then using PCR to isolate the DNA that flanks it. If the PCR product is then run through a high-resolution gel, individuals with the shorter DNA (who have the deletion) will have their DNA sample run further than individuals with the longer DNA (who have the insertion). Indels represent inserted or deleted bases, unlike RFLPs, which are generally caused by SNPs (i.e., single-base changes). Like RFLPs, however, indels act as codominant systems when typed for autosomal chromosomes. Heterozygotes will have two bands, while each of the homozygotes will have a solitary band.

One of the best-studied indels is a **9 bp deletion** that is found between the coding sequence for cytochrome *c* oxidase II (COII) and for lysine tRNA (K) in the mitochondrial genome. Because mtDNA is haploid, the 9 bp deletion is simply expressed as present or absent. In the case of the 9 bp deletion, this indel is referred to as a deletion (rather than as a lack of an insertion) because the original complete mtDNA sequence (Anderson et al. 1981) and its revision (Andrews et al. 1999) include two stretches of nine noncoding bases between COII and K and because most individuals have these two stretches as follows:

COII – GCA – CCCCCTCTA – CCCCCTCTA – GAG – tRNA[Lys]

In individuals with the deletion the sequence is as follows:

COII – GCA – CCCCCTCTA – GAG – tRNA[Lys]

There are a number of other known variations (Handoko et al. 2001) in this small region (such as three repetitions of the nine bases), but by and large individuals have either the first motif (with two repeats) or the deletion motif. Because the 9 bp deletion was initially described among only Asians and peoples with Asian ancestry (Native Americans and at least some Pacific Island populations), it was viewed as an "Asian marker." More recently, the deletion also has been described for some sub-Saharan African populations (Soodyall et al. 1996) and for populations from other parts of the globe, indicating that the deletion arose multiple times in human prehistory. Thus, the 9 bp deletion probably is not a UEP.

Table 8.2 contains frequencies of the 9 bp deletion for a few Indonesian populations (Handoko et al. 2001), which we will examine in greater detail here. As in Chapter 3 (see equation 3.14), we can use these frequencies to calculate an overall F_{ST}. To find the expected heterozygosity (H_e), we first total the sample sizes and allele counts, finding that 403 of 532 individuals lack the deletion, for an overall frequency of 0.7575. The heterozygosity is therefore $2 \times 0.7575 \times (1 - 0.7575)$, or 0.3674 (*heterozygosity* is a bit of a misnomer here because mtDNA is haploid). To get the observed heterozygosity, we calculate $2pq$ within each population (see Table 8.2) and average these values across populations for an observed heterozygosity of 0.3610. Now the estimate of F_{ST} is $(H_e - H_o)/H_e$, or $(0.3674 - 0.3610)/0.3674$, which is equal to 0.0174. This F_{ST} value is quite small, suggesting that the observed heterozygosity is nearly equal to what we would expect if the total population is mating randomly with respect to the individual population divisions.

Table 8.2 Frequency of the mitochondrial DNA COII–K 9 bp deletion in five Indonesian populations

Population	Location	Sample Size	+	p	–	q	$2pq$
Sumbanese	Lesser Sunda	98	82	0.8367	16	0.1633	0.2732
Batak	North Sumatra	125	100	0.8000	25	0.2000	0.3200
Kaili	Central Sulawesi	105	79	0.7524	26	0.2476	0.3726
Javanese	Central Java	105	78	0.7429	27	0.2571	0.3820
Dayak	Central Kalimantan	99	64	0.6465	35	0.3535	0.4571
Average							0.3610
Total		532	403	0.7575	129	0.2425	0.3674

+ represents the number of individuals with two repeats of 9 bases, while – represents the number with one set of 9 bases; p and q are the respective frequencies. Data are from Handoko et al. (2001) and ignore two Sumbanese, a Batak, a Kaili, and a Javanese individual listed there, whose genotypes do not fit into these two categories.

Table 8.3 Pair-wise F_{ST} based on the 9 bp deletion frequencies shown in Table 8.2

	Sumbanese	Batak	Kaili	Javanese	Dayak
Sumbanese	0	−0.00466	0.01170	0.01635	0.08076
Batak	−0.00466	0	−0.00223	0.00053	0.04973
Kaili	0.01170	−0.00223	0	−0.00937	0.01676
Javanese	0.01635	0.00053	−0.00937	0	0.01203
Dayak	0.08076	0.04973	0.01676	0.01203	0

An intuitively clearer way to look at F_{ST} values is to use a statistical tool called **analysis of variance** (ANOVA). Because ANOVA has been adapted for use in molecular genetic analysis, Excoffier et al. (1992) refer to the method as "analysis of molecular variance," or "AMOVA." The gist of the method is to partition the total genetic variation into two additive parts, a part between populations and a part within populations. F_{ST} is then defined as the proportion of the total variance that is due to between-population variation. Viewed as a percentage (i.e., multiplied by 100), the F_{ST} we have just calculated indicates that 1.74% of the total genetic variation for these Indonesian populations is between populations, while 98.3% of the total genetic variation is within populations.

The overall F_{ST} we have calculated tells us about the overall level of genetic differentiation between these five Indonesian populations, but it does not allow us to identify which populations are more similar to one another. A quick glance at Table 8.2 suggests that the Sumbanese (deletion frequency of 0.1633) and Batak (deletion frequency of 0.2000) might be grouped together, the Kaili (deletion frequency of 0.2476) and Javanese (deletion frequency of 0.2571) might also be grouped, and the Dayak (deletion frequency of 0.3535) are the most dissimilar from the other four populations. In later examples we will use more populations and more loci/alleles, but we first need a general way to look at genetic relationships among populations.

There are many different ways to measure genetic similarity/dissimilarity between two groups, but we will use a simple one based on F_{ST} here. Instead of calculating F_{ST} for the entire group of five Indonesian populations, we calculate the value for each pair of populations. We use AMOVA because it is well suited for the relatively small samples available from these populations. Because hand calculation of all of the pairwise F_{ST} values is a tedious proposition, we instead use the program Arlequin, (see supplemental resources for Chapter 3), the output from which is shown in Table 8.3. The F_{ST} for a population compared with itself is 0 by definition. Ordinarily, all of the other pair-wise F_{ST} values should be positive, and if negative values do arise, they are disregarded and considered to be equal to 0. The pair-wise F_{ST} values can be interpreted as the proportion of total genetic variance for the two samples that is due to between-population variation. Thus, larger pair-wise F_{ST} values indicate greater genetic dissimilarity, or divergence, between the two populations. From Table 8.3, the largest genetic divergence is between the Sumbanese and the Dayak, with a pair-wise F_{ST} of 0.08706.

Interspersed Nuclear Elements: SINEs and LINEs

In the case of indels, it is not possible to tell, in the absence of further information, whether the original mutation was an insertion or a deletion. In the case of **short**

Table 8.4 Alu insert in the sixteenth intron of the angiotensin I–converting enzyme (ACE) gene

14041	ctg<u>CTGGAGA</u>	CCACTCCCAT	<u>CCTTT</u>CTccc	atttctctag	acctgctgcc	tat**ACAGTCA**
14101	**CTTTTTTTTT**	**TTTTTTGAGA**	**CGGAGTCTCG**	**CTCTGTCGCC**	**CAGGCTGGAG**	**TGCAGTGGCG**
14161	**GGATCTCGGC**	**TCACTGCA<u>AG</u>**	<u>CT</u>**CCGCCTCC**	**CGGGTTCACG**	**CCATTCTCCT**	**GCCTCAGCCT**
14221	**CCCAAGTAGC**	**TGGGACCACA**	**GGCGCCCGCC**	**ACTACGCCCG**	**GCTAATTTTT**	**TGTATTTTTA**
14281	**GTAGAGACGG**	**GGTTTCACCG**	**TTTTAGCCGG**	**GATGGTCTCG**	**ATCTCCTGAC**	**CTCGTGATCC**
14341	**GCCCGCCTCG**	**GCCTCCCAAA**	**GTGCTGGGAT**	**TACAGGCGTG**	**A**tacagtcac	ttttatgtgg
14401	tttcgccaat	tttattccag	ctctgaaatt	ctctgagctc	cccttacaag	cagaggtgag
14461	ctaagggctg	gagctcaagg	cattcaaacc	cctaccag<u>AT</u>	<u>CTGACGAATG</u>	<u>TGATGGCCAC</u>
14521	<u>ATC</u>ccggaaa	tatgaagacc	tgttatgggc	atgggagggc	tggcgagaca	aggcggggag

The Alu insert is shown in bold upper case, and the *Alu*I restriction sites (from which the Alu insert draws its name) are marked with double underlines. The upper-case single-underlined sections are the primer sites usually used in polymerase chain reaction assays for this polymorphism. The second primer site (14491–14523) is also the location for the start of the sixteenth exon (at 14491). The ACE Alu shown is listed in the SNP consortium database under NCBI Assay Id 4729.

interspersed nuclear elements (SINEs) and **long interspersed nuclear elements** (LINEs), we know that all mutational events are insertions. Both types of DNA sequence are capable of **transposition**, which means that they can copy and insert at new locations within the genome. SINEs are on the order of 150–300 bp long. The best studied of the SINEs is *Alu* (Watkins et al. 2001). Alu sequences are about 300 bp long and contain one or more *Alu*I restriction sites, for which they are named. Table 8.4 shows an Alu sequence within an intron for the angiotensin I–converting enzyme (ACE) gene on 17q23. There may be as many as a million Alu inserts interspersed throughout the genome, and collectively they comprise about 10% of the genome. Because these sequences appear to have no function, they are not subject to direct selection (provided that they do not disrupt the functioning of genes by inserting into an exon). As a consequence, there are considerable sequence variations for Alu elements, there are places where the Alu insertions are only fragments, and there are even cases of Alu insertions within Alu insertions (Comas et al. 2001).

Because the Alu sequences have been actively transposing (i.e., inserting) throughout human history, the presence of an Alu element at a particular site in the genome may be polymorphic. In other words, some individuals may have an insert at a particular location, while others may not. Further, the genome is so vast that the probability of an Alu element inserting more than once at exactly the same spot is vanishingly small (the proverbial "lightning striking twice"). Consequently, site-specific Alu elements are UEPs, which makes them very useful in population genetics analyses. Alu insertions are particularly easy to type using PCR and gel electrophoresis. To do this type of analysis, a researcher will pick primers to either side of an Alu insertion site and then run the PCR product on a gel. Individuals with the Alu insertion will produce a longer PCR product (hence, the DNA will not run as far in the gel), while individuals lacking the insertion will produce a shorter PCR product. Because of the specificity of the primers (usually about 25 bases long), an Alu PCR test is targeted for one and only one insertion site. Some Alu polymorphisms are listed in the SNP database (see supplemental resources), even though this is a misnomer since an Alu is not a single nucleotide. For sites that are autosomal or on the X chromosome when typing females, heterozygotes can be distinguished from homozygotes. One extensively studied Alu insertion is on the Y chromosome, referred to as the Y Alu polymorphism (YAP), which is either present or absent in males (who are haploid with respect to the Y chromosome). Together with 92R7 discussed above, YAP is one of the two UEPs used to define the Cohen modal haplotype, discussed later.

LINEs are much longer than SINEs, generally about 6,000 bp, and present in excess of 100,000 copies, most of which are incomplete copies of the LINE sequence. Together, these LINEs comprise about 15% of the genome. Sheen et al. (2000) describe a method they call "L1 display," which can be used to identify polymorphic sites for LINEs. The "L1" is a reference to the primary human LINE, known as "long interspersed nuclear element—1." The L1 display method can be used to type individuals as being homozygous for the insertion, heterozygous, or homozygous for lacking the insert. Myers et al. (2002), Ovchinnikov et al. (2002), and Salem et al. (2003) describe particular classes of polymorphic L1 variants that may prove particularly useful in human population genetics studies.

DNA Sequencing

The determination of DNA sequences has now become a routine matter, with the process so highly automated that a draft of the entire human genomic sequence was released in 2001 (International Human Genome Sequencing Consortium 2001). The most popular method for determining sequences is the **chain termination reaction**. The method begins with many copies of single-stranded DNA obtained either by asymmetric PCR, or by tagging one of the primers so that after the DNA is denatured the targeted strand can be isolated. Once single-stranded DNA is obtained, the chain termination reaction is run. This is run more or less like a normal PCR but uses only one primer and limited amounts of dideoxy-ATP, -GTP, -CTP, and -TTP (referred to generically as "ddNTPs") in addition to the dNTPs. These dideoxy molecules lack a part of the regular dNTPs; consequently, if they are incorporated into a growing chain during extension, they will terminate the growth of the DNA chain. The ddNTPs can be labeled so that they are differentiated. In automated sequencing, each type of ddNTP has a different-colored fluorescent marker. When the chain termination product is run through a high-resolution gel, a laser detects the position and color of the dye and converts these automatically into the sequence (see Supplemental Resources).

Although sequencing has become a very routine and automated molecular genetic technique, it is still time-consuming. In human population genetics studies, which typically focus on a large number of individuals, it is completely impractical to sequence long stretches of DNA. As a consequence, anthropological geneticists have typically used relatively short sequencing runs and, thus, may have missed many informative DNA sites. However, new techniques have recently been developed that allow specific mutation sites to be located. Once these sites are located, a short sequencing run surrounding the site can be executed in order to characterize the nature of the mutation. This saves the time and expense of sequencing thousands of invariant DNA bases. In their article "The Mitochondrial Gene Tree Comes of Age," Richards and Macaulay (2001) suggest that whole-sequence mitochondrial genome analyses are a very real possibility for the future. In their words (p 1319), "It seems there may be life in this old molecule yet."

As with indels, we can analyze DNA sequence variation using F_{ST}. F_{ST} is again defined (equation 3.14) as the decrease in observed heterozygosity relative to the expected heterozygosity. In the following example, we use AMOVA to calculate F_{st} (Excoffier et

Table 8.5 Mitochondrial DNA (mtDNA) sequence data for the Haida and Bella Coola

	1 1111111112 22222 1234567890 1234567890 12345			Haida	Bella Coola
1	TTTTACCCCT	TTTCTTTTCT	ATCCC	1	8
2	....G.....	....	.....	20	3
3	..C.G.....	.......C...	GC...	2	5
4	.CC.G.....	.......C...	GC..T	0	2
5	....G.....	....	...T.	10	0
6	....G.....	.......T.	...T.	1	0
7	....G...T.	....	...T.	1	0
8	....G....C	....	.....	1	0
9	..CG.....	....	.....	1	0
10	C.C.G.....	...T..CC..	GCT.T	3	3
11	..C.G.....	..C...C..C	G...T	1	0
12	......TT.C	....	.....	0	5
13	....GT....	....C.....	.....	0	2
14	....G.....	.....C....	.....	0	1
15	....G....C	....	...T.	0	6
16	..C.G.....	C.....C...	GC...	0	3
17	....G....C	.CC...C...	G...T	0	2

Dots indicate bases that are the same as in the first line, numbers in the top row give the position (see Ward et al. 1993 for original mtDNA positions), and the counts under "Haida" and "Bella Coola" give the number of individuals with the listed sequence.

After Ward et al. 1993.

al. 1992) for Ward et al.'s (1993) sequence data for mtDNA from the Haida and the Bella Coola, two Pacific Northwest coast Native American tribes. The sequence data for polymorphic sites are listed in Table 8.5.

We again use Arlequin, this time to form all of the AMOVA tables by polymorphic sites, and then sum across these 25 sites, giving a final F_{ST} value of 0.08407. This value means that about 8% of the genetic variation is between the two tribal samples and the remaining 92% is within the tribes. The F_{ST} value of 0.08407 that we just found is based on sequence divergence. For DNA sequence data, this is usually the quantity that we want to estimate. However, we also can treat the sequences as individual haplotypes (or alleles), in which case F_{ST} is equal to 0.14889. Notice that this haplotype F_{ST} is higher than the sequence F_{ST} value of 0.08407. Lynch and Crease (1990, 386) note that "estimates of population subdivision that are based on nucleotide divergence need not be the same as those based on haplotype frequencies," as is indeed the case in this example. Yet another way to calculate F_{ST} makes specific use of a model known as the "coalescent." To understand the basis for these calculations, we need to discuss this model in some detail.

The Coalescent Model

Coalescent refers to the "coming together" of descendants into ancestral DNA. Thus, the **coalescent model** gives us a "family tree" for DNA. The coalescent is relatively easy to understand by using a simple heuristic diagram, as in Figure 8.3. In this figure, we start in the current time period with, for example, 10 mtDNA molecules from 10 different

Past

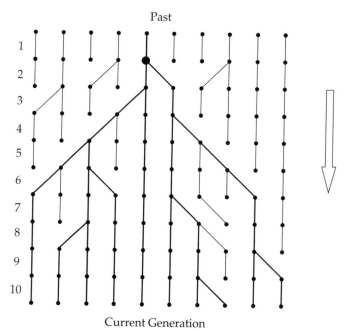

Current Generation

Figure 8.3. Schematic diagram representing the coalescent model. The diagram shows the coalescence of 10 alleles. The *arrow on the right* shows the "flow of time." The *numbers on the left* show the number of lineages in existence that give rise to the 10 current alleles at the bottom of the diagram.

individuals. As we go back to a previous time, there can only be as many molecules or some lesser number because of coalescent events. Moving against the flow of time, the molecules coalesce because we are witnessing meioses in reverse. In this example, we start with 10 mtDNA molecules that have coalesced to nine within the first time period, then eight, then seven, and so on until finally there is a single lineage. The individual shown at the top of this lineage is the **most recent common ancestor (MRCA)** for the 10 descendants in the current generation. What about the other lineages that existed at the time the MRCA was alive? The figure shows that there have always been 10 mtDNA molecules alive at any given point in time, so at the time of the MRCA there are nine other lineages, none of which extends into the current generation. This particular coalescent model is thus one with a constant population size.

Table 8.6 lists all of the probabilities of coalescence for 20 mtDNA genomes down to one and for (female) population sizes of 100 and 1,000 (these were found using the "and" and "or" rules of probability). Looking first at the probabilities of coalescence for $n = 100$ and starting at the bottom of the table, we can see that the probability of coalescence decreases with each coalescent event. As a heuristic, we think of the mtDNA genomes as undergoing "Brownian motion" and as being "sticky." Consequently, in recent history, when there are more distinct mtDNA molecules, they are more likely to "bump into" each other and "stick together" (coalesce). As the number of mtDNA lineages decreases, the further back in time we go, they become less and less likely to coalesce. The rate of coalescence consequently decreases as we move backward in time. Comparing the

Table 8.6 Probabilities of mitochondria DNA coalescence for population sizes of 100 and 1,000 females

	n = 100		n = 1000	
Number of alleles	Stay	Move Up	Stay	Move Up
1	1.0000	0.0000	1.0000	0.0000
2	0.9900	0.0100	0.9990	0.0010
3	0.9702	0.0298	0.9970	0.0030
4	0.9411	0.0589	0.9940	0.0060
5	0.9035	0.0965	0.9900	0.0100
6	0.8583	0.1417	0.9851	0.0149
7	0.8068	0.1932	0.9792	0.0208
8	0.7503	0.2497	0.9723	0.0277
9	0.6903	0.3097	0.9645	0.0355
10	0.6282	0.3718	0.9559	0.0441
11	0.5653	0.4347	0.9463	0.0537
12	0.5032	0.4968	0.9359	0.0641
13	0.4428	0.5572	0.9247	0.0753
14	0.3852	0.6148	0.9126	0.0874
15	0.3313	0.6687	0.8999	0.1001
16	0.2816	0.7184	0.8864	0.1136
17	0.2365	0.7635	0.8722	0.1278
18	0.1963	0.8037	0.8574	0.1426
19	0.1610	0.8390	0.8419	0.1581
20	0.1304	0.8696	0.8259	0.1741

The probability of coalescence is given in the "Move Up" column. For example, the value of 0.8696 in the bottom row, third column is the probability of going up the first column from 20 alleles to 19 alleles (a coalescent event). The columns labeled "Stay" give the probability of no coalescence.

probabilities of coalescence for populations of $n = 100$ and $n = 1,000$, it is clear that the probability of coalescence decreases with increasing population size. This is just like genetic drift, where the influence of random sampling events (drift) decreases with increasing population size. Indeed, the coalescent model is another way of looking at genetic drift, and this is why we can use the model to "motivate" calculation of F_{ST}.

While Table 8.6 allows us to see a few interesting properties of the coalescent, it is far easier to use computer simulation to explore properties of this model (Strobeck 1987, Hudson 1990). Figure 8.4 shows one such simulation with 20 mtDNA alleles and a (female) population size of 100 individuals (as in the first part of Table 8.6). Figure 8.5 shows another simulation, also for 20 mtDNA alleles, but this time with a (female) population size of 1,000 (as in the second part of Table 8.6). The two simulations are broadly similar; both show that most coalescent events occur within recent population history. As in Table 8.6, the coalescent process "slows down" as the number of lineages decreases back through coalescent events. Figure 8.6 shows both simulations together drawn on the same scale of 2,000 generations. Note how much more rapidly the smaller population (top of figure) coalesces.

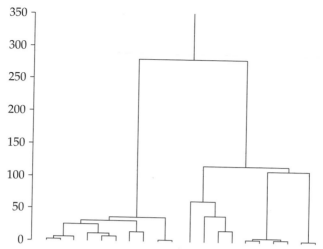

Figure 8.4. Simulation of the coalescent for 20 mtDNA alleles with a female population size of 100. The axis on the left shows number of generations.

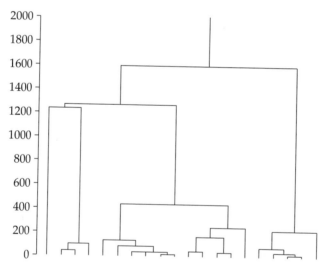

Figure 8.5. Simulation of the coalescent for 20 mitochondrial DNA alleles with a female population size of 1,000. The axis on the left shows number of generations.

Mutation, Mismatch Distributions, and F_{ST} in the Colaescent Model

It is a simple matter to include mutational events in our simulated histories. To simulate mutations in the coalescent model, we say that there is a constant low rate of mutation and then "expose" the vertical (descending) lines in the simulated tree to this mutation rate. The longer the descending line, the more mutations we expect to occur along the path. Shorter vertical paths are less likely to have mutational events, while longer paths

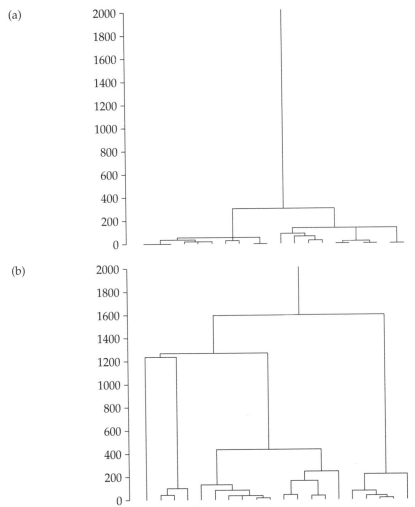

Figure 8.6. Simulations from Figure 8.4 (a) and 8.5 (b) shown together on the same scale of 2,000 generations.

may "see" one or more mutational events. Once we have simulated mutational events, we need to consider how these events affect the DNA history we are trying to construct. In this model, we assume that each mutation is to a new UEP. Figure 8.7 shows a simulation with 36 mtDNA molecules. We have removed the scaling on the vertical axis because usually the scale when there is mutation is in units which are a function of both time and the mutation rate (Rogers 1995, Harpending et al. 1998). Rogers (1992) discusses the specifics of this model, though there are still some points of contention (Rogers et al. 1996, Schneider and Excoffier 1999).

Now, we can apply a graphical method to summarize the data. The particular graphical device we use is known as a **mismatch distribution**, which is just a bar graph (histogram) showing the number of sites at which individuals differ for all pairs

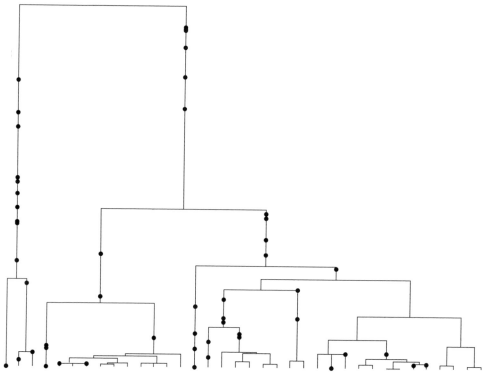

Figure 8.7. Simulation of the coalescent model with mutation. Each dot on the gene tree represents a unique mutation.

of individuals. With 36 mtDNA molecules, there are 630 pair-wise comparisons (we compare individual 1 against individual 2, 1 against 3, 1 against 4, and so on). To compare individual 1 against 2, we count the number of mutations (black balls in the figure) that occur as we go up the tree to a common ancestor. On the left ascending branch from individual 1 there is one mutation, while on the right ascending branch from individual 2 there are two mutations. Consequently, for this pair-wise comparison, the individuals differ at three sites. We simply need to count up the number of mutations for all pair-wise comparisons as we ascend to the MRCA for the two individuals. Figure 8.8 shows the mismatch distributions for two simulations with a higher mutation rate than in Figure 8.7. We show two simulations so that it is clear that each is characteristic of the process and not an individual fluke. Both of the mismatch distributions are "peaky," "ragged," or "bumpy" in appearance due to the fact that a few deep branches disrupt the distribution.

Mismatch distributions from sequence data can be used to estimate F_{ST} (Harpending et al. 1996), which is why we discuss the coalescent model at this point. If we find the average mismatch within the Haida and within the Bella Coola (Figs. 8.9 and 8.10) and then average the two groups, we get an average within-group mismatch of about 3.72. In other words, on average, two individuals who are randomly chosen within the same tribe will differ at about three and three-quarters bases. If we pool the Haida and Bella Coola and then find the average mismatch, it is a bit higher, at about 3.91. Using equation 3.8,

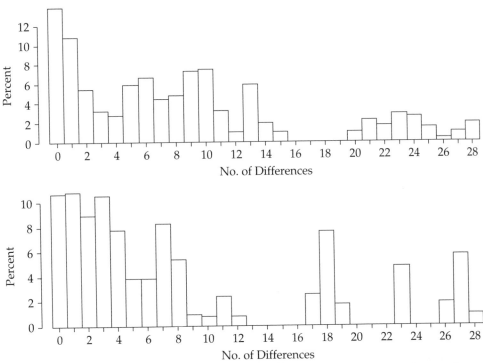

Figure 8.8. The mismatch distributions from two simulations of coalescence with mutation.

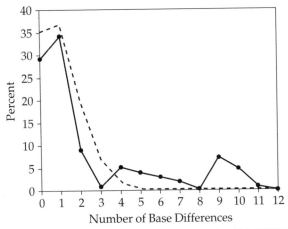

Figure 8.9. The mismatch distribution from the Haida mitochondrial DNA sequence data. The dots connected by the *solid line* are the observed percentages, while the *dashed line* represents a fitted model.

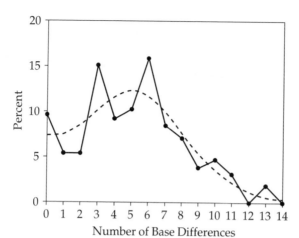

Figure 8.10. The mismatch distribution from the Bella Coola mitochondrial DNA sequence data. The dots connected by the *solid line* are the observed percentages, while the *dashed line* represents a fitted model.

we can substitute the average mismatch within tribes for H_o and the average mismatch for the combined tribes for H_e. This gives $(3.91 - 3.72)/3.91$, or an F_{ST} of about 0.048. While it may be disturbing to have three different ways to calculate F_{ST} from sequence data (averaging across polymorphic sites, treating as haplotypes, and using the mismatch distribution) which lead to three different results, it is important to understand that different methods of calculation exist. One has to be extremely careful in comparing DNA sequence F_{ST} estimates from different studies because they may have used different calculation methods.

Demography and the Coalescent Model

In addition to providing information about F_{ST}, the mismatch distribution can provide information about demographic processes. We saw in Figures 8.8–8.10 that the mismatch distributions were "bumpy" or "ragged"-looking. In Figure 8.8, this is a result of assuming a constant population size through time. We can only guess why the Haida and Bella Coola mismatch distributions in Figures 8.9 and 8.10 are ragged-looking. In Figure 8.11, we compare a simulation where population size increased in the past with a simulation where population has always been a constant small size. As we move back in time to the top of the "graph," the probability of coalescence drastically increases right before the population explosion (as in Table 8.6, where the probability of coalescence is greater at a size of 100 than at 1,000). As a result, most of the coalescent events in the expanding population occur just prior to the population explosion so that the gene tree has long tips that descend from the point of population increase. Under constant population size, in contrast, the coalescent events are at first very rapid and then slow down, producing a tree that has a few deep bifurcations.

Different tree shapes can have profound effects on the final genetic outcome of mutational events, and this affects the shape of the mismatch distribution. Figure 8.12 shows a simulation of the coalescent model with mutation where at some time in the past the population underwent an instantaneous increase in size. Figure 8.13 shows the mismatch distributions from two simulations with population expansion. Because coalescent

(a)

(b)

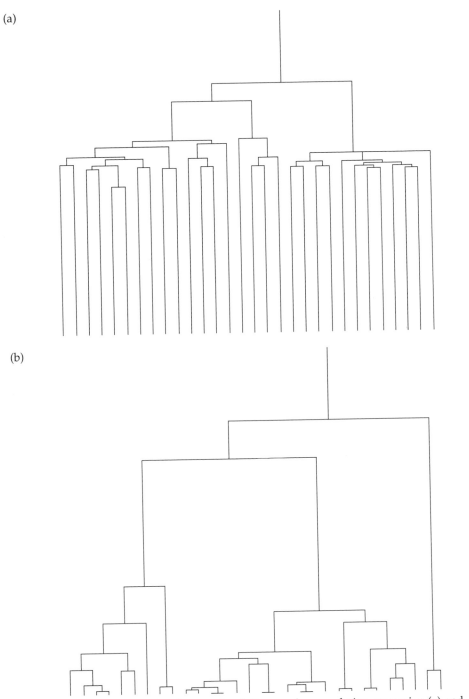

Figure 8.11. Comparison of coalescent simulations under population expansion (a) and constant population size (b). In the top simulation, the population increased by a factor of 10, while in the bottom simulation the population remained a constant size.

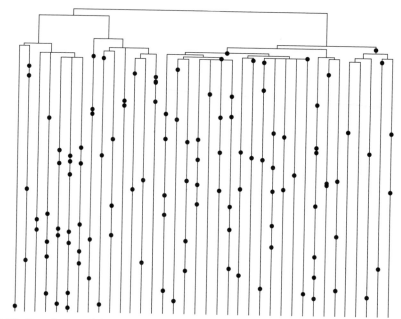

Figure 8.12. Simulation of the coalescent model with population expansion. Each dot on the gene tree represents a unique mutation.

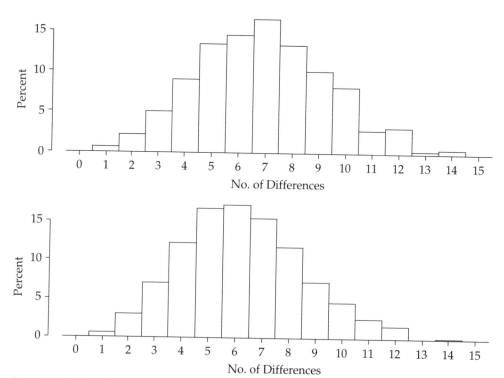

Figure 8.13. The mismatch distributions from two simulations of coalescence. Unlike Figure 8.8, which shows the mismatches from populations of constant size, this figure is from simulations where there was a population explosion in the past.

events are concentrated at a point in time just prior to the expansion, the gene tree has many deep branches (see Fig. 8.12), which results in smooth mismatch distributions. These mismatch distributions are considerably different from the ones we previously examined from simulations of stationary populations (see Fig. 8.8).

The mismatch distributions for the Haida and Bella Coola (Figs. 8.9 and 8.10) do not follow the smooth pattern expected for populations that have undergone population expansion. Figures 8.9 and 8.10 show the expected distributions (from Arlequin) together with the rather "bumpy" observed distributions. Stone and Stoneking (1998, 1167) comment on this fact for the Haida: "The Haida do exhibit reduced diversity and a ragged mismatch distribution, which may indicate recent population restriction or long-term constant size." In an analysis of Choctaw and Menominee mtDNA sequences, Baker (2001) finds that these Native American tribes did show the smooth mismatch distributions expected for populations that have undergone expansion or, alternatively, a "selective sweep" (see Rogers 2001). Baker's work is particularly interesting in that she extracted mtDNA from hair clippings that had been collected as part of Franz Boas's large-scale anthropometric project for the World Columbian Exposition in 1894 (Jantz 1995). Stone and Stoneking's (1998) analysis of prehistoric (circa 1300 A.D.) Oneota mtDNA sequences from Illinois also provides a smooth mismatch distribution. Some of the "bumpiness" observed in extant Native American mtDNA mismatch distributions (such as the Haida and Bella Coola) could well be due to the extreme population decimation attendant with the European arrival in the "New World." For other regions of the world, there have been intense arguments in the literature over how to interpret bumpy mismatch distributions (see, e.g., Bandelt and Forster 1997).

Selection can produce some of the same patterns that demographic factors produce in the coalescent model. For example, a smooth unimodal mismatch distribution can arise from either selection or a population expansion. There is no particular way to differentiate these two explanations for the same pattern, unless we have multiple lines of evidence. If smooth unimodal mismatch distributions are observed for a number of different genetic systems within the same population, then this suggests population expansion. If some mismatches are unimodal while others are bimodal, selection may have occurred for the traits represented in the unimodal mismatches. The difference between demographic processes and selection is that the prior are expected to affect all genetic systems, while the latter should affect only systems that are not selectively neutral. Because mtDNA represents a single genetic locus, we cannot differentiate among these explanations when only mtDNA data are available.

Tandem Repeats

Unlike the interspersed elements (SINEs and LINEs) which are present in single copies dispersed throughout the genome, tandem repeats are repetitive sequences that occur at specific locations in the genome. These repetitive elements are also called "satellites" since DNA separated out by density (buoyancy) in a centrifuge shows single-copy DNA as a main dense band and tandemly repeated DNA as lighter "satellite bands." **Classical satellite DNA** forms long sections that range from 100,000 to 1,000,000 total bases. Alphoid DNA, one of the main classical satellites, is found at the centromere of all chromosomes, is composed of a 171-base "motif" that is repeated many times, and can comprise as much as 5% of the DNA in a chromosome. Santos et al. (1995) describe a PCR-based assay for variation in Y-chromosome alphoid DNA.

Smaller minisatellites have a motif of about 8–100 bases, while microsatellites have an even shorter motif. For historical reasons, minisatellites and microsatellites each have an acronym associated with them: minisatellites are often referred to as "variable number of tandem repeats" (VNTRs), while microsatellites are referred to as "short tandem repeats" (STRs). We will look at the STRs first as they are simpler to understand as well as to assay. The STRs have become the system of choice in much of forensic DNA analysis (Butler 2001) and have also been used extensively in population genetics analyses. STRs are tandem repeats of 2 to 6 bases, such as CA as a two-base motif or AGAT as a four-base motif. To assay an STR, geneticists use flanking primers for a specific STR locus and then some form of electrophoresis to resolve the STR and count the number of repeats. There is extensive information on the web (see Supplemental Resources) for the STRs typically used in forensic work, so we do not repeat that information here. STRs allow one to distinguish between heterozygotes and homozygotes for autosomal loci.

VNTRs are generally more complicated to assay as well as to analyze. While in theory they can be treated just like STRs, VNTR motifs are long enough that they are more subject to sequence variation within individual motifs. As a consequence, the more useful assays indicate not only the repeat number but also the variable types of motif within the VNTR. Jobling et al. (1998) describe the Y-chromosome minisatellite MSY1. This locus has repeat units (motifs) that are 25 bp long, and there are five main types that differ from each other at one or two bases. Jobling and co-workers (1998) designed a PCR procedure that can be used to count the number of repeats of each type and

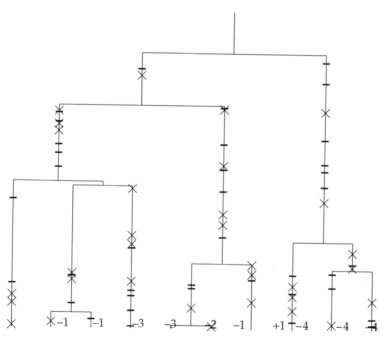

Figure 8.14. Simulation of the coalescent model under a microsatellite mutation model for a population of constant size. X's represent the gain of an additional repeat in the short tandem repeat, while *minus signs* represent the loss of a repeat. The numbers at the bottom give the net loss or gain of repeat units relative to the ancestral state.

determine the order in which the types occur. As we will see in a later section, genetic evidence from the MSY1 locus has figured heavily in ongoing discussions about whether Thomas Jefferson fathered children with his slave Sally Hemings. In the remainder of this section, we will focus only on microsatellites because analysis of minisatellites is more complicated.

The analytical methods for microsatellites are related to those we have already discussed for sequence data but differ in that we must account for the number of repeat differences between two individuals. With microsatellite data, we must measure precisely how much individuals differ at a locus. For example, two individuals with repeat lengths of 15 and 16 are considered to be more similar than two individuals with repeat lengths of 15 and 18. While the basic AMOVA approach still works, it requires modifications (Slatkin 1995, Goldstein et al. 1995). The microsatellite calculations are usually made using specialized software available on the internet (MICROSAT or RSTCALC, see Supplemental Resources) and are too complicated to deal with here.

Microsatellites and the Coalescent Model

Microsatellite variation can also be discussed within the coalescent model. The tree process for the coalescent model (Figs. 8.3–8.6 and 8.11) is the same for microsatellites as it was for sequence variation. However, the mutational process for microsatellites is different, generally involving the loss or gain of one repeat unit. Figure 8.14 shows a simulation where loss of a repeat is marked with a minus sign and gain of a repeat with an X. Listed at the bottom of the tree are the net increases or decreases in repeat lengths relative to the MRCA. This particular simulation was done using the assumption that there is a constant population size through time, whereas in Figure 8.15 we show a

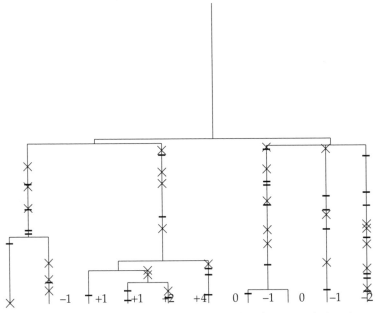

Figure 8.15. A simulation identical to that in Figure 8.14 but for a population that underwent a 100-fold expansion.

similar simulation that includes population expansion. The values for both of these simulations were taken from King et al. (2000). Similar simulations can be found in Reich and Goldstein (1998).

In an actual genetic analysis, we cannot observe the lineages and instead need to characterize the alleles in the current generation. A simple way to do this is to make a histogram of the allele sizes. Figure 8.16 shows such histograms for four simulations, where the top two are for stationary (constant population size) cases and the bottom two are for expanding populations. As can be seen from Figure 8.16, stationary populations produce markedly different histograms from expanding populations. Stationary populations produce histograms that are more spread out and uneven, with a number of spikes. This is because there are only a few deep branches in the **gene tree** that provide lineages that have evolved independently for many generations. In contrast, in an expanding population, there are many deep branches, so there are many independent lines; consequently, the mutation process tends to average out. Adding population expansion to a coalescent model with stepwise mutation of STRs (loss or gain of single repeats) produces a histogram that is much less dispersed, typically with a single peak.

The coalescent model has figured heavily in interpretations of the timing of population events based on microsatellite and other DNA variations. This is a complicated area, and one for which the math is entirely beyond the level we have adopted in this text. Walsh (2001) gives a method for determining the time to the MRCA for nonrecombining DNA sequences found in two individuals. Slatkin (1995) and Zhivotovsky (2001) give such methods for determining times to coalescent events from microsatellite data (see Zhivotovsky et al. 2003 for a recent application). The usual model adopted is one of stationary populations with growth rate of 0. More complicated models that incorporate population growth generally require computer simulation methods in order to estimate times to coalescent events (see, e.g., Thomson et al. 2000), though Zhivotovsky's microsatellite model does allow for population expansion.

TWO DNA VIGNETTES FROM ACROSS THE GLOBE

We have discussed DNA markers by the type of variation, be they RFLPs, indels, sequence variation, or tandem repeats. Another way to organize the material in this chapter is by portions of the genome. For example, both mtDNA (Richards and Macaulay 2001) and the Y chromosome have recently been reviewed (Jobling and Tyler-Smith 2003). A third way to organize this material is by looking at applications of various DNA markers to regional analyses. Jobling and Tyler-Smith (2003) refer to this as "phylogeography," and this is the path we follow in this section.

The literature on DNA variation in local and regional populations is now so voluminous that we must choose from among a myriad of possible examples. In this section, we will look at only two cases, the first being the region of Oceania and the second being the Middle East. We look at Oceania first because it should, in theory, form a less complicated example. The Middle East example is particularly complicated because we will focus on one religious group (Jews) that has undergone a number of diasporas. Our reason for picking this second example is the comparatively short time lapse between oral and written histories of this group. Jewish oral history is some of the earliest to be placed in writing. Oral traditions for the founding of the "New World" are probably more ancient

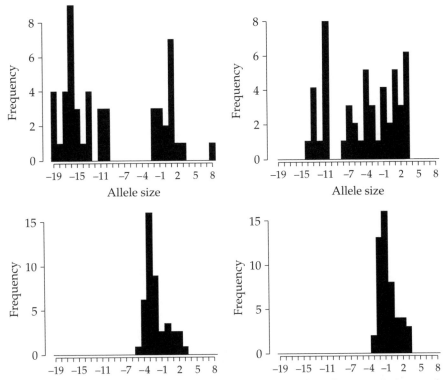

Figure 8.16. Histograms of relative short tandem repeat sizes from four simulations under stepwise mutation. The *top two* simulations are for populations of constant size, while the *bottom two* simulations are for populations that have undergone expansion.

but have been written down so recently that we would have difficulty finding clear examples of where oral history and genetics coincide.

Even though we are giving only two examples, there are many interesting and complicated questions to be addressed on the regional level across the world. For example, the suggestion by some anthropologists that the "New World" was settled in a number of discrete "waves" of migration from Asia makes for interesting studies that attempt to (1) document the number of waves, (2) identify ancestral sources, and (3) document pre- and postsettlement demographic processes (Forster et al. 1996, Karafet et al. 1999, Derenko et al. 2001). The picture from Europe is also vastly complicated as we must contend not only with whether or not Neandertals were replaced by "modern humans" (or whether they admixed) (Hawks et al. 2000; Krings et al. 2000; Labuda et al. 2000; Wall 2000; Adcock et al. 2001; Relethford 2001a,b; Takahata et al. 2001) but also with the complicated history and prehistory of Europe (Lell and Wallace 2000, Rosser et al. 2000, Barbujani and Bertorelle 2001, Helgason et al. 2001, Nasidze and Stoneking 2001, Nasidze et al. 2001, Pereira et al. 2001, Renfrew 2001, Zerjal et al. 2001, Capelli et al. 2003). Africa, as one of the largest and longest populated continents, also has a complicated genetic history, much of which may have been distorted by the way in which geneticists have chosen to partition variation (MacEachern 2000).

Human Colonization of Oceania: "Express Trains," "Slow Boats," "Entangled Banks," and Embittered Battles

Archaeological evidence indicates that the Pacific Islands constituting "far Oceania" (the islands east of Papua New Guinea and the Solomon Islands) are one of the more recently settled parts of the globe. Because colonization required open ocean voyages, it might be presumed that the genetic history of Oceania would be a simple one punctuated by a few long-range migrations. Thus, the Pacific is often viewed by geneticists as a Panglossian "lab experiment" in human population genetics that we should be able to easily read and interpret. In contradiction to this view, studies of the Pacific have been slow to yield any obvious and unambiguous signatures of human population genetic history. Terrell (2000) argues that this may be as much a result of overreliance on such simple metaphors as the "express trains" and "slow boats" of the section heading as anything else and that the history of science and of publication in science has driven much of the debate in interpreting the genetic history of Oceania (the "embittered battles" in the heading).

Some of the background we need for this discussion unfortunately comes from physical anthropology's early interest in racial classification. In the ineluctable drive to classify humanity, early anthropologists typically divided populations from the Pacific Islands into three "varieties": Polynesians, Melanesians, and Micronesians. *Polynesians* were those peoples living on islands bounded by and including the triangle formed by New Zealand, the Hawaiian Islands, and Easter Island. *Melanesians* were those peoples living to the west of Polynesia, principally on Papua New Guinea, the Solomon Islands, Fiji, Vanuatu, and New Caledonia. *Micronesians* were those peoples living on the many small islands north of Melanesia. Houghton (1996, 20) gives a map of Oceania with these divisions, appropriately labeled "The Parceling of the Pacific into Melanesia, Micronesia and Polynesia." While these three regions represent a geographic demarcation, the parceling also was originally intended to apply to the peoples who inhabited these vast regions. More recently, Micronesians and Polynesians have been viewed as a single geographically dispersed group, referred to as "Austronesians," while Melanesians are often referred to as "Papuans." Terrell et al. (2001, 97) argue that anthropology's long fixation with neat, tidy classification and oversimplification of the past has essentially produced an approach that focuses on "two peoples and two periods: an elegantly simple paradigm or model of the past." In this "elegantly simple paradigm," Papuans were viewed as the earliest wave to arrive in the Pacific, followed by Austronesians as a second wave. However, as Terrell et al. (2001, 97) point out, this view "is not just simple, it is too simple."

Within the context of this paradigm, geneticists have tended to focus (some might say "fixate") on two diametrically opposed models. One model, termed the "express train" or "speedboat" model by the physiologist Jared Diamond (1988), posits that Austronesians settled Micronesia and Polynesia as a rapid and late migration that "skipped across" Melanesia without any genetic admixture. Opposed to this model is what Redd et al. (1995, 605) term the " 'Melanesian' model," where "Polynesian origins involved a founder event from a genetically diverse source population(s) from Melanesia." Kayser et al. (2000, 1237) refer to this latter model as the "entangled banks hypothesis," where the settlement of Polynesia was characterized by a "long and more complex interaction between Polynesia, Melanesia and Southeast Asia." Their choice of this catchphrase is a bit odd as they cite Terrell's 1986 book, *Prehistory in the Pacific Islands*, where he does not use this phrase, but fail to cite his 1988 article "History as a Family Tree, History as an Entangled Bank: Constructing Images and Interpretations of Prehistory in the South Pacific." Intermediate between the "express train" and the "entangled bank" is the "slow boat" model, where "Polynesian ancestors did originate from Asia/Taiwan" (an assumption in the "express train" model) "but did not move rapidly through Melanesia; rather, they interacted with and mixed extensively with Melanesians, leaving behind their genes before colonizing the Pacific" (Kayser et al. 2000, 1237). This "slow boat" model differs from the "entangled bank" model only in the specification of a source population (Asia/Taiwan).

Kayser et al. (2000) find support for the "slow boat" model in their analysis of Y-chromosome markers, but other researchers have not in analyses of mtDNA. This points up additional problems in DNA analyses. First, both mtDNA and the non-recombining portions of the Y chromosome are, technically speaking, single and highly polymorphic loci. As such, each provides a history of a very small fraction of the genome. Second, mutational rates differ between different genetic systems (e.g., STRs mutate much more rapidly than SNPs) and within different parts of the genome. Consequently, "genetic clocks" run at different speeds, with some too slow to register historical events and others so fast that they quickly are overwritten. Third, the whole process of colonization of formerly uninhabited areas is presumably complex enough that we may never fully understand it or be able to differentiate between stereotypical models.

Thus, at the end of a Y-chromosome study by Hurles et al. (2002, 301), we read the equivocal statement that "this study, while not strongly supporting the hypothesis of a rapid Austronesian expansion from Taiwan, is not necessarily incompatible with it." Capelli et al. (2001, 433), however, are less equivocal about their Y-chromosome data, writing that "the genetic heritage of Austronesian agriculturalists throughout Southeast Asia and Melanesia has a conspicuous indigenous origin and ... in Melanesia in particular, the dispersal of the Austronesian languages was mainly a cultural process, in contradiction to the express-train model." Su et al. (2000, 8225) write that "the Y-chromosome data do not lend support to either of the prevailing hypotheses" for Polynesian origins. Kayser et al. (2001) expand their own analyses in the Pacific to include indigenous Australians and suggested that Y-chromosome markers differentiate Australians from Melanesians. These groups are presumed to have a relatively recent common history that ended with the termination of the last glacial period. In the case of Australians and Melanesians, both the Y-chromosome and mtDNA markers indicate divergence between the two groups, while autosomal markers do not. Kayser et al. (2001, 184) use a "different clock rate" argument to explain these results: "We therefore hypothesize that the Y-chromosome (and mtDNA) difference between PNG [Papua New Guinea] and Australia arose after the separation and isolation of Australia and New Guinea, whereas autosomal markers reflect their preceding shared history."

An additional complexity in Oceanian studies is that there has been admixture from Europeans and Native Americans into the island populations. Hurles et al. (2003) report on Native American Y-chromosome haplotypes from the Polynesian island of Rapa Iti.

Rapa Iti, not to be confused with the farther east Rapa Nui or Easter Island, is in the east end of the Austral Islands, which are the southernmost group of French Polynesia, south of the Society Islands and Tahiti. Rapa Iti is about equidistant from the Chilean and Papua New Guinean coasts, and as such Native American genes at Rapa Iti could be used to argue for a South American origin for Polynesians, a now largely discredited hypothesis due to Heyerdahl (1950). Hurles et al. (2003) identify a more plausible argument, which is that the Native American chromosomes found on Rapa Iti are a relict of the historic Polynesian slave trade from Peru. Of the many slave ships that visited Polynesia from Peru, one was captured in 1863, and five of the crew members remained on Rapa Iti. Hurles et al. suggest that these crew members are the source of the Native American Y-chromosomal haplotypes found on Rapa Iti. In an ingenious study, Matisoo-Smith et al. (1998) remove this additional complexity of historical European and Native American admixture in Polynesia by studying the mtDNA of extant Pacific rats. Because this species of rat can swim only a very limited distance in the ocean, any Pacific rats in Polynesia must have come in the outrigger canoes used to explore and settle the region. European rats, which are also found in Polynesia as a result of historical European exploration, belong to two different species that cannot breed with the Pacific rat. Using the mtDNA of Pacific rats across Polynesia, Matisoo-Smith et al. studied the colonization of Polynesia free of the effects of later historical intrusions into the area.

On the finer scale for the Pacific, Murray-McIntosh et al. (1998) studied mtDNA data specifically for the Maori (Polynesians) from Aotearoa (New Zealand) and compared their results to mtDNA studies from other Polynesians and Melanesians. They find that Melanesian populations have the smooth unimodal mtDNA mismatch distribution typical of nonequilibrium populations (i.e., populations that have been through an explosion in size), while Polynesian populations tend to have bimodal mismatch distributions typical of a stable population. Rather than conclude that Polynesian populations have not undergone a population explosion, they make the more reasonable suggestion that two formerly isolated groups (one Austronesian and one Melanesian) fused during the founding of Polynesia.

Murray-McIntosh et al. (1998) also examine the mtDNA data for the Maori using a computer simulation. Their simulation results suggest that the founding number of women (remember that mtDNA provide only a maternal history) for the Maori was between 50 and 100. They evelute these results within two diametrically opposed models, one in which Aotearoa was founded by one or a few "accidental" landings versus one in which a number of canoe voyages were made with the intention of colonizing the new land. The relatively large number of founding females suggested in their simulations is incompatible with the first alternative of a few accidental landings, and Underhill et al. (2001, 276) write that "Y-chromosome data also suggest multiple arrivals." Murray-McIntosh et al. (1998) noted that a pattern of intentional colonization is in synchrony with Maori oral history. This is an interesting point because, with the exception of Europe, oral histories are often blithely ignored by geneticists when reconstructing human history and prehistory. In the case of Aotearoa, the islands were colonized recently enough (about 1150 A.D.) that oral history should be a good guide. Indeed, Murray-McIntosh et al.'s (1998, 9052) study is one of the few genetic analyses which acknowledges the broader interpretive web: "we note that genetic data, including interpretation and analysis of DNA sequences, are just one aspect of the multifaceted and profound issues of relationship, kinship, and differences between human groups, their cultures, and perceptions of self-identity."

THE COHEN MODAL HAPLOTYPE AND THE DIASPORA

In 1997, Skorecki et al. published a scientific correspondence in *Nature* titled "Y Chromosomes of Jewish Priests"; and in 1998, Thomas et al. published (again in *Nature*) a correspondence on Y-chromosome variation titled "Origins of Old Testament Priests." The logic applied in these two correspondences is relatively simple. In the story of Exodus, the second of the five books of Moses, Aharon HaCohain (usually anglicized as "Aaron the Priest"), brother of Moses, is established as the priest (a literal translation of the Hebrew *HaCohain*). This position was to pass from generation to generation through the male descendants of Aaron. Among Jews today, the surnames *Cohn, Cohen, Kahn, Katz,* and *Kahner* are viewed as representing descent from the "priestly line." However, this picture is made more confusing by the fact that Jews have undergone a number of diasporas, including, the Babylonian exile and the expulsion from Spain in 1492. Today, Jews are often broadly classified as being either Ashkenazic (European) or Sephardic (Mediterranean). These terms, as in any other classificatory scheme, are difficult to apply in practice. Zoossmann-Diskin (2000, 158) comments in reference to both Skorecki et al. (1997) and Thomas et al. (1998) that "the way the authors define the term Sephardic is probably related to its use in Israeli religious politics; it has nothing to do with science." In any event, both Skorecki et al. and Thomas et al. present Y-chromosome data on Cohens and non-Cohens for Ashkenazim and Sephardim. Thomas et al. also break down the non-Cohens into Levites and Israelites. According to Exodus, Levites are descendants of the "tribe of Levi," one of the 12 tribes of Israel, accorded special tasks with the priests. Again, these tasks were performed by male descendants, and particular surnames (e.g., *Levi, Levy,* and *Lavi*) are found among Levites. Israelites form the remainder of the "tribes of Israel."

Skorecki et al. (1997) examine the frequency of the Y-chromosome Alu insertion polymorphism (YAP) that we previously discussed as well as the frequencies of a 4 bp motif microsatellite known as DYS19. This microsatellite is typically present in repeat lengths of 13–17. Skorecki et al. (1997, 32) find that YAP and DYS19 frequencies were nonrandomly distributed across Cohanim versus Israelites and, in particular, that the absence of the YAP insert combined with a 14-count DYS19 was common among both Ashkenazic and Sephardic Cohanim, "suggesting that this may have been the founding modal haplotype of the Jewish priesthood." However, YAP in and of itself is not informative enough to define a modal haplotype, and microsatellites mutate too rapidly to be of much use in defining haplotype groups. As a consequence, Thomas et al. (1998) expand the original study by adding a number of Y-chromosome SNPs and microsatellites to Skorecki et al.'s YAP and DYS19. Ignoring for the moment their microsatellite data, Thomas et al. define three haplotypes on the basis of YAP and 92R7 (the C-T SNP that we previously mentioned can be typed by PCR followed by RFLP analysis with *Hind*III). Thomas et al.'s haplotype 1 is defined as YAP⁻ and 92R7(C), haplotype 2 is YAP⁻ and 92R7(T), and haplotype 3 is YAP⁺ and 92R7(C) (where the superscript plus and minus signs for YAP represent the presence/absence of the insert and the parenthetical terms with 92R7 represent the base found at the SNP position). Thomas et al. (1998) refer to the type 1 haplotype as the "Cohen modal haplotype" because it is common among Ashkenazic and Sephardic Cohanim (97 of 106 individuals, or 91.5%) and less common among Ashkenazic and Sephardic Levites and Israelites (109 of 200 individuals, or 54.5%). They also note that the type 1 haplotype is commonly associated with a particular array of counts for six Y-chromosome microsatellites.

A few years later, Thomas et al. (2000, 674) report that the Cohen modal haplotype was present in high frequency among the Lemba, "commonly referred to as the 'black Jews' of South Africa" (see Supplemental Resources for information on a *Nova* episode on the Lemba). The Lemba are Bantu speakers, and they maintain in their oral traditions that they are descended from Jews who migrated to Africa from "Sena." The exact location of "Sena" is unclear, though one possibility from Lemba oral tradition is that it was in Yemen. Thomas et al. (2000) analyzed Y-chromosome SNPs and microsatellites from Ashkenazic Israelites, Sephardic Israelites, Yemenites (non-Sena), Sena (Yemenites from the town of Sena), Lemba, and Bantu. Some of the Yemenites and Lemba, as well as most of the Bantu speakers, have a fourth haplotype not seen among any Ashkenazim or Sephardim. The fourth haplotype is defined by the SNP sY81, which represents a mutation from A to G. Table 8.7 shows these four haplotypes and their frequencies among Israeli and Palestinian Arabs, northern Welsh (Nebel et al. 2000), Yemen, Sena, Lemba, Bantu (Thomas et al. 2000), and Ashkenazic and Sephardic Cohanim, Levites, and Israelites (Thomas et al. 1998). Table 8.7 shows that while the Lemba have a relatively high frequency of haplotype 4, which is a common haplotype among Bantu speakers, they also have a high frequency of haplotype 1, the Cohen modal haplotype. Thomas et al. (2000) suggest that the Lemba oral tradition of a Jewish origin is supported by these results.

Table 8.7 also shows us that the term *Cohen modal haplotype* is a misnomer as this haplotype is found in all of the samples considered in the table, including Israeli and Palestinian Arabs, Northern Welsh, South African Bantu speakers, and non-Cohen Jews. Zoossmann-Diskin (2000, 158) makes a similar point, writing in reference to the original papers that defined the Cohen modal haplotype (Skorecki et al. 1997, Thomas et al. 1998): "Had it been found that the other [non-Cohen] Jewish chromosomes carry various haplotypes while the priestly ones are essentially monomorphic, it could have been a strong argument to substantiate the two papers." Further, all of the Sena chromosomes are the Cohen modal haplotype, yet all of these individuals are Hadramaut (Arab Muslims). It would appear then that the Cohen modal haplotype is really a Semitic haplotype (Semites are peoples associated with the Hebrew and Arabic languages), which inexplicably is also found at low frequency among the Northern Welsh and South African Bantu speakers. Zoossmann-Diskin (2000) also cites studies showing that the Cohen modal haplotype is relatively common among Iraqi Kurds, southern and central Italians, and Hungarians.

Table 8.7 Percentages and sample sizes for four haplotypes among Israeli and Palestinian Arabs, Northern Welsh (Nebel et al. 2000), Yemen, Sena, Lemba, Bantu (Thomas et al. 2000), and Ashkenazic and Sephardic Cohanim, Levites, and Israelites (Thomas et al. 1998)

Haplotype	Arabs	Northern Welsh	Yemen	Sena	Lemba	Bantu	Ashkenazic Cohen	Sephardic Cohen	Ashkenazic Levite	Sephardic Levite	Ashkenazic Israelite	Sephardic Israelite
−, A, C	69.93	7.45	73.47	100.00	65.44	16.88	95.92	87.72	30.19	67.86	61.76	62.75
−, A, T	9.79	89.36	16.33	0	1.47	0	4.08	10.53	56.60	17.86	17.65	23.53
+, A, C	20.28	3.19	6.12	0	2.94	2.60	0	1.75	13.21	14.29	20.59	13.73
+, G, C	0	0	4.08	0	30.15	80.52	0	0	0	0	0	0
Number	143	94	49	27	136	77	49	57	53	28	68	51

All classificatory terms for groups are from the original articles. The haplotypes are defined by three polymorphisms (YAP −/+, sY81 A/G, and 92R7 C/T, listed in that order).

The suggestion that the Cohen modal haplotype is really just a Semitic haplotype is supported by two recent studies (Hammer et al. 2000, Rosenberg et al. 2001). Although these studies do not specifically examine the Cohen modal haplotype or differentiate between Cohanim and Israelites, both show that Jews and Middle East Arabs are genetically very similar. The Rosenberg et al. (2001) study is particularly interesting in that they examined autosomal microsatellites and used a relatively new approach to admixture estimation from Pritchard et al. (2000). While previous methods for admixture estimation from molecular data required an a priori assignment of individuals to groups (either two ancestral and one hybrid group, as in Bertorelle and Excoffier 1998, or multiple ancestral and one hybrid group, as in Dupanloup and Bertorelle 2001), Pritchard et al.'s (2000) method fits within a particular statistical method called a "mixture model" or "model-based clustering." In *model-based clustering*, individuals are not initially assigned to groups but are instead allowed to classify into their own groupings (clusters). Further, it is possible to test whether the number of groups should be increased or decreased in order to best fit the data. Rosenberg et al. (2001) find that their autosomal microsatellite data from Ashkenazi Jews (from Poland), Druze, Ethiopian Jews, Iraqi Jews, Libyan Jews, Moroccan Jews, Palestinian Arabs, and Yemenite Jews were best fit with a three-group cluster model. Their cluster model essentially partitions the sample into a cluster that represents Ethiopian Jews, a cluster that represents Libyan Jews, and a large cluster that contains everyone else (Ashkenazi Jews, Druze, Iraqi Jews, Moroccan Jews, Palestinian Arabs, and Yemenite Jews). Nebel et al. (2001, 1105) suggest from Y-chromosome data that Palestinian Arabs and Bedouin represent, to a large extent, early lineages derived from Neolithic inhabitants and that "the early lineages are part of the common chromosomal pool shared with Jews."

It could be argued that the fact that Jews and Arabs are found in the same cluster based on autosomal variation does not "sink" the Cohen modal haplotype because Cohen status is paternally defined. However, Hammer et al.'s (2000) study, which uses only Y-chomosome variation, also finds that "Jewish and Middle Eastern non-Jewish populations share a common [gene] pool." As Judaism, like any other religion, is not genetically inherited, we should not be particularly surprised by Hammer et al.'s results. Since Talmudic times, Judaism has been traced through maternal lines, so technically it should sort with mtDNA. Indeed, in Thomas et al.'s (2002, 1417) study of mtDNA, they write that "Jewish populations . . . appear to represent an example in which cultural practice—in this case, female-defined ethnicity—has had a profound effect on patterns of genetic variation." However, regardless of whether we are looking at paternal or maternal descent, Judaism has always recognized the status of converts. In fact, Richards et al. (2003) note from mtDNA data that for eastern Jewish communities (e.g., Iraqi, Iranian, and Yemenite Jews) "it seems likely that substantial conversion of indigenous women took place at the time of the founding of these communities." Cruciani et al. (2002, 1203), on the basis of Y-chromosome data, "suggest that the Ethiopian Jews acquired their religion without substantial genetic admixture from Middle Eastern peoples and that they can be considered an ethnic group with essentially a continental African genetic composition." As a consequence of cultural diffusion and conversion, it would be naive to assume that there are genetic markers for Judaism or any other religion.

There was, at one point, another study that suggested a close genetic relationship among Mediterranean populations. This study, by Arnaiz-Villena and colleagues, was published in the journal *Human Immunology* on September 9, 2001. In the next month's copy of the journal, the editor-in-chief published a brief editorial in which she

(Suciu-Foca 2001, 1063) states in reference to Arnaiz-Villena et al.'s study "that the authors . . . confounded the elegant analysis of the historic basis of the people of the Mediterranean Basin with a political viewpoint representing only one side of a complex political and historical issue." Suciu-Foca then closes her editorial by noting: "This paper has been deleted from the scientific literature." The article was removed from the online copy of the journal, and subscribers were sent a letter suggesting that they "physically remove the pages" containing the article (see Shashok 2003 for further discussion). Ultimately, the journal's retraction of a published article was not viewed favorably by the scientific community (Krimsky 2002, Shashok 2003, Smith 2003).

HOW MANY RACES ARE DOCUMENTED IN OUR DNA?

We have seen from the two extended examples above that DNA rarely provides cut-and-dry answers to questions of local or regional population history and prehistory. A simpler question is whether or not we can partition DNA variation along racial lines and, if so, whether we can potentially determine the race of individuals from DNA markers. We are careful at this point to use the term *race* here, with all its implicit definitional imperfections (see Chapter 1). The use of the term *ethnicity* as a more politically correct synonym for *race* is inappropriate as people can have multiple ethnicities and can change ethnicities over time. Ethnicities are implicitly cultural constructions, and as such, the term *ethnicity* should not be substituted for *race*. In forensic anthropology, it has become common to use the term *ancestry* in place of *race*, but the ancestries discussed are usually the typical racial terms discussed in Chapter 1. The current consensus on DNA variation and "racial variation" is that a recent replacement model for human origins obviates the possibility of there being biologically meaningful races (Disotell 2000). In fact, when partitioned by continents, F_{ST} values from DNA are rarely much more than 15% and are generally in the vicinity of 10%, indicating that about 90% of the genetic variation is within continents. This high level of within-continent variation (and correspondingly low levels between) is often used to argue that racial classifications have no veracity. In point of fact, *race* is such a vague word in biology that we cannot know what level of between-group variation defines it. Cooper et al. (2003, 1168) make this point when they write that "although everyone, from geneticists to laypersons, tends to use 'race' as if it were a scientific category; with rare exceptions, no one offers a quantifiable definition of what a race is in genetic terms."

Before we dispatch this topic, we need to ask why we should care whether or not DNA variation is "racially" patterned. The first answer to this question is a practical one, which is that the "race concept" has not died in the medical and legal communities. Typically, "race" is one of the attributes used by law-enforcement officers in making identifications. MacEachern (2000, 359) acerbically, but perhaps accurately, notes that one of forensic anthropology's research goals is "the substantiation of law-enforcement folk taxonomies." Racial attributions are frequently made by witnesses on the basis of phenotypic observations, and these attributions generally are admissible in a court of law. "Race" is also often seen as an important variable in the health professions, although the intertwining of "race" (however defined) and culture or ethnicity makes any classificatory scheme suspect from the start and the issues surrounding race and medical care/research are exceedingly complex (Foster and Sharp 2002, Burchard et al. 2003,

Cooper et al. 2003). Also, the history of, at least, Western medical genetics' relationship with the "race" concept has not been a particularly pretty one, though matters are slowly improving (Fisher 1996). However, the demands of legal practice and medical science continue to shape and reshape how we look at human variation, whether or not we would choose to study this variation.

On the less practical side, there is an inherent human fascination with "where we came from" (and maybe less with "where we are going"). The worldwide interest in genealogy has not gone unnoticed by those who work with DNA markers. Indeed, Sykes and Irven (2000) have looked at Y-chromosome variation and defined a "Sykes haplotype" common in individuals with the *Sykes* surname. There are now commercial companies that will type mtDNA or Y chromosomes to aid in genealogical projects (see Supplemental Resources), and there are some large ongoing surname studies (see the Mumma DNA Surname Project in Supplemental Resources). While the genealogy projects do not necessarily touch on the "race" concept, in some instances individuals have apparently been interested in determining their "ethnic background" using DNA data. As a consequence, a number of commercial companies now offer genetic tests to help people determine their ancestral background.

We do not appear to be in a good position to answer the question heading this section ("How Many Races Are Documented in Our DNA?"). This is not to say, however, that DNA variation is negligible or that it is random (unstructured). On the contrary, we saw in our discussion of the Cohen modal haplotype that model-based clustering was capable of separating regional populations using autosomal microsatellite data (Rosenberg et al. 2001). Similarly, Pritchard et al. (2000) show that RFLPs can partition African populations from European populations. A particularly interesting aspect of their work is that they were also able to consider model-based clusters with more than two groups and found some support for there being more than two groups. They note that (p 952) "This may reflect the presence of population structure within the continental groupings, although in this case the additional populations do not form discrete clusters and so are difficult to interpret." For North American populations, Melton et al. (2001) show that about 19% of mtDNA variation is between groups defined as African American, European American, and Hispanic. To date, though, model-based clustering has not been applied to non-Native American data. Rosenberg et al. (2002, 2381) apply model-based clustering to "377 autosomal microsatellite loci in 1,056 individuals from 52 populations" and "without using prior information about the origins of individuals . . . identified six main genetic clusters, five of which correspond to major geographic regions, and subclusters that often correspond to individual populations." Bamshad et al. (2003, 584) apply model-based clustering to microsatellite and Alu loci and find that "when data from all 160 loci were used, the mean correct assignment to the continent of origin increased to 99%–100%."

DNA MARKERS AND INDIVIDUAL VARIATION

DNA markers have often been in the news over the past 10 years or so, with the O. J. Simpson trial bringing DNA testing to the general U.S. public's attention, continuing through to nightly news features on Monica Lewinsky and her blue dress, and more recently the identification of Laci Peterson and her unborn son's remains. Butler (2001) gives an excellent discussion of such "high-profile cases" in the last chapter of his book.

Here, we will look at two such cases, one involving the third president of the United States and the other a notorious outlaw (Jesse James). These examples are interesting not only for what they tell us about some of the modern uses for DNA markers but also for what they tell us about the importance of nongenetic information in assessing DNA.

Thomas Jefferson and the Descendants of Sally Hemings

Foster et al. (1998) published an article in the journal *Nature* with a title worthy of the tabloid presses: "Jefferson Fathered Slave's Last Child." In the "News and Views" section of that journal, Lander and Ellis (1998) introduce Foster et al.'s work under the more restrained and ambiguous title of "Founding Father," though they clearly support Foster et al.'s conclusion that Thomas Jefferson, the third president of the United States, fathered the last son of his slave Sally Hemings. Subsequent letters to *Nature* (Abbey 1999, Davis 1999) critiqued the conclusion from Foster et al., and the Thomas Jefferson Foundation, Inc., has put up a website addressing Jefferson–Hemings DNA testing (see Supplemental Resources). This has become a highly politicized and emotionally charged issue, with the Thomas Jefferson Foundation, Inc., essentially supporting Foster et al.'s claim and the Thomas Jefferson Heritage Society (see Supplemental Sources) raising extensive objections. At issue is the paternity of the last child born to Sally Hemings, an enslaved African American owned by Jefferson. Since the early 1800s Jefferson has been variously implicated as the father of a number of Hemings's children. Specifically, Thomas Woodson (named after a later owner) was suggested as a son of Jefferson and Hemings, as was Eston Hemings Jefferson. It has also been suggested that Eston Hemings Jefferson's father was Samuel or Peter Carr, nephews of Jefferson (sons of his sister).

To sort out these competing claims, Foster et al. obtained DNA samples from living male descendants of the parties in question. Further, they sampled DNA only from males who were uninterrupted patrilineal descendants—in other words, individuals whose Y chromosomes should trace back to the Jefferson line if Thomas Jefferson or one of his patrilineal relatives were one of their ancestors. Foster et al. type seven Y-chromosome SNPs (including YAP and 92R7), 11 microsatellites (STRs), and the minisatellite system MSY1. Table 8.8 shows the results from the SNP and STR analyses, and Figure 8.17 shows in graphical form the results from the minisatellite analysis. There are a few single-step

Table 8.8 Bi-allelic markers and Short Tandem Repeats (STRs) from male descendants in the Jefferson paternity case

Male Descendants of	Number Typed[1]	Bi-allelic Markers	Microsatellites (STRs)
Field Jefferson	5	0 0 0 0 0 0 1	15 12 4 11 3 9 11 10 15 13 7
Eston Hemings Jefferson	1	0 0 0 0 0 0 1	15 12 4 11 3 9 11 10 15 13 7
John Carr	3	0 0 0 0 0 1 1	14 12 5 12 3 10 11 10 13 13 7
Thomas Woodson	4	0 0 0 0 0 1 1	14 12 5 11 3 10 11 10 13 13 7
Thomas Woodson	1	1 1 1 0 0 0 1	17 12 6 11 3 11 8 10 11 14 6

[1]Number of descendants with the given genotype, and in the case of Thomas Woodson one descendant has a different genotype from the other four descendants.

From Foster et al. 1998.

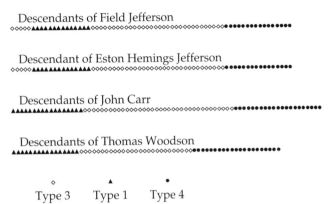

Figure 8.17. Schematic representation of the MSY1 variable number of tandem repeats on the Y chromosome for descendants of various males involved in the Jefferson paternity article (Foster et al. 1998). The pattern shown for descendants of Thomas Woodson is for three of the four individuals in Table 8.8 with the same Y-chromosome haplotype.

mutations for the satellite data that we have deleted to simplify the table and figure. The table and figure show that what we might call the "Jefferson haplotype" is present in the descendant of Eston Hemings Jefferson, the last child of Sally Hemings. The four descendants of Thomas Woodson (an earlier child of Sally Hemings) in Foster et al.'s study do not have the Jefferson haplotype and instead have a haplotype that is only a few mutational steps away from John Carr, Jefferson's brother-in-law. Kenneth Kidd (1999, 16), a geneticist from Yale, comments on this in his report to the Thomas Jefferson Memorial Foundation:

> I notice that the pattern for the Woodson descendants is very similar to the pattern for the Carrs. It would not take many mutations to convert one into the other. That makes it possible that John Carr and Thomas Woodson had fathers who were male-line relatives within a few generations. That could just mean that their fathers came from the same town or county back in Europe since I do not know how common these particular allelic combinations are in Europe (England?).

There is also one descendant of Thomas Woodson with a haplotype that Foster et al. (1998, 27) note is "most often seen in sub-Saharan Africans" (see Table 8.8).

How should these results be interpreted? The data seem to suggest that Thomas Jefferson, or for that matter any male Jefferson, is an unlikely candidate for having been the father of Thomas Woodson and that instead Woodson's paternity lies somewhere in the line of Thomas Jefferson's brother-in-law John Carr. The Woodson family maintains that their oral tradition ties them to descent from Jefferson (http://www.woodson.org/), though the genetic data suggest otherwise. Lanier and Feldman (2000) interviewed many of the descendants of Jefferson and of Sally Hemings (Lanier is a descendant of Madison Hemings, one of Sally Hemings's sons) including descendants of Thomas Woodson, and their book provides a very balanced view of competing claims on ancestry. In particular, they quote an interview with Michele Cooley Quille (Lanier and Feldman 2000, 49), a descendant of Thomas Woodson:

> the DNA stuff doesn't worry me, because we know who we are. . . . We know five people [descendants of Thomas Woodson] volunteered [for the Foster et al. 1998 study]. Out of

fifteen hundred Woodson family members, five were chosen. We can't even vouch for the people who volunteered, especially since their identities are being kept secret. . . . If they had taken DNA from my brother, then I would believe it as representing me, but they didn't. The point is, *we know who we are.* [emphasis in original]

Quille is quite correct in questioning the documentation for the five Woodson family members in the Foster et al. (1998) study. The fact that of the five Woodson family members one has a substantially different Y haplotype indicates that not all can be biological descendants of Thomas Woodson.

While the Woodson family still claims their link to Thomas Jefferson, there have been many who have argued vociferously against the claim that "Jefferson fathered slave's last child." The DNA data from Y chromosomes fairly clearly tie Eston Hemings Jefferson into the Jefferson line as by inference from Eston's living male descendant Eston must have had the Jefferson haplotype. It is not particularly likely that the Jefferson haplotype entered the Eston Hemings Jefferson line at a later time because Eston was freed on Thomas Jefferson's death and consequently left the state of Virginia. It is much more likely that Eston received the Jefferson haplotype from his father, but who was his father?

As Daniel Jordan (1999, 14), director of the Thomas Jefferson Memorial Foundation, acknowledges: "After the initial rush to conclusions ('Jefferson Fathered Slave's Last Child,' read the original misleading headline) came another round of articles explaining that the study's results were less conclusive than had earlier been reported." Along with these articles were two letters (Abbey 1999, Davis 1999) and a reply from Foster et al. (1999) published in *Nature* under the heading "The Thomas Jefferson Paternity Case." The two letters noted that there were Jeffersons other than Thomas who could have been Eston's father. As Davis (1999, 32) notes, "any male ancestor in Thomas Jefferson's line, white or black, could have fathered Eston Hemings." The phrase "white or black" may seem odd, but Davis continues that "As slave families were passed as property to the owner's offspring along with land and other property, it is possible that Thomas Jefferson's father, grandfather or paternal uncles fathered a male slave whose line later impregnated another slave, in this case Sally Hemings." This suggestion is made less likely by the observation that there is a strong correlation between Jefferson's visits to Monticello while he was vice president and president and the birth of Sally Hemings's children 9 months later (Neiman 2000). In the "Scholar's Commission on the Jefferson–Hemings Matter," it is noted that this argument of a correlation between Jefferson's visitations to Monticello and Hemings's conceptions could also be used as evidence that a (white) Jefferson other than Thomas was Eston's father. The logic here is that Jefferson's relatives were much more likely to visit Monticello when the master of the plantation was at home.

The primary problem is that a paternity case would never be assessed on the basis of Y-chromosome markers. Y-chromosome markers are inherited through paternal lines (Jobling and Tyler-Smith 1995, Jobling 2001, Sykes and Irven 2000), so all paternally related males should have the same Y-chromosome markers. Paternity suits would typically be based on autosomal markers, which because of independent assortment are generally unique to individuals (if enough markers are typed). However, this increased information is of virtually no use in the Jefferson paternity case because we do not have DNA from Thomas Jefferson, Sally Hemings, and Eston Hemings Jefferson. Instead, we have DNA from living patrilineal descendants, so the probability of identity by descent for any given autosomal locus is remarkably low. Consequently, the Y-chromosome analysis is the best that can be hoped for at present, and one must rely on non-genetic evidence to reach a conclusion as to who transmitted the Jefferson haplotype to Eston Hemings Jefferson.

Who Is Buried in Jesse James's Grave?

One of the most recent high-profile cases to be published (Stone et al. 2001) is that of the nineteenth-century outlaw Jesse James. As is common with those living "beyond the law," James is reported to have been living under an alias (Thomas Howard). As a result, when one Thomas Howard was killed in 1882 and interred on the James family farm, some argued that the buried remains were not those of Jesse James and that Jesse was in fact still "on the lamb." Later, the remains were transferred from the family farm to a plot in a local cemetery, adding uncertainty to the identification as Jesse James. To settle the identity issue, the cemetery remains were exhumed in 1995 under a court order. mtDNA was successfully extracted from two molars from the cemetery grave and from two hairs from the original grave on the James family farm. The DNA extractions from the two molars and two hairs were amplified by PCR and sequenced between bases 16,055 and 16,356. The sequences were identical and bore five base substitutions from the Cambridge reference sequence (Anderson et al. 1981). These results suggested that the remains buried on the James family farm and later moved to a local cemetery were indeed from the same individual. Blood samples were obtained from two individuals maternally related to Jesse James, and the mtDNA sequences from those two individuals matched the extractions from Thomas Howard. Stone et al. (2001) indicate that a database search of sequences from 2,426 individuals did not reveal the same sequence as in their study. Assuming a "hit" on the 2,427th try, this gives a frequency of 1/2,427, meaning that the sequence obtained from the cemetery remains is about 2,500 more likely to be from Jesse James or a maternally related relative than from the population at large.

It would seem that the mtDNA evidence, taken together with other evidence, is strong enough that there should be virtually no doubt that Jesse James is indeed buried in Kearney, Missouri (near the James family farm). However, Duke (1998) argues that the DNA testing is in error and that, in point of fact, her great-grandfather, James L. Courtney, was the real Jesse James and died in Marlin, Texas, in 1943, a few days past what would have been his 96th birthday if he were indeed Jesse James. That one of the West's most notorious gun-slingers would live to such an advanced age seems dubious, but there is a similar claim that a J. Frank Dalton, who is buried in Granbury, Texas, was the real Jesse James. If true, Dalton would have been 103 years old at his death (given the date of Jesse James's birth). On May 30, 2000, Dalton's grave was exhumed in order to extract DNA from his remains and compare the results with DNA from known relatives of Jesse James. Unfortunately, the wrong grave was opened, and a one-armed man was disinterred (later identified as William Henry Holland; Dalton was not missing any limbs when he was buried). One of Holland's descendants later commented "Before you go dig somebody up, do a little research" (Hanna 2001).

CHAPTER SUMMARY

This chapter has covered the genetic markers that have become the tools of choice within the last few decades for analyzing human variation. Many of the techniques start with the polymerase chain reaction (PCR) to make many copies of highly specific portions of the DNA. PCR is designed so that specific loci or parts of loci can be amplified into many copies. Beginning with the PCR products it is then possible to assay a wide variety of DNA markers. One of the simpler assays is referred to as an RFLP (for restriction fragment length polymorphism). In RFLP analyses the DNA is exposed to particular enzymes that will cut at specific sequence sites, provided that those sequences are present. The "digested" DNA is then run on an electrophoretic gel (much as is done in detecting protein polymorphisms) and those DNA samples with the particular sequence will show two bands on the gel (for two fragments of DNA) while those DNA samples lacking the sequence will show one band. Gel electrophoresis can also be used to assay various insertions and deletions in a particular DNA segment isolated by PCR. Many of these insertions or deletions are quite short, such as the 9 base-pair deletion found in mitochondrial DNA, both other insertions can be rather long and quite frequent across the genome. For example, the alu insert is about 300 bases long and is found in approximately a million copies spread throughout the human genome. Alu inserts can be very simply assayed by picking PCR primers to either side of a known alu insertion site, amplifying the DNA, and then using gel electrophoresis to determine whether or not the alu insert is present. It is also possible using what is known as a "chain termination reaction" to determine the complete DNA sequence for PCR products, so that variation in the sequence can be determined.

Many of the DNA markers are analyzed within the framework of what is known as the "coalescent model." This model, which is another way of looking at genetic drift, examines the probability that two allelic lineages were maintained as two separate lineages in the past, as versus merging (coalescing) into a single lineage. This probability is a function of the number of existent lineages and the population size. When the number of lineages is high the probability of merging lineages (viewed against the flow of time) is also high. When the population size is high the probability of merging lineages is low. More elaborate versions of the coalescent model that include population growth can be formed, and these ultimately provide expectations for how DNA variation should be structured in existing samples. The coalescent model can also be applied in analyzing the evolution of tandemly repeated DNA sequences found between genes.

In order to give some tangible examples of how DNA markers have been used in recent studies of human variation this chapter focused on two particular applications. The first example was drawn from studies of the peopling of the Pacific Islands, while the second was a more circumscribed example of a particular Y chromosome haplotype found in high frequency among some mid-East groups and particularly among males with particular surnames. After a brief discussion of geographic patterns of DNA marker variation, the chapter closed with two examples of how DNA markers have recently been used in interesting historical analyses.

SUPPLEMENTAL RESOURCES

http://helios.bto.ed.ac.uk/evolgen/rst/rst.html. Page for RSTCALC 2.2, another program for analysis of microsatellite (STR) data. Also contains links to other STR analysis programs, but some of the links are out of date.

http://hpgl.stanford.edu/projects/microsat/. Page for MICROSAT, a program to analyze microsatellite (STR) data.

http://j.webring.com/hub?ring=dnasurnameprojec&id=3&hub. A "webring" for "DNA Surname Projects : Genetics for Genealogy."

http://snp.cshl.org/. This site is the home for the SNP Consortium. The site lists the currently known human single nucleotide polymorphisms (SNPs), which can be searched by chromosomes.

https://www.agenus.com/ClassA/ClassB/ychrome.cfm. Home to the "Y chromosome ethnicity calculator."

http://www.anthro.utah.edu/popgen/programs/TreeToy/. This page is home to a Java applet that simulates the coalescent model with or without change in population size. Produces a simulated tree, as well as the mismatch distribution and frequency spectrum. A lot of fun, and a good way to explore the coalescent model.

http://www.codoncode.com/TraceViewer/docs/Readme.htm. Software for viewing DNA sequence data. The web page gives some good visual examples of what automated sequence output looks like.

http://www.cstl.nist.gov/biotech/strbase. Extensive information on short tandem repeats (microsatellites).

http://www.familytreedna.com/. Another commercial site for DNA genealogy, which includes Y-chromosome short tandem repeat analyses "yielding the world's tightest parameters to the most recent common ancestor (MRCA)."

http://www.genetree.com/compreltesting.htm. Another commercial site for DNA genealogy.

http://www.monticello.org/plantation/hemingscontro/hemings_resource.html. A website maintained by the Thomas Jefferson Foundation, Inc., that provides links to information about the ongoing debate over DNA testing and the Hemings–Jefferson controversy.

http://www.mumma.org/DNA.htm. Home to the "Mumma DNA Surname Project," one of the larger and older DNA surname projects (not listed in the webring above).

http://www.oxfordancestors.com/. A commercial site that will do Y-chromosome typing for individuals and provide a "Y-line certificate, suitable for framing."

http://www.pbs.org/wgbh/nova/israel/familylemba.html. A *Nova* special on the Lemba that aired on February 22, 2000. The website contains a transcript of the broadcast and additional information.

http://www.pbs.org/wgbh/pages/frontline/shows/jefferson/. A PBS site on "Jefferson's Blood" that contains some links under "Is It True?" and a transcript of the program that aired on May 2, 2000.

http://www.protocol-online.org/prot/Molecular_Biology/PCR/index.html. Polymerase chain reaction protocols.

http://www.tjheritage.org/. A web-site maintained by the Thomas Jefferson Heritage Society, whose purpose, among others, is "to stand always in opposition to those who would seek to undermine the integrity of the name of Thomas Jefferson." Contains links that are primarily critical of studies suggesting a Jefferson–Hemings affair.

Section 3
Variation in Complex Traits

9

Quantitative Variation

In Chapter 1, we saw that there is a long history of subclassifying humanity, often using phenotypes that are quantitative. Whenever there is a reference to the size or shape of some aspect of human morphology or to the amount of something, we are dealing with a quantitative trait. There are many possible examples of quantitative traits, but some of the more familiar ones are stature, weight, **intelligence quotient (IQ)**, and skin color (in Chapter 11 we will find that "skin reflectance" is a more accurate descriptor). However, having established what we mean by "quantitative variation," we reach a dilemma; we have taken great pains to lay out the genetic basis for qualitative traits (Chapters 2 and 3) such as blood group types (Chapter 4), blood proteins and enzymes (Chapter 5), hemoglobin (Chapter 6), human leukocyte antigen (HLA, Chapter 7), and DNA markers (Chapter 8). All of these types of traits follow the simple rules of transmission genetics that we described in Chapter 2, but how can we describe quantitative trait variation within the framework of these rules? This chapter explores what is known as the **quantitative genetic** or **polygenic** model, which posits a large number of loci that affect a quantitative trait.

THE MENDELIAN BASIS FOR QUANTITATIVE TRAIT VARIATION

The first detailed analyses of quantitative (continuous) variation in humans were made for variable stature, so in this section we consider the problem of how to explain genetic variation in this trait. Early researchers were well aware of the fact that tall men and women tend to have children who grow to be taller than average adults, while short men and women tend to have children who grow to be shorter than average adults. However, it is not immediately clear how Mendelian laws should apply since there is continuous variation in stature. Many of the traits that anthropologists have traditionally studied vary continuously, instead of having discrete categories. Body measurements (see Chapter 10), of which stature is but one, have a continuous distribution, as do skin pigmentation (see Chapter 11) and many physiological traits. R. A. Fisher (1918, 400) first gave an explicit explanation of the Mendelian basis for quantitative trait variation by positing that many loci contributed to the trait. In Fisher's words: "The simplest hypothesis, and the one which we shall examine, is that such features as stature are determined by a large number of Mendelian factors, and that the large variance among children of the same parents is due to the segregation of those factors in respect to which the parents are heterozygous." Fisher's explanation provides the basis for continuous variation and indicates why siblings often vary widely in their genotypes.

To demonstrate Fisher's model, we first pretend that stature is controlled by a single Mendelian locus with two alleles, *A* and *B*, where each *B* allele adds 1 inch to an individual's adult stature, while an *A* allele adds nothing. We will assume that the baseline stature is 5 feet and 11 inches (71 inches) so that someone with the *AA* genotype will be 71 inches tall, someone with the *AB* genotype will be 72 inches tall, and someone with the *BB* genotype will be 73 inches tall. Now, if the allele frequencies are 0.5, then by Hardy-Weinberg equilibrium the expected frequency of people who are 71 inches tall is 0.25, the expected frequency of people who are 72 inches tall is 0.5, and the expected frequency of people who are 73 inches tall is 0.25; or on a percentage basis, we expect 25% to be 71 inches tall, 50% to be 72 inches tall, and 25% to be 73 inches tall. Now, assume that there are two loci that act jointly so that a person with the *AA*, *AA* genotype has an increment of 0 inches added to the baseline height, while someone with the *BB*, *BB* genotype has 4 inches added to the baseline height. With a baseline stature of 5 feet and 10 inches (70 inches), people's statures will now range between 70 and 74 inches. If both loci have

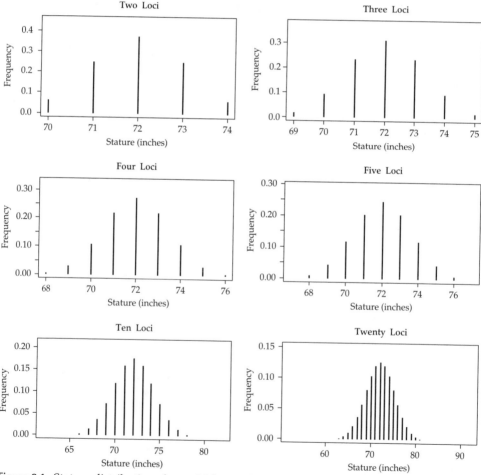

Figure 9.1. Stature distributions for variable numbers of loci with additive alleles at each locus.

allele frequencies of 0.5, then by the "and" rule, the probability that both loci are homozygous for the A allele is 0.25×0.25, or 0.0625. Consequently, the expected frequency of individuals who are 70 inches tall is 0.0625, as is the frequency of individuals who are 74 inches tall. The frequencies of intermediate statures are more complicated, but ultimately we find that 6.25% of people are expected to have no B alleles and will be 70 inches tall, 25% are expected to have one B allele and will be 71 inches tall, 37.5% are expected to have two B alleles and will be 72 inches tall, 25% are expected to have three B alleles and will be 73 inches tall, and 6.25% are expected to have four B alleles and will be 74 inches tall. For more than two loci, the record keeping gets rather tedious, so we instead show the expected frequencies graphically in Figure 9.1. Note that with increasing numbers of loci, the distribution of stature tends toward a normal, or Gaussian, distribution. This distribution is symmetric and has the appearance of a "bell" in shape.

The above description may seem overly simplistic, and indeed there are other situations we could consider. For example, can we get this bell-shaped distribution for stature if there is complete dominance at each of the loci? The answer is yes, as Figure 9.2 shows.

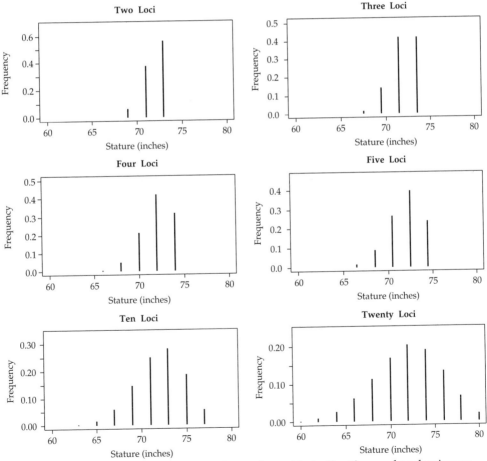

Figure 9.2. Stature distributions for variable numbers of loci, all with complete dominance.

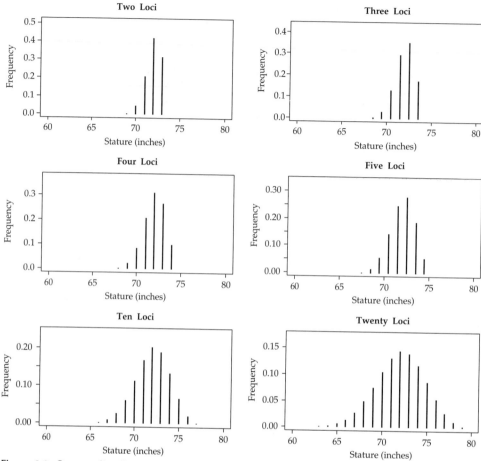

Figure 9.3. Stature distributions for variable numbers of loci, all with additive alleles and allele frequencies of 0.75 and 0.25.

Can we get a bell-shaped distribution if the alleles at each of the loci are additive but, rather than an equal frequency of *A* and *B* alleles, the frequencies are 0.25 and 0.75 at each locus? Again, the answer is yes, as shown in Figure 9.3. We can even think of an example with complete dominance and a disparate allele frequency (such as 0.25 and 0.75) and find that if the number of loci is very large we will get a bell-shaped distribution. This tendency to normality (the bell-shaped distribution) is a demonstration of a classic theorem from probability theory called the **central limit theorem**. The central limit theorem says that even if the contributing variables themselves (in this case, the one-locus distributions) do not follow a bell-shaped distribution, the overall distribution (in this case, the stature distribution) will tend toward the bell shape.

The above description covers only genetic effects on quantitative traits, while we know that nutrition and other environmental variables affect human growth and development. To make the model more realistic, we should account for these environmental

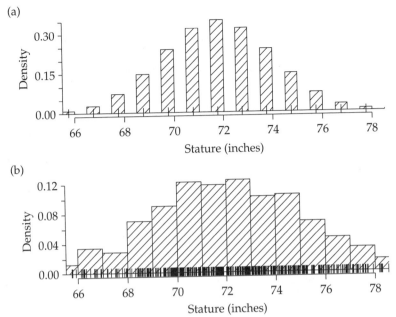

Figure 9.4. Histogram of stature distribution for 10 loci with additive alleles (a) and for the same loci with the addition of environmental effects (b). The vertical lines on the horizontal axis (known as a "rug") show the actual values.

effects on variables such as stature. The top part of Figure 9.4 shows the same 10-locus example with allele frequencies of 0.5 and additive effects that we examined in Figure 9.1. The bottom part of Figure 9.4 shows the same example but with random environmental noise added to each of 500 individuals. Note how the simulated environmental variation "smears out" the measurements so that the discrete distribution in the top panel becomes a continuous distribution in the bottom panel.

Components of Variation for a Quantitative Trait

Now that we see how continuous variation can arise from the influence of many Mendelian loci and environmental effects (such as nutrition), we need a language for describing the relative contribution of genetics and environment to human variation. All of the terminology we will use is based on the concept of variance. As we saw in Chapter 3, the variance of a trait is the average squared deviation around the average. For example, if we had measured the statures of five brothers and found them to be 73.6, 71.6, 73.8, 75.2, and 68.8 inches tall, then their average stature would be 72.6. The sum of the squared deviations around the mean would be $(73.6 - 72.6)^2 + (71.6 - 72.6)^2 + (73.8 - 72.6)^2 + (75.2 - 72.6)^2 + (68.8 - 72.6)^2 = 24.64$. The average squared deviation is then $24.64 \div 5 = 4.928$. Many textbooks would correctly note that the variance is actually $24.64 \div 4 = 6.16$, because one "degree of freedom" is lost in calculating the average, but we will not make this distinction here. With 500 individuals represented in Figure 9.4, it would be far too tedious to show the calculation of the variances; but in the top panel (with only the effect of 10 loci with additive alleles) the variance is 5.0, while in the bottom panel (with the effect of 10 loci and random environmental deviations) the variance is 10.0. The variance from the bottom panel is called the **phenotypic variance** (often symbolized as V_p) of the trait since it includes both the genetic variance from the top panel and the variance due to environmental effects. The variance from the top panel is called the **additive genetic variance** (often symbolized as V_a) since it is due to the additive effects of alleles at each locus. When there is dominance at loci, the variances become more complex, so for the moment we will consider only alleles with additive effects at loci. The difference between the phenotypic variance and the additive genetic variance is know as the **environmental variance** (often symbolized as V_e) since it indicates the amount of variation in the phenotype not attributable to additive genetic effects (i.e., due to the environment).

Because the additive genetic and environmental variances are additive, we can write the simple equation $V_a + V_e = V_p$. The proportion V_a/V_p is a measure of how heritable the trait is since it measures the relative contribution of the additive genetic variance to the total phenotypic variance. Indeed, this proportion is known as the **narrow sense heritability** of the trait and is often symbolized as h^2. For those familiar with the squared correlation coefficient as a measure of how much variance is explained by a particular variable, there is a basic parallel here such that h^2 is a measure of how much phenotypic variance is explained by additive genetic variation. If there is "narrow sense" heritability, then it would make sense that there would also be **broad sense heritability**. The concept of broad sense heritability is rather difficult to understand, so we need to develop it carefully.

When there is dominance at loci, the **total genetic variance** (usually symbolized as V_g) can be partitioned into two additive parts, one due to the additive effects at loci (this again is the additive genetic variance) and one due to the interaction of alleles at loci, known as the dominance variance (usually symbolized as V_d). Now we have $V_a + V_d + V_e = V_p$, or simply $V_g + V_e = V_p$. The broad sense heritability is the proportion V_g/V_p, or $(V_a + V_d)/V_p$. Broad sense heritability is usually denoted with the symbol H^2 and is more inclusive than narrow sense heritability. Narrow sense heritability gives the proportion of phenotypic variance explained by transmissible genetic effects, while broad sense heritability gives the proportion of phenotypic variance explained by all genetic effects. Broad sense heritability consequently includes effects due to dominance at loci, even though these effects cannot be passed on to children because a parent transmits only one allele per locus. For either heritability, the range of possible values is 0.0–1.0. For a trait with narrow sense heritability of 0.0, none of the phenotypic variation is due to transmissible genetic effects (additive effects of alleles). For a trait with narrow sense heritability of 1.0, all of the phenotypic variation is due to additive genetic effects. We will focus in the remainder of this chapter on narrow sense heritability since it is the more relevant parameter for discussing human evolution and variation.

ESTIMATION OF NARROW SENSE HERITABILITY

It is one thing to describe the quantitative genetic model and quite another to worry about how to estimate narrow sense heritability. In animal breeding studies, there is a classic equation which relates the proportion of parents "culled" from the herd to the amount of economic gain expected given the narrow sense heritability. For example, if dairy cattle are bred on the basis of milk production (with the males being bred at random) such that only those cows above the average are allowed to mate and if the heritability is 1.0, then there will be a large response to the selective breeding program (where response refers to the change in milk production in the next generation relative to the preselection mean in the parental generation). However, if the heritability is 0, there will be no response to selection. Consequently, if the heritability is unknown but the proportion "culled" (not allowed to breed) is known and the response to selection can be measured, then the heritability can be estimated.

While the animal breeding model is very useful for trying to improve agricultural yields, it is of no use when studying humans. Instead, human geneticists have typically used information on relatives to estimate heritabilities. The simplest type of study, indeed one of the first used in the literature (Galton 1889), relates measurements of parents to the measurements of their grown children. Logically, if a trait like stature has a high heritability (near 1.0), then we would expect tall parents to produce tall children and short parents to produce short children. The higher the heritability, the stronger the relationship. If we plot offsprings' statures against their parents' statures, we would expect to see this relationship in the form of clustering along a diagonal line. Figure 9.5 shows six simulations each of 500 daughters' adult statures plotted against their mothers' statures. These six simulations were generated with a heritability of 1.0. On each graph there is a dashed line which represents the case where each daughter's stature is exactly equal to her mother's stature. Even though the heritability is 1.0, the daughters' statures do not fall on this line. This is for two reasons. First, we have not accounted for the influence of the fathers' statures on their daughters. Second, even if we did account for the fathers' statures, the allele at each maternal locus is randomly selected for transmission to the daughter. As a consequence, there is still considerable uncertainty when predicting the daughter's stature.

Figure 9.5 also contains what are known as **linear regression** lines. These represent the best-fitting lines, where "best-fitting" means that they minimize the squared distances of daughters' statures to the line (measured parallel to the vertical axis of the graph). For the statistical cognoscenti, the distances of daughters' statures to the regression line are known as "residuals." The term *regression* is a more modern version of what Galton referred to as "reversion." He noted that there was "reversion to the mean" in the data he obtained. In other words, in Figure 9.5, short mothers tend to have children taller than themselves, while tall mothers tend to have children shorter than themselves. This reversion to the mean (or now more properly "regression" to the mean) is because the process of randomly selecting among alleles in the mothers (as well as in the fathers) has an averaging effect. Further, we assume that there is no assortative mating (see Chapter 3) so that, on average, a short mother will have mated with a father of average stature and the daughter's stature will be greater than the mother's. Similarly, on average, a tall mother will have mated with a father of average stature and the daughter's stature will be less than the mother's.

Figure 9.6 shows the same types of simulation as in Figure 9.5, but now the heritability is 0.0. Note that the regression lines are nearly parallel to the horizontal axes of the graphs. As a consequence, there is no relationship between daughters' statures and their mothers' statures in these latter simulations. While some traits (in particular the number of dermal ridges on fingers) do have heritabilities that are nearly 1.0 and others have heritabilities near 0, a more typical value for heritability is around 0.35. Figure 9.7 shows such a case for 500 pairs of daughters with their "mid-parents" (the average of the fathers' and mothers' statures).

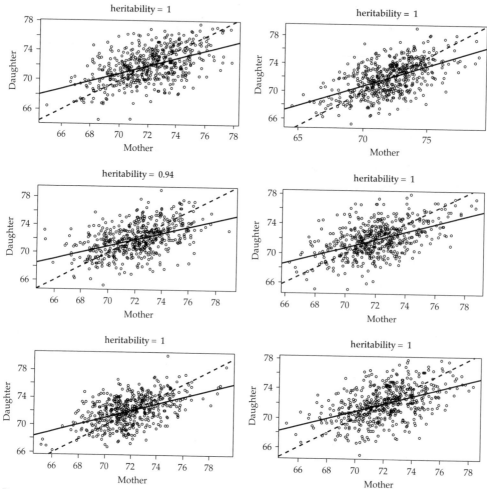

Figure 9.5. Six simulations of 500 mother/daughter pairs for stature under a model where heritability is equal to 1.0. The plot titles give the realized heritability for each simulation. The *dashed line* gives the case where a daughter's stature (in centimeters) equals her mother's stature, while the *solid line* is the observed linear regression of daughters on mothers.

The regression method we have used is but one of the ways to estimate narrow sense heritability. The method uses the slope of the regression line to estimate the heritability. For the regression of children's values on one of their parent's values, the heritability is estimated by two times the slope of the line, while for the regression of children's values on the average of the parental values, the slope is directly equal to the heritability. However, we could also use similarity among siblings (brothers and sisters) to estimate the heritability. In cases of high heritability, we would expect brothers and sisters to not vary much between themselves, while between-sibships the variance should be higher. In cases of low heritability, we would expect brothers and sisters to be as variable among themselves as between sibships. Consequently, the relationship between the average

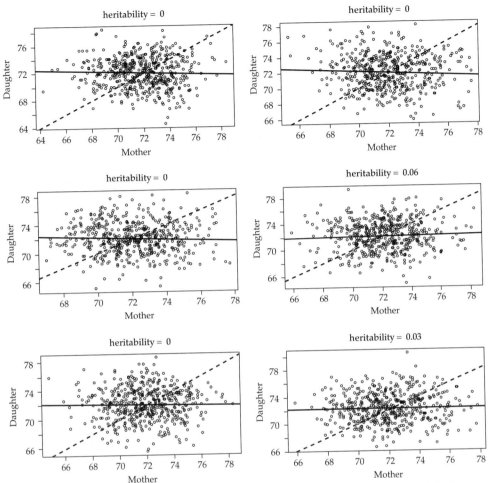

Figure 9.6. Six simulations of 500 mother/daughter pairs for stature under a model where heritability is equal to 0.0. The plot titles give the realized heritability for each simulation. The *dashed line* gives the case where a daughter's stature (in centimeters) equals her mother's stature, while the *solid line* is the observed linear regression of daughters on mothers.

within-sibship variances and the between-sibship variance can be used to estimate heritability. This is known as analysis of variance, which we considered in Chapter 8 when we discussed AMOVA.

Generally speaking, most data from humans are collected opportunistically, so we might have stature data from parents, children, brothers, sisters, uncles, and aunts (as well as grandparents and cousins). The data consequently exist within pedigrees. We would expect that if the heritability is high, then those relatives who share more alleles would be phenotypically more similar than those relatives who share few alleles. If the heritability for a given trait is 0, then even monozygotic twins (who share all their alleles) would be no more similar than completely unrelated individuals. The similarity

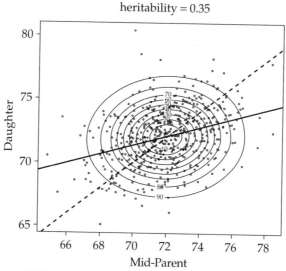

Figure 9.7. Simulation of 500 mid-parent/daughter pairs under a heritability of 0.35. The ellipses enclose approximately 90% (outer ellipse) down to 10% (innermost ellipse) of the data in 10% increments.

between various types of relative's measurements can be used to estimate narrow sense heritability, but to do this we need a formal way of measuring how genetically similar pairs of relatives are. For this, we use a **coefficient of relationship**, which is the expected proportion of shared alleles for two individuals. The coefficient of relationship between two individuals is equal to two times the probability of identity by descent for a hypothetical child with the two individuals as parents. Table 9.1 gives the coefficients of relationship for a number of relationships. It shows, for example, that monozygotic twins share all of their alleles, a parent and child share half of their alleles (as half of a child's alleles come from one parent and half from the other), and two unrelated individuals share no alleles. The coefficients of relationship and the measurements on individuals can

Table 9.1 Proportion of alleles shared for various pairs of relatives

Relationship	Proportion of Alleles Shared
Monozygotic twins	1.0000
Full sibs	0.5000
Half-sibs	0.2500
Parent–child	0.5000
Grandparent–grandchild	0.2500
Great-grandparent–great-grandchild	0.1250
First cousins	0.1250
Second cousins	0.0312
Double first cousins	0.2500
Unrelated individuals	0.0000

Table 9.2 Narrow sense heritability estimates for morphological and physiological traits

Trait	Heritability
Stature	0.65[1]
Head length	0.54[2]
	0.37[3]
Head breadth	0.58[2]
	0.46[3]
Face height	0.38[3]
Face breadth	0.50[2]
	0.47[3]
Nose height	0.49[3]
Nose breadth	0.50[3]
Left thumb, radial ridge count	0.69[4]
Systolic blood pressure	0.34, 0.41, 0.42, 0.32, 0.13, 0.24, 0.34, 0.22[5]
Log total serum immunoglobulin E	0.47[6]
Serum immunoglobulin E to house dust mites and Timothy grass	0.34[6]

[1]Roberts et al. 1978.
[2]Sparks and Jantz 2002.
[3]Konigsberg and Ousley 1995.
[4]Ousley 1997.
[5]Weiss 1993.
[6]Palmer et al. 2000.

be used to estimate heritabilities using a method known as **maximum likelihood esti-mation**. This method is too complicated to discuss here, but students wanting a fairly gentle introduction can find one in Konigsberg (2000).

Table 9.2 lists narrow sense heritabilities for a number of different morphological and physiological traits. There are a few salient points which we should make about heritabilities, based both on the values listed and on other studies not reported in the table. First, narrow sense heritability measures only the proportion of variation in a trait explained by polygenic inheritance. As such, heritabilities should not be viewed as invariant values. For example, if environmental variance decreases, heritability will increase. A clear example of this can be seen among populations that have recently undergone exposure to a new and fairly uniform environment. Young et al. (1969) report a heritability of myopia based on parent–offspring correlation that is only 0.10, while the heritability is 0.98 based on analysis of sibs. Their particular study was done among Inuit, who had recently undergone exposure "to compulsory education and a 'westernized' environment during their childhood" (Guggenheim et al. 2000). Second, it is often difficult to extrapolate from a heritability study made in one sample to heritability for another sample. To some extent, this is due to differing environmental variance, but it can also be due to different allele frequencies at loci. Third, heritabilities may vary over the life span. For example, Towne et al. (1993) found a heritability for length at birth that was 0.83, which is considerably higher than most estimated heritabilities for adult stature. This higher heritability is presumably because the newborns have not yet been subjected to the many environmental factors that may affect postnatal growth. Finally, there is no necessary relationship between the level of heritability and the evolutionary "importance" of a trait. In fact, those traits that have the most to do with relative fitness typically have the lowest heritability. This follows from a principle known as Fisher's fundamental theorem, which says (among other things) that selection will remove additive genetic variation for traits that are "close" to fitness. The decreased additive genetic variation decreases the narrow sense heritability. Houle (1992) also suggests that traits which are close to fitness tend to have large nongenetic (environmental) components, which would also decrease the heritability. In contrast to traits such as fecundity and survivorship, which are close to fitness, morphological and physiological traits tend to have higher heritabilities because they have not been exposed to extreme selection.

Absent from Table 9.2 are any heritability estimates for behavior or IQ. We take up the difficult issues of genetics, behavior, and human variation in the final chapter of this text. A few comments are probably in order here, though, regarding the heritability of behavior. Genetic studies of neural function or behavior in humans are a very dicey proposition, particularly because their results are subject to interpretation in the sociopolitical arena. Kearsey and Pooni (1996, 192) aptly make this point:

> No one would be upset to hear that the heritability of fat content of cow's milk was 70%, but if a similar value were published, say, for intelligence or personality in humans the political correctness of the scientist would immediately be questioned, as would be the relevance of the character.

Behavioral genetic studies are also greatly complicated by the problem that behaviors have low **repeatability**. Repeatability refers to whether or not a trait varies when it is measured multiple times. A morphological trait such as stature does not vary greatly from time to time, though there is stature reduction during the day due to intervertebral disc compression and subsequent regain of stature after a night off of one's feet. Aside from possible measurement errors, then, stature is a highly repeatable trait, particular if measured immediately following waking in the morning. In contrast, behaviors tend to be much less repeatable; and as a result, heritability estimates for behaviors tend to be underestimated. Hoffmann (2000) discusses this problem in studies of fruit fly mating and courtship speed, where heritabilities are very low unless multiple measurements are averaged for each individual. In humans, it is rare to have such multiple observations on related individuals. There is the further problem that a variable such as intelligence is very amorphously defined. Even if something labeled as "intelligence" can be measured with high repeatability, this does not mean that the gloss actually measures intelligence (however it might be defined).

NARROW SENSE HERITABILITY OF THE CEPHALIC INDEX

The **cephalic index** (head breadth divided by head length and multiplied by 100 to convert to a percentage) is one of the most intensively studied quantitative traits in physical anthropology. A cephalic index of 50 would indicate a head half as wide as long, while a cephalic index of 100 would indicate a round head. In actuality, about 99% of human cephalic indices range between 72 and 93, so the examples of 50 and 100 are outside of the real range. We discuss the cephalic index here as an example of a single measurement, while in the next section we discuss the study of multiple quantitative traits. In fact, the cephalic index is really composed of two traits, head breadth and length, but one is divided by the other in order to simply describe the shape of the head independent of head size. Physical anthropology's fascination with the cephalic index has deep historical roots centering on the misguided notion that the index could be used to distinguish between "races" (see Chapter 1). In fact, this idea had largely fallen out of favor by the mid-1960s, as seen in the following quote from J. Michael Crichton (of *Jurassic Park* fame; 1966, 63):

> For discrimination of race, measurements of the face are more efficient than those of the cranial vault. This implies that, if it can be avoided, investigations of racial history should not be based on the cranial index and similarly insensitive indicators.

A further problem with the cephalic index is that it is not a particularly complete summary of cranial vault shape. The index cannot distinguish where the maximum breadth falls relative to the anterior/posterior axis of the vault, and it loses a considerable amount of information about the actual shape of the vault. These problems can be addressed by using geometric coordinates instead of caliper-based measurements (a point presaged by Franz Boas in the early 1900s, see Cole 1996 for a historical account). As an example of the problems with the cephalic index, Figure 9.8 shows an idealized

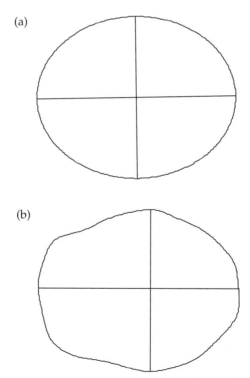

Figure 9.8. Comparison of cephalic indices. Part (a) shows the cranial vault as an ellipse so that the maximum length and breadth pass through the center of the ellipse. Part (b) shows these measurements on an actual vault (with the nose pointing to the left). Besides losing much of the information on the actual contour of the vault, the top figure places the maximum breadth too close to the nose.

drawing of a transverse slice through the top of a skull (the general location in which cranial length and breadth would be measured). This idealized drawing is shown as an ellipse because we do not know from the measurements alone where the two measurements cross (for an ellipse the two measurements must cross in the center, as shown in the top panel of Fig. 9.8). Below this idealized drawing is the actual cranial vault with maximum length and breadth marked. The vault is drawn such that the nose would be to the left. Note how the maximum cranial breadth falls farther away from the nose in the actual skull than in the ellipse and how there is a great deal of shape information that would be lost by summarizing the vault by the ratio of two measurements.

Because of physical anthropology's fascination with the cephalic index, there are reams of data for this trait, including a very large study conducted by Boas of late nineteenth- and early twentieth-century immigrants (and their children) to New York City. As Boas (1928) published all of the raw data from this study, the data set has now been the focus of a number of computer-intensive reanalyses (Sparks and Jantz 2002, 2003; Gravlee et al. 2003a,b) that would have been impossible in Boas's time. In this section, we briefly reanalyze the data, which was downloaded from Clarence Gravlee's website (see Supplemental Resources). We use data on the 13,732 individuals in Boas's study who have cephalic index and age recorded. Because data were recorded on nuclear families, we can use the method of maximum likelihood to estimate the narrow sense heritability of the cephalic index. From this analysis, we find that on average the cephalic index is about 82.3 and the narrow sense heritability is about 70% with a **standard error** of about 1%. The standard error is a single number that gives us an idea of the level of certainty we should place on an estimate. Typically, ±2 standard errors around an estimate is expected to contain the true value of the estimate about 95% of the time. Consequently, we can say that we are about 95% certain that the real narrow sense heritability for the cephalic index is between 68% and 72%.

This is a rather crude analysis as it does not take into account any possible **fixed effects**. Fixed effects are any known factors that could affect the quantitative trait. For example, an individual's age may influence the cephalic index (via growth), so age would be an environmental fixed effect. In this example, age is a fixed effect because it is a factor that we can measure and which is fixed (invariant) for the individual at any given point in time. Age is an environmental effect because it is not genetically transmissible to offspring. Other possible environmental fixed effects in Boas's study are an individual's sex and whether he or she was born in Europe or the United States (i.e., the effect of immigration). A genetic fixed effect that we have ignored is ancestry as Boas's (1928) data included individuals classified as Sicilians, Central Italians, Bohemians, Hungarians and Slovaks, Poles, Scots, and Hebrews. These terms are the ones used in Boas's original report and, with the exception of *Hebrews*, refer to geographic regions. *Hebrews* is used by Boas to refer to European Jews, specifically those from "Western Russia, Poland, Germany, Austria, Switzerland, and Romania" (Sparks and Jantz 2002, 14636). We refer here to the classification into Sicilian, Central Italian, Bohemian, Hungarian and Slovak, Pole, Scotch, and Hebrew as the fixed genetic effect ancestry because it most accurately captures the structure of the data. While it is tempting to refer to the classification as "ethnicity," this cannot be a fixed genetic effect because ethnicity is neither genetically transmitted nor immutable. It is also not possible to refer to the classification as being geographically based because individuals from Poland are found both in the group "Poles" and in the group "Hebrews."

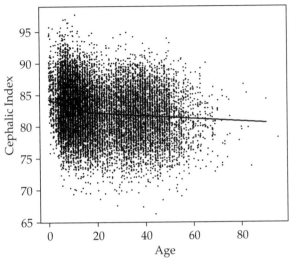

Figure 9.9. Cephalic index plotted against age for the 13,732 individuals in the Boas (1928) study (age has been randomly "jittered" by a small amount to decrease the overlap of points). The two lines show a "change point" regression, demonstrating the decrease of the cephalic index throughout life.

Starting with age as an environmental fixed effect, Figure 9.9 shows a plot of cephalic index against age for the 13,732 individuals with recorded ages in the Boas data. Also shown is what is known as a "change point regression," which in this case is just two connected straight lines that fit best through the data. The method of maximum likelihood is used to find both the equations for these lines and the age at which one line stops and the other starts, which in this case is 17 years. As can be seen in the figure, the cephalic index decreases throughout life, though the rate of this decrease slows by age 17. Turning again to estimation of the narrow sense heritability, we find that the estimated heritability has increased from 70% to 74%. Why did this happen? The reason is that narrow sense heritability is the proportion of the additive genetic variance out of the sum of the additive genetic and unexplained environmental variance. By accounting for age in the analysis, we have decreased the unexplained environmental variance (some is now explained by age) so that the heritability increases. To this analysis we can now add the environmental fixed effect of immigration by allowing European and U.S.-born individuals to have their own mean cephalic indices. In this analysis, we still model the effect of age as shown in Figure 9.9. In this case, the estimated heritability remains at 74%, so it does not appear that allowing for the effect of immigration has lowered the amount of unexplained environmental variance in the cephalic index. However, this last analysis is extremely misleading because Boas was able to demonstrate that the environmental effect of immigration differed across ancestral groups. Clearly, we need to take a fine-grained approach to analyzing these data.

The situation discussed above where groups with different ancestry show different responses to an environmental effect (in this case immigration) can be referred to as **genotype by environment interaction** (Falconer and Mackay 1996, Lynch and Walsh 1998), or more properly "polygenotype by environment interaction" (Blangero 1995). In polygenotype by environment interaction, individuals or groups with different

Table 9.3 Quantitative genetic analysis of the cephalic index from Boas's (1928) immigrant data

Ancestral Group	Foreign-born	U.S.-born	Heritability
Sicilians	80.0	81.6	54.5 (47.5–61.4)
Central Italians	82.8	84.0	56.8 (50.1–63.4)
Bohemians	86.9	85.9	46.0 (38.5–53.5)
Hungarians and Slovaks	87.1		58.2 (42.4–73.9)
Poles	86.4		50.4 (31.3–69.5)
Scots	78.1		53.2 (25.0–81.3)
Hebrews	86.1	84.7	52.6 (47.8–57.5)

The average cephalic indices (adjusted to birth) for foreign-born and U.S.-born individuals are shown separately when they are significantly different. The narrow sense heritability is given as a percentage, with its 95% confidence interval shown in parentheses.

polygenotypes (additive genetic values) react to the environment in different ways. In Boas's data, he noted that when comparing "American-born descendants of immigrants" to "their foreign-born parents. . . . The changes which occur among various European types are not all in the same direction" (Boas 1940, 60). This suggests that we need to analyze the ancestral groups separately if we are to understand the effects of immigration on the cephalic index and, ultimately, the effects on our calculation of the heritability. Table 9.3 contains the mean cephalic indices by ancestral group and by the foreign versus U.S.-born distinction. Although not shown in the table, each ancestral group was allowed to have its own particular age progression for the cephalic index. As a consequence, the mean values are shown adjusted to age 0 (i.e., at birth). In this table, only one mean value is listed for a group if the foreign and U.S.-born means were not significantly different, which is true for the Hungarians and Slovaks, the Poles, and the Scots. Here, "significance" means that there is a <0.1% chance of randomly getting foreign and U.S.-born means as different as those observed if the means were actually equal. Where only one mean is listed, the chance of getting the observed difference was arbitrarily set at >20%. For the Sicilians and Central Italians, the cephalic index increased for U.S.-born individuals relative to foreign-born, while for the Bohemians and Hebrews, the index decreased.

It is interesting now to look at the calculation of within-group narrow sense heritabilities for the seven ancestral groups. All of the calculated heritabilities are substantially lower than the 74% heritability we calculated with the ancestral groupings ignored, clustering around about 55%. The reason the heritability is so much lower is that the previous analysis ignored the presence of "group structure." The existence of partially isolated ancestral groups led to genetic drift between these groups. As a consequence, the heritability around the grand mean for all of the groups considered together includes both a within-group and a between-group component. Allowing for the seven individual groups having their own means accounts for the fixed genetic effect of ancestry and, thus, lowers the heritability. This is an important point because it means that if there is undetected genetic heterogeneity within a study, then we will overestimate the narrow sense heritability for traits. An additional point from Table 9.3 is that the confidence intervals on heritability are much broader than from our previous calculations. This is because we are estimating heritability from seven smaller samples versus our original analysis with all seven groups considered together.

MULTIPLE QUANTITATIVE TRAITS

Just as we can estimate the heritability for single traits, we can also estimate the heritability for multiple traits in a **multivariate quantitative genetic** model. The multivariate quantitative genetic model is more complicated than the single-trait model. Now we must account not only for transmission and environmental effects on single traits but also for their common effects on two or more traits. Common effects due to the environment are easy enough to understand. Nutritional variation clearly can affect growth, and this can in turn cause **environmental correlation** among traits. For example, we might expect arm and leg length to be correlated because of environmental effects such that poor nutrition could lead to short arms and legs and better nutrition could lead to long arms and legs. This is a positive correlation in that long is found with long and short with short. Correlations can take values anywhere between negative 1 (complete negative correlation in which, e.g., short arms are found with long legs and long arms are found with short legs) and positive 1 (complete positive correlation). Zero correlations represent traits with no relationship (e.g., arm lengths and leg lengths vary independently of one another).

Correlations between traits can also arise because of the common effects of alleles at genetic loci. These common effects are referred to as pleiotropy and can be either positive or negative. In positive pleiotropy, alleles cause small and small to co-occur and big and big to co-occur. In negative pleiotropy "small" and "big" co-occur. Because of both pleiotropy and environmental correlations, it is quite common for morphological and physiological variables to be correlated at the phenotypic level. Indeed, a vast amount of biological anthropological research has focused on these correlations. For example, there are scores of articles written about the phenotypic correlation between stature and femur length (primarily because of the practical application of being able to estimate stature from femur length). There have, however, been very few studies that have resolved phenotypic correlations into their constituent (additive) genetic and environmental correlations. This is because such studies typically require multiple measurements on related individuals, and the methods for calculating environmental and additive genetic correlations between traits are not particularly easy.

It is unfortunate that there have been so few multivariate quantitative genetic studies of anthropometric data among humans. In lieu of additive genetic correlations from quantitative genetic studies, anthropologists and human biologists have typically substituted phenotypic correlations. These correlations are readily available because they can be calculated without the need for pedigree data; however, phenotypic correlations are not necessarily a good guide to either additive genetic correlations or environmental correlations. The relationship among the correlations is as follows (Falconer and Mackay 1996, 314):

$$r_P = h_X h_Y r_A + e_X e_Y r_E$$

where the r values are correlations (with the subscripts P, A, and E referring to phenotypic, additive genetic, and environmental, respectively), h is the square root of the narrow sense heritability for traits X and Y, and e is the square root of 1 minus the narrow sense heritability for traits X and Y. As an example, Konigsberg and Ousley (1995) estimated all of these values for head length and breadth using data collected by Boas from five Native American tribes. The heritabilities from their study can be found in Table 9.2, and the additive genetic correlation between the two traits was 0.377 and the environmental correlation was 0.199. This leads to a phenotypic correlation of 0.272, which overestimates the actual environmental correlation and underestimates the actual additive genetic correlation.

EVOLUTION AND QUANTITATIVE TRAITS

In Chapter 3, we discussed the role of evolutionary forces in shaping human variation for simple Mendelian loci. In the remaining section of this chapter, we discuss the comparable effects of mutation, selection, genetic drift, and migration on quantitative traits. Unfortunately, this is a highly technical area to discuss, so we refer readers to the supplemental material for detailed accounts. Such accounts are typically very mathematically challenging. In this section, we will of necessity eschew all math and just give the basic results. There is an additional complication that many traits appearing to be discrete in nature may actually be explained by the polygenic model with a threshold acting on an underlying unobservable distribution. For example, heart attacks are decidedly discrete events; but rather than there being a Mendelian locus controlling attacks, all evidence points to a polygenic model for underlying risk factors. We will not discuss the evolution of such **threshold traits** here, though it is important to understand that many presence/absence traits found in the human skeleton and dentition (and hence of interest to paleoanthropologists) likely evolve as polygenic traits. We also will not consider an additional and important complication, which is that many continuous traits may be influenced both by *polygenes* (many loci with small effects) and by **major genes** (genes at a few loci that have big effects on the phenotype). These major genes are of considerable biomedical import as many laboratories are currently exploring what is known as **quantitative trait linkage** (QTL) analysis. In QTL analysis, researchers try to establish statistically whether a putative (and not located) major gene locus is linked to a marker locus with a known position in the genome. Such analyses can lead to the identification of loci that have large effects on quantitative traits (referred to as **quantitative trait loci**, also [confusingly] abbreviated as QTL).

Mutation

As with simple Mendelian loci, mutation is a necessary force in order to create genetic variation influencing quantitative traits. Without mutation, the additive genetic variance would be 0 (because there would be no variation in the genes) and the heritability (broad or narrow) would be 0. Again, as with simple Mendelian loci, quantitative traits would evolve incredibly slowly if mutation were the sole force driving evolution.

Selection

Selection can be a potent evolutionary force affecting quantitative traits. As a result of differential fitness, selection can change both the average value for a quantitative trait as well as the additive genetic variance. Specifically, selection can act to increase or decrease the average value, while at the same time decreasing the additive genetic variance. The fact that differential fitness for the quantitative trait results in decreased additive genetic variance is an important point. As a result of this decrease, the narrow sense heritability decreases so that selection becomes less effective at changing the average.

This relationship is so important that it is often referred to as Fisher's "fundamental" theorem (after R. A. Fisher). As a consequence of this, traits that are under heavy selection will eventually have their additive genetic variance depleted so that there is little left to select. Animal breeders have long noted this effect, where they can obtain a large response to selection in the first few generations and then the response diminishes. However, as Simm (1998) notes, usually the economic goals of selective breeding can be reached before the additive genetic variation is exhausted. Barton and Keightley (2002) review the literature on diminished response to generations of artificial selection and find that in general species show a remarkable level of potential genetic adaptability.

Another important aspect of selection on quantitative traits is that there can be both direct and indirect selection. In direct selection, we are just concerned with the fitness values for a single quantitative trait and how the trait responds to selection. With indirect selection, we are concerned with what are known as correlated responses to selection. A correlated response means that direct selection is operating on one trait and that another quantitative trait is "along for the ride." Correlated responses occur because of pleiotropy so that if certain alleles are under selective pressure for one trait, then those alleles can also affect the evolution of another (genetically correlated) trait. As an example from the animal breeding literature, wool fiber diameter generally has a positive genetic correlation with fleece weight in sheep (reviewed in Adams and Cronjé 2003). Consequently, if one selected for increased wool yield, an expected correlated response would be an increase in fiber diameter. The increased fiber diameter would be an indirect response to the selection on increased fleece weight. As thicker wool fibers increase the "prickle factor" (itching and scratchiness), selection for increased yield could have the unintentional consequence of producing wool that has a much lower economic value.

Because of possible correlated responses, it is difficult to predict how selection will act on morphology. Without knowing the additive genetic correlations between traits, we can make only an uneducated guess about responses to selective regimes. This point has apparently been lost on a number of biological anthropologists. For example, Brace and Hunt (1990, 344), in a discussion of "nonadaptive traits," write the following:

> What possible difference in selective value might there be in having a cranial contour that is ovoid as opposed to pentagonal—or any other of the myriad named shapes? Why on Earth would it be better to have high rounded orbits under some circumstances and low rectangular ones under others?

The problem with this statement is that even if we were clever enough to figure out direct selective effects for individual morphological traits, we cannot intuit our way through an understanding of the correlated responses to direct selection.

Genetic Drift and Migration

Unwittingly, we have already covered genetic drift in our discussion of the Boas cephalic index data. As with single loci, genetic drift causes homogenization of (heritable) quantitative traits within demes and increasing variance between demes. Such was the case in the Boas data, where the ancestral groups did not all have the same mean cephalic index. In fact, statistically "forcing" the groups to have a common mean caused the heritability of the cephalic index to be overestimated because it was including a between-group component of variation. For quantitative traits, migration has an opposite effect to genetic drift, just as we saw in Chapter 3 for single-locus traits and as we will see in Chapter 12.

CHAPTER SUMMARY

This chapter covered the inheritance and population genetic basis for variation of quantitative traits in humans. First the chapter worked through the Mendelian basis for the inheritance of polygenic (many locus) traits and then distinguished between the concepts of narrow and broad sense heritability. Broad sense heritability was shown to be the proportion of phenotypic variation due to the effects of genes, while narrow sense heritability was shown to be only that proportion of phenotypic variation due to the additive effects of genes. The broad sense heritability gives an overall idea of how "genetic" a trait is, but because broad sense heritability includes dominance effects it is also measuring effects that cannot be transmitted from parents to children. Narrow sense heritability is consequently the measure of greater relevance in discussing human evolution and variation. Narrow sense heritability can be assessed in a variety of ways including looking at the relationship between parent's and children's phenotypes, looking at the relationships of phenotypes within as versus between sibships, and using pedigree data from larger samples. As an example of how narrow sense heritability studies are done, this chapter reanalyzed a large collection of data on cephalic index from Franz Boas' U.S. immigrant study. Following on this, the chapter discussed extensions to multiple quantitative traits and closed with a brief discussion of how evolution shapes quantitative trait variation.

SUPPLEMENTAL RESOURCES

Franz Boas's immigrant study: http://lance.qualquant.net/boas/data.htm. Clarence Gravlee's website contains the Boas data, which can be downloaded in various formats.
GENUP: http://www-personal.une.edu.au/~bkinghor/genup.htm. Home to Brian Kinghorn's GENUP package, which is "designed to help you master concepts in Quantitative Genetics . . ."
Marks J (1995) *Human Biodiversity: Genes, Race, and History*. New York: Aldine de Gruyter. Good coverage of the history of anthropological thought on human variation. See also http://www.uncc.edu/jmarks/pubs/main.html for more recent work by Marks.

10

Anthropometric Variation

Although many may not have ever had DNA testing or analysis of the biochemical variants described in earlier chapters, most have likely had some anthropometric measurements taken at various times. The word *anthropometric* literally translates as "human measures," but specifically it refers to the measurement of external dimensions of the human body, including the face and skull. Throughout our lives, we have all had some of the more common anthropometric measurements taken, such as height and weight. There are many other types of anthropometric measure used by anthropologists to assess variation in physical size and shape of the body and the skull. Analysis of anthropometric data provides clues about the causes of variation. Anthropometric traits are typical quantitative ones; as discussed in the previous chapter, variation in such traits reflects the joint influences of both genetic inheritance and environment. In addition, anthropometric measures change during the life cycle, adding a developmental component to observed variation. This chapter reviews some of the more commonly used anthropometric measures and their patterns of variation both within populations as well as between populations living in different parts of the world.

ANTHROPOMETRIC MEASURES

Many of us have had the experience of our parents standing us up next to a door frame and marking off our height with a pencil. By doing so, they were getting an estimate of our stature, one of the most commonly taken anthropometric measures. However, most people do not measure their child's height in a precise, systematic way. Height might change from one measuring session to another purely by error introduced by standing in a slightly different manner, wearing or not wearing shoes, or even the time of day. To be useful scientifically, measurements have to be taken in a precise, clearly defined manner. The anthropometric measurements routinely collected by anthropologists fit these criteria, having precise definitions and measuring protocols. There are many different types of anthropometric measure, using different types of measuring equipment and each with a specific standardized method of measurement. As used in this chapter, the word *anthropometrics* refers to measures taken on living subjects, although many of the same measures are also taken on skeletal material, a field known as **osteometrics**.

Body Measures

A number of anthropometric measures focus on the size and shape of the entire body and different body segments. Some of these measures focus on overall size (weight, stature), while others focus on a specific part of the body, such as the breadth across the shoulders or hips or the length of the arms and legs.

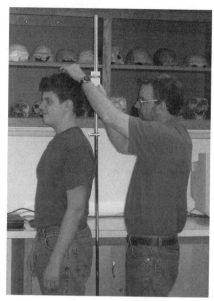

Figure 10.1. One of the authors (J. H. R.) measuring stature. (Photo by Hollie Jaffe.)

Weight and Stature

One of the most common measures is weight, measured in kilograms using standard scales. Weight is a composite measure of body size, made up of a number of components, including bone, fat, and muscle. **Stature** (height) is the other most common measurement, measured using an **anthropometer**, a long rod with movable ends (Fig. 10.1), and recorded using a metric scale, in centimeters or millimeters. As with weight, stature is a composite measure made up of a number of components, each of which varies within and between populations, including leg length, upper body length, and head height.

Body Mass Index

Weight and stature can be combined to compute the **body mass index (BMI)**, an approximate measure of obesity used in epidemiological studies as a risk factor for many diseases. BMI is computed as

$$BMI = \frac{Weight}{Stature^2}$$

where weight is measured in kilograms and stature is measured in meters. For example, if one is 1.75 m tall (slightly less than 5 feet 9 inches) and weighs 70 kg (slightly more than 154 pounds), one has a BMI of

$$BMI = \frac{70}{1.75^2} = \frac{70}{3.0625} = 22.9$$

BMI is often used as a rough index of relative weight; values between 19 and 24.9 are considered normal, values between 25 and 29.9 are considered overweight, and values equal to or greater than 30.0 are considered obese.

Although BMI is easily computed, it is not without its limitations. For example, in people with athletic builds, more of the weight reflects muscle mass rather than fat. As a consequence, BMI will tend to overestimate the amount of body fat. More precise measures of body fat and body composition are available using elaborate laboratory methods based on processes such as differences in electrical flow through tissues, the amount of air displaced by a person when sitting in a sealed chamber, differences in absorption of X-rays in various tissues, and other such methods.

Sitting Height

There are several body measures that provide information on relative body length. The most common of these is **sitting height**, measured by having the subject sit on a table and then measuring the height from the table surface to the top of the head. **Subischial length** is an estimate of lower body length, computed by subtracting sitting height from stature. Sitting height can also be used to compute a measure of relative sitting height known as the **cormic index**:

$$\text{Cormic index} = \frac{\text{Sitting height}}{\text{Stature}} \times 100$$

For example, let us assume we have measured someone whose sitting height is 940 cm and whose stature is 1,816 cm. This person's cormic index is

$$\text{Cormic index} = \frac{940}{1,816} \times 100 = 51.8\%$$

This ratio means that 51.8% of this person's height is due to upper body length (sitting height); therefore, $100 - 51.8 = 48.2\%$ of the height is in the lower body. As such, this person's upper body is slightly longer than the lower body. In general terms, a value of 50 indicates that 50% of someone's total height is due to lower body length (sitting height). A value higher than 50 indicates someone who has a relatively long torso, while a value less than 50 indicates someone with relatively long legs. The cormic index varies across our species: Australian Aborigines and some Africans typically have short trunks and cormic indices less than 50, while some Asian and Native American populations have relatively long torsos and cormic indices greater than 50 (Harrison et al. 1988). As discussed later in this chapter, much of the difference among populations relates to climate. In addition, we must remember that for this trait (and most traits in general) there is considerable variation around the mean value. For example, the average cormic index for 2,222 adult Irish men between the ages of 35 and 49 is 53.2 (J. H. R., unpublished data). This figure shows that Irish men have relatively long torsos. However, there is considerable variation around this mean. The standard deviation is 1.32, meaning that 95% of all Irish men have cormic indices that range from 50.6 to 55.8. Some individual subjects had even more extreme values; the smallest value observed was 47.7, and the largest was 58.7. It is important to always remember that a mean does not describe everyone within a sample.

Body Breadths

Anthropometric measures are also made on various body breadths. For example, **biacromial breadth** is a measure of body width across the shoulders, and **bi-iliac breadth** is a measure of body width at the hips. **Chest breadth** and **chest depth** measure the size of the thoracic region from side to side and from front to back, respectively. Although many other measurements are available, these are the most widely used.

Circumferences

A tape measure is used to measure circumferences of the body and limbs. Some common measures of circumference include chest circumference and waist circumference. In addition, limb circumferences (arms and legs) are often measured, sometimes at different parts of the limb.

Skinfolds

Weight provides a measurement of total body mass, which includes components made up of bone, muscle, fat, and internal organs. For many purposes, including studies of nutrition, sports physiology, and disease risk, it is useful to have estimates of the relative contribution of these components to body mass. In practice, most methods of analyzing body composition provide a means of distinguishing between fat and fat-free (lean) body mass. One method of assessing body composition is the use of skinfold measures. Much of the body's fat is **subcutaneous fat**, lying just below the skin. The thickness of subcutaneous fat is a valuable measure of nutritional status and is easily measured (Himes 1980). Subcutaneous fat is assessed by measuring the **skinfold thickness** at different parts of the body. Skinfold thickness is measured by grabbing a fold of skin and fat and measuring the double fold with a skinfold caliper (Fig. 10.2). Since only skin and fat can be pulled loose from the body in this manner and since skin is very thin, these measures provide an estimate of the total amount of subcutaneous fat at a given part of the body. Three body sites are typically measured in many anthropometric studies: (1) subscapular skinfold, (2) triceps skinfold, and (3) calf skinfold. Subscapular skinfold is measured by pinching the skin and subcutaneous fat immediately below the bottom margin of the

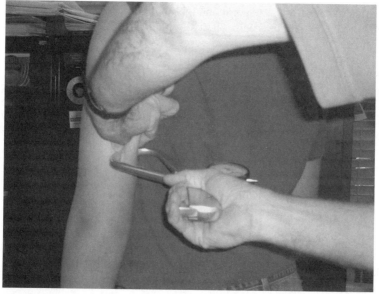

Figure 10.2. One of the authors (J. H. R.) measuring triceps skinfold using a skinfold caliper. (Photo by Hollie Jaffe.)

shoulder blade (scapula). Triceps skinfold is measured on the back of the arm (above the triceps muscle), halfway from the shoulder to the elbow. Calf skinfold is measured on the back of the leg, halfway from the knee to the ankle. Skinfold measures can be used to predict body density and fat proportion using equations developed on reference populations where both skinfolds and other measures of body density and fat have been observed.

Head Measures

Many anthropometric measures are taken on the head, including measures of overall head size, cranial and facial breadths, and nasal measures. Head measures have often been found to be reliable markers of population history and genetic affinities between populations (e.g., Friedlaender 1975, Relethford and Crawford 1995). Head measures also provide valuable information on climatic adaptation (e.g., Beals et al. 1984, Roberts 1978), as discussed later in this chapter. Most head measures are taken using a spreading caliper (Fig. 10.3) or a sliding caliper (see later, under Nasal Measures).

Head Size and Shape

Three of the most common head measures are head length, head breadth, and head height. **Head length** measures the overall length of the skull between standardized locations on the front and rear of the skull (Fig. 10.3). **Head breadth** is measured as the greatest width across the back of the skull, usually above and somewhat behind the ears. **Head height** is generally measured using an anthropometer and is the distance from the ear hole to the top of the head. These three measures provide data on the overall size of the upper part of the skull and can be used together to roughly estimate cranial volume.

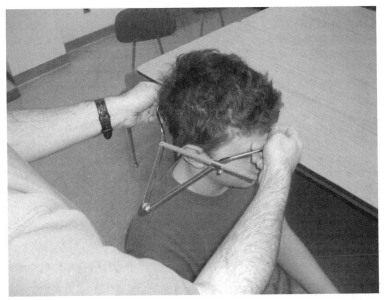

Figure 10.3. One of the authors (J. H. R.) measuring head length using a spreading caliper. (Photo by Hollie Jaffe.)

As noted in the previous chapter, head length and head breadth can be used to construct a simple index of cranial shape known as the cephalic index (CI). The CI, as measured on living humans, is equivalent to the cranial index discussed in the previous chapter. This index, invented by the Swedish anatomist Anders Retzius, is a measure of the relative width of the head, computed as follows:

$$CI = \frac{\text{Head breadth}}{\text{Head length}} \times 100$$

For example, assume we have just measured someone's head and obtained a value of 186 cm for head length and 156 cm for head breadth. This person's CI is

$$CI = \frac{156}{186} \times 100 = 83.9\%$$

The width of this person's skull is roughly 84% of the length of the person's head. As noted in Chapter 9, higher CI values indicate heads that are relatively broad, whereas low values indicate relatively narrow heads.

The CI was once thought to provide an objective means by which to classify individuals into different races (Klass and Hellman 1971). At first glance, this might seem to make some sense because there is a rough correspondence of CI and the geographic origin of a population. For example, on average, sub-Saharan Africans tend to have relatively narrower heads (a lower CI) than Europeans, who tend on average to have relatively broader heads. However, closer inspection reveals a problem with analyzing CI variation in this manner. There is considerable variation within geographic regions such that the ranges of variation overlap considerably (see Fig. 10.4 for an example). This makes assignment of any given CI to a particular geographic region difficult at best. As

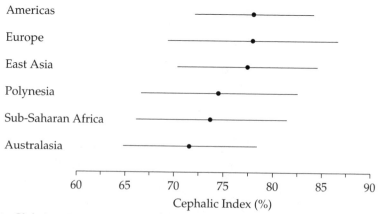

Figure 10.4. Global variation in cephalic index based on the crania of 907 adult males from six different geographic regions. These data come from the skulls of skeletal remains and not from living people. The dots correspond to the mean values within each region, and the lines show the range within which 95% of the crania belong (±1.96 standard deviations). Although there is a difference in the average cephalic index between regions, there is also considerable overlap between regions. (Data from Howells 1996.)

a consequence, there are cases where two populations in different parts of the world have the same CI. One example is Koreans and Germans, both have the same mean CI of 83% (Harrison et al. 1988). Although there is some overall relationship between mean CI and geographic location, it does not correspond to a view of discrete and readily identifiable races. Similarity between populations is in part a reflection of shared gene flow but is also affected by adaptation to different environments. As discussed later, mean CIs of humans across the world are correlated with climate to a certain extent, with a tendency toward higher values in colder climates and lower values in warmer climates. However, as is clear from Figure 10.4, there is also variation within populations. Although Europeans tend to have broader heads than Africans, this does not mean that all Europeans have broad heads or that all Europeans have broader heads than all Africans.

Facial Breadth

A variety of measures have been developed to measure the width of the skull at various points. Differences between these measures provide estimates of craniofacial shape that can frequently be used to distinguish populations. Three craniofacial breadths are often measured in anthropometric studies (Fig. 10.5): (*1*) **minimum frontal diameter** (cranial width between the indentations on the skull above the eye orbits), (2) **bizygomatic breadth** (facial width between the outermost parts of the cheek bones), and (3) **bigonial breadth** (width across the rear portion of the lower jaw). A number of ratios have been devised to use these measures to express craniofacial shape; for example, the ratio of bizygomatic breadth to bigonial breadth provides an index of lower facial shape. High values of this index indicate a more triangular lower face.

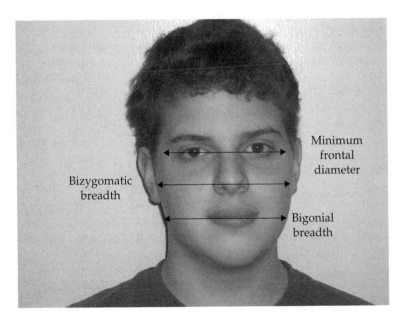

Figure 10.5. Three different measurements of facial breadth commonly used in anthropological studies. (Photo by Hollie Jaffe.)

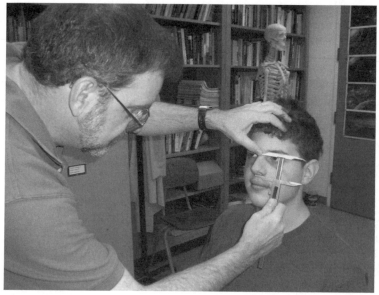

Figure 10.6. One of the authors (J. H. R.) measuring nose height using a sliding caliper. (Photo by Hollie Jaffe.)

Nasal Measures

There are two common measures of nose size, **nose height** and **nose breadth**. Both are measured with a sliding caliper (Fig. 10.6). Nose height (also often called "nose length") is the distance from the top of the nose to its base (or, on skeletal remains, the height of the nasal opening). Nose breadth is measured as the width of the nose at the base. When measured on living people, care must be taken not to compress the soft tissue of the nose and get a false reading.

Nose height and nose breadth are often used together to compute the **nasal index** (NI), a measure of relative nasal breadth:

$$NI = \frac{\text{Nose breadth}}{\text{Nose height}} \times 100$$

A high NI indicates a relatively broad nose, while a low NI indicates a relatively narrow nose. As discussed below, several studies have found a correlation between NI and climate.

VARIATION WITHIN POPULATIONS

Variation in anthropometric traits is due to the joint influences of genetic ancestry, environmental modifications, and the growth process.

Genetics

Anthropometric traits are typical quantitative ones (see Chapter 9), whose distribution reflects inheritance of different genes combined with the influence of environmental factors. Many studies have been performed on anthropometric traits in order to determine the relative roles of genetic and nongenetic factors. As always, estimates of heritability are often specific to particular populations, although some general inferences can be made.

Anthropometric traits in general tend to show moderate heritability, often in the range of 0.3–0.6. For example, in their analysis of Native American anthropometrics, Konigsberg and Ousley (1995) found an average heritability of 0.4 using a multivariate method that takes correlation between traits into account. Heritability estimates for anthropometrics tend to vary by trait, generally showing lower values for measures of body fat and mass than for lengths or craniofacial measures. In their study of Mennonites, Devor et al. (1986) derived estimates of "transmissibility" for 34 anthropometric traits. *Transmissibility* is an estimate of parental influence on a trait, including both genetic inheritance and aspects of the shared environment. As such, it is an upper estimate of actual heritability. Devor et al. (1986) found average transmissibilities of 0.63 for measures of body and limb length, 0.43 for body breadth measures, 0.45 for head measures, and 0.34 for measures of body fat and circumference. In a review of studies of craniofacial inheritance, Devor (1987) found that most head measures have transmissibilities between 0.45 and 0.60. The finding of lower heritabilities for measures of body fat and circumference makes sense because body fat is also strongly influenced by diet, exercise, and other nongenetic factors.

As noted in Chapter 9, heritability is a relative estimate of the proportion of total variation due to genetic variation. As environmental variation changes, so does heritability: the lower the environmental variation, the higher the heritability. A good example of this comes from an analysis of stature performed by Towne et al. (1993), who showed higher heritability among infants than among adults. As noted in Chapter 9, this age-related difference appears to reflect the fact that infant body length has not yet been affected by environmental factors, whereas adult stature reflects lifelong nongenetic influences (see also Konigsberg 2000).

As noted in the previous chapter, new methods of quantitative genetic analysis offer great promise for unraveling the specific genetic factors affecting anthropometrics and other quantitative traits. For example, several studies have found evidence of a major gene affecting BMI (e.g., Ferrell 1993). More recent analysis using the quantitative trait loci (QTL) method has found evidence of significant linkage of body fat with a marker on chromosome 2 (Comuzzie et al. 1997). Analysis of baboons has found evidence of a significant QTL for body length (Rogers et al. 1999) and offers a glimpse of possibilities that may soon be available for anthropometric traits in humans. The technology is thus far rather expensive, but as noted by Rogers et al. (1999), the necessary data might be collected as part of larger biomedical projects, typically better funded than pure anthropological research. We may soon be at the stage where we can relate the genetic component of anthropometric variation to specific genes, which will also improve our ability to analyze and understand nongenetic influences and the interaction between genotype and phenotype.

Human Growth

Everyone can easily tell that body size and shape changes during growth. This observation may seem too obvious to warrant much further consideration. However, the reason for such changes may be less clear. Much of growth research focuses on describing and analyzing patterns of human growth and relating these patterns to genetic, hormonal, nutritional, and other environmental influences.

Human growth is subdivided into two phases, **prenatal** growth (before birth) and **postnatal** growth (after birth). This section focuses on postnatal growth in anthropometric measurements. There are two main types of study for postnatal human growth. One is a **cross-sectional study**, which compares individuals at different ages, all measured at the same point in time. For example, if we went to a local school and measured the heights of girls aged 6–16, we would be conducting a cross-sectional study and the comparisons would be between girls at different ages as an approximation of the changes that take place as a child grows. For example, if we looked at the average height of 6-year-old girls and compared it to average height of 7-year-old girls, we would have an idea of how much change typically takes place from age 6 to age 7. However, we are not actually observing this change over time and instead are inferring it from the average height in two different samples that presumably are similar except for age.

If we wanted more precise data on growth from age 6 to age 7, we might take a different approach by starting with a sample of 6-year-old girls and then measuring these same girls again a year later when they were 7 years old. This second approach is known as a **longitudinal study**, where we compare the same individuals at different points in their lives. Ideally, we would track the same individuals over as much of their postnatal growth as possible by returning to measure them year after year. For certain types of analysis, notably those that look at changes in the velocity of growth (see below), longitudinal studies are preferred (Tanner 1989). However, they take more time since we would have to continue measurements throughout a person's life. As such, we would also run into the problem that some of the original subjects may move or die over the long course of the study.

Growth Curves

Human physical growth is not simply a matter of the body becoming larger over time. Different parts of the body grow at different rates, and the rates themselves change over time. Following the old adage that a picture is worth a thousand words, the process of human growth is most easily visualized and analyzed using growth curves. One type of growth curve is a **distance curve**, which is simply a plot of size measured over time. Figure 10.7 presents a typical distance curve for stature, showing the obvious point that people generally get taller as they age, at least up through adulthood. Closer inspection of this curve shows that growth is not constant, which would result in a straight line rather than one that has some curvature. More height is gained at certain points during an individual's life than others.

Changes in height gain are more easily shown using a **velocity curve**, which measures the rate of growth over time. The velocity curve for human stature is shown in Figure 10.8, which plots the amount of height gained per year as a function of age. This curve shows that the greatest amount of gain in stature occurs during infancy; at no other point during the life span do we grow so quickly. Immediately after birth, the rate of growth decreases quickly (deceleration) and levels off to a relatively constant value during much of childhood. During adolescence, the rate of growth increases briefly, leading to the **adolescent growth spurt**, part of the entire biological transition of puberty. By the time we reach adulthood, the velocity of growth has declined to close to 0. Keep in mind

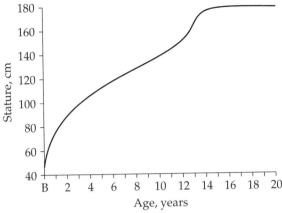

Figure 10.7. Typical distance curve for human height. (Courtesy of Robert Malina.)

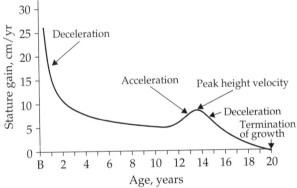

Figure 10.8. Typical velocity curve for human height. (Courtesy of Robert Malina.)

that Figure 10.8 shows the velocity of growth. People generally continue to grow during the first 20 years of life but not at the same rate each year.

One useful way of understanding the difference between a distance curve and a velocity curve is to consider similar quantities when driving. When we are driving on the highway, we keep track of two different measurements: one is distance, how far we have traveled as recorded on the car's odometer, and the other is velocity, how fast we are traveling. Note that velocity can change up and down, but as long as the car is moving, we are still going forward, although at different speeds, and the total distance will increase. Likewise, for a measurement of body size such as stature, we continue to grow during most of the first 20 years of life but at different rates. Sometimes we are growing more quickly than at other times.

The velocity curve shown in Figure 10.8 demonstrates two important characteristics of human growth: (1) an extended childhood and (2) an adolescent growth spurt. An extended childhood separates infancy and the time of sexual maturation. In our species, the extended childhood is an evolutionary adaptation to our cultural way of life and our emphasis on learned behavior as an adaptive strategy. Our brain growth is complete

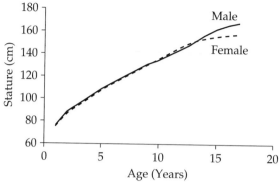

Figure 10.9. Distance curves for stature of male and female urban Chinese from 1 to 17 years of age. (Data from Eveleth and Tanner 1990.)

during childhood, but we still need time to learn the physical and social skills needed to survive as adults and raise children. The evolutionary significance of the adolescent growth spurt is still debated. While some argue that this spurt allows us to "catch up" and become adults quickly after we have passed childhood, others suggest that the situation is more complicated (Bogin and Smith 1996). Although many have argued that the adolescent growth spurt is unique to humans, it turns out that it is also seen in nonhuman primates, although the human growth spurt does occur relatively later in life, following the extended childhood (Leigh 1996).

Growth curves are useful because they can be easily compared between different samples. An example of comparing distance curves for stature is shown in Figure 10.9, which compares male and female growth in stature by plotting the average height of urban Chinese males and females from 1 to 17 years of age. During infancy and childhood, the two curves are almost exactly the same, showing very little difference between the average height of Chinese boys and girls. Between roughly 11 and 13 years of age, the situation changes and girls are slightly taller than boys during this time. After 13 years of age, the situation changes again, and males are now larger than females, a trend which continues through adulthood. This graph summarizes what is widely known about sex differences in body size. Females begin adolescence roughly 2 years earlier than males and, thus, have their growth spurt earlier, giving them slightly greater stature for these 2 years. Males catch up later, and since they have had 2 additional years of growth prior to their growth spurt, they wind up larger as adults. Figure 10.9 represents the average values for each age group, but there is variation in overall height in both sexes and across all ages. For example, not all males are taller than all females in adulthood. Still, the use of averages represented by growth curves does provide a useful summary of general patterns of growth.

Another example of comparison using distance curves is shown in Figure 10.10, which compares groups based on different levels of modernization. Here, female stature is plotted from ages 6 to 11 for two Samoan populations: Western Samoa, which has a more "traditional" lifestyle, and American Samoa, which is more modernized. The sample from American Samoa is slightly taller at all ages than the sample from Western Samoa, clearly showing the biological impact of modernization. Increased height reflects changing environmental conditions. Such change is not always healthy; it is typical for

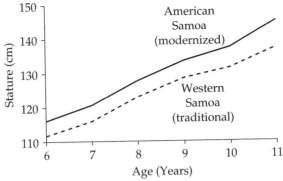

Figure 10.10. Distance curves for stature for girls from two Samoan populations: American Samoa (more modernized) and Western Samoa (more traditional). (Data from Bindon and Zansky 1986.)

modernizing populations to also show an increase in obesity due to changes in diet and a more sedentary lifestyle (Harrison et al. 1988).

The above discussion has focused specifically on stature as a measure of human growth. The basic distance and velocity curves also apply to other anthropometric measures, such as weight and limb length. However, not all measurements follow the same growth curves. In general, head measures and brain size increase more rapidly during infancy and early childhood, such that adult brain size is reached by roughly 7 years of age (Cabana et al. 1993). In addition, sex organs and reproductive maturity lag in growth, showing little change until adolescence. The basic pattern of human growth is that head/brain size reaches adult levels earliest, followed by body size and sexual maturation. Again, these differences reflect the evolution of the human growth pattern, where brain size is attained first and the rest of growth lags behind in order to provide ample time for children to learn the social and physical skills needed for life.

Genetic and Hormonal Influences

Comparison of growth curves among relatives shows that human growth is partially affected by genetic factors. The body size of adults tends to be close to the average of their parents' heights, so tall parents generally have tall children. However, we also know that environment (particularly diet and disease) can strongly affect human growth, and changes in these factors can cause deviations from the expected pattern. In general, measures of length (e.g., stature, limb length) tend to show a fairly close resemblance between parents and children (as adults). The degree of correspondence is somewhat less for measures of body width (e.g., biacromial breadth) and even less for measures that include fat, which can be strongly affected by diet and exercise.

Hormones also affect growth. Hormones are chemicals released by the endocrine glands that travel in the bloodstream to body tissues and stimulate and regulate biological processes such as growth and development. Hormone secretion is regulated by the hypothalamus in the brain, which releases hormone-releasing chemicals that travel to the pituitary gland, which in turn secretes hormones that travel to different sites throughout the body or to other glands that produce other hormones.

Many hormones influence the process of human growth. One, human growth hormone, is secreted by the pituitary gland and influences growth by affecting protein synthesis and the rate of cell multiplication. Another hormone, thyroxine, is produced by the thyroid gland and influences oxygen consumption by bodily tissues. Various sex hormones, such as the androgens (male sex hormones) and the estrogens (female sex hormones) stimulate the development of secondary sex characteristics and affect patterns of body growth differently between the sexes (e.g., increased muscle mass in males).

Nutrition and Disease

Genetics is not destiny but sets limits to growth that are also affected by many environmental factors. Proper nutrition is critical in determining whether a person reaches his or her genetic potential for growth. Ingested nutrients provide the energy necessary for proper growth and development as well as basic maintenance of the body. The proportion of nutrients needed for growth is greatest during infancy, childhood, and puberty; and too little nutrient energy can result in delayed maturation and smaller adult body size. If too much is ingested, then the excess turns to fat, which in turn can affect predisposition to a number of health problems. In addition to having the appropriate level of total caloric intake, the human body requires protein, fat, vitamins, and minerals. Lack of one or more of these can seriously affect human growth (and health). Adequate nutrition is thus defined in terms of both quantity and diversity.

Poor nutrition is a problem that unfortunately affects much of the world today. Poor nutrition (often a lack of adequate calories) acts to slow down the growth process, leading to later maturation and smaller adult body size. If nutritional levels are too low, the consequences are even more severe, including increased susceptibility to disease, death, and mental retardation. Although it has been suggested that the small body size found in malnourished populations represents an adaptation to limited nutrition (Seckler 1980), this view assumes that reduced body size is without any biological cost, which is clearly not the case.

Infectious disease can also affect human growth, particularly if the child is poorly nourished, again leading to delayed growth and smaller adult size. In undernourished populations, there is often a vicious cycle between poor nutrition and infectious disease. Poor nutrition acts to lower the body's resistance to infectious disease, and in turn, infection affects nutritional status by reducing the efficient absorption of nutrients and causing a loss of appetite. The onset of infectious disease thus leads to a reduction in nutritional status, which in turn leaves the child even more susceptible to the effects of infectious disease (Martorell 1980).

Socioeconomic Influences

A variety of socioeconomic factors have been found to affect human growth, primarily by affecting nutrition, prevalence of disease, and access to health care. In many societies, there is a relationship between social class and growth, where children in upper-class families tend to be larger and grow more quickly, a reflection of class differences in available nutrition and/or health care. An exception is obesity, which sometimes is greater in lower social classes (Bogin 1988, Tanner 1989).

A number of studies of economically developed societies have found that children growing up in urban environments are often larger and grow more quickly than those in rural environments. Tanner (1989) argues that the primary cause of this difference is that rural children have less caloric intake and more physical activity. Tanner also notes that urban–rural contrasts are not as great in societies that are less economically developed. Another social correlation with growth is the number of children in a family and the order of birth. In general, children from larger families tend to grow more slowly, perhaps due to less nutrition per family member (especially since family size is often correlated with socioeconomic class, such that poorer families have more children).

Secular Change

When one is taller than the same-sexed grandparents, this is evidence of **secular change**, a change in the average pattern of growth from generation to generation. During the past century, average height and weight have increased from one generation to the next in economically developed populations. The secular increase in body size reflects changing environmental conditions, especially improvements in nutrition and reduced incidence of childhood infectious disease (Malina 1979). These environmental changes mean that individuals in recent generations are more likely to reach their genetic potential since the nutritional and infectious disease stresses have been reduced. While secular increases in body size continue, there is some suggestion that it is slowing down (Eveleth and Tanner 1990).

Figure 10.11 shows an example of a dramatic secular change in the height of Japanese children from 1900 through 1971 (Oiso 1975). This particular graph shows the average height of all 10-year-old Japanese boys as an example, but the pattern shown here is also found at other ages and for Japanese girls. The average height of the 10-year-old boys has increased from 124 cm (roughly 4 feet, 1 inch) in 1900 to 136 cm (roughly 4 feet, 5.5 inches) in 1971. That is, the average 10-year-old boy in 1971 was 12 cm (almost 5 inches) taller than the average 10-year-old boy in 1900. Figure 10.11 also shows an interesting reversal in the general trend of secular increase: between 1940 and 1948 the average height declined. This period of time includes the Second World War, during which there was widespread nutritional deficiency in Japan. The nutritional status in Japan is known to have improved since then, and the secular increase after the Second World War is even greater than that which took place before, reflecting rapid and intensive improvements in the nutritional status of Japanese children, as measured by national nutritional surveys (Oiso 1975).

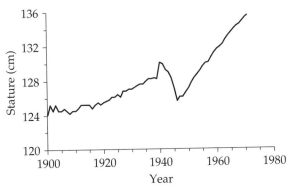

Figure 10.11. Stature of 10-year-old Japanese boys from 1900 to 1971. (Data from Oiso 1975.)

Developmental Plasticity

The influence of growth on anthropometric measurements has also played a role in debates regarding **developmental plasticity**, a change in morphology that occurs when individuals grow up in different environments. The classic example of developmental plasticity is the pioneering study conducted by Franz Boas (1912) noted in Chapter 9. Between 1908 and 1910, Boas took a number of anthropometric measurements on thousands of European immigrants and their children. One of his goals was to examine the impact that growing up in a different environment had on cranial measurements. To this end, he compared cranial measurements of the children of European immigrants by placing each child into one of two different groups, those born in Europe and those born in the United States. This comparison allowed the effects of two different environments during growth to be assessed. He analyzed these differences in seven different ethnic groups: Bohemians, Central Italians, Hebrews, Hungarians, Poles, Scots, and Sicilians (the group names given here are those used by Boas). Boas noted that cranial measures of the U.S.-born and foreign-born children were different, suggesting that growing up in different environments affected cranial measures. His findings were quite influential in the further development of American anthropology as they came at a time when the dominant ideology focused on presumed fixity of cranial measures that would reflect only genetics and not environment (Harris 1968).

Although Boas clearly demonstrated that cranial measures were not solely affected by genetics, it was also clear that environmental factors were not solely responsible for cranial variation. As is most often the case for complex traits, variation reflects the interaction of both genetic and environmental factors. The relative influence of these factors has recently been brought into question by two groups of researchers who have reanalyzed Boas's original data (Sparks and Jantz 2002, Gravlee et al. 2003b). Although both studies used the same data, they reached rather different conclusions regarding the influence of developmental plasticity on cranial measures. Sparks and Jantz (2002) found only slight evidence for cranial plasticity, primarily in head shape (as measured by the CI). Gravlee et al. (2003b) found a more substantial effect; in four out of seven immigrant groups analyzed, there was a significant difference between the CI of U.S.-born and foreign-born immigrant children. While Sparks and Jantz asserted that Boas's original claim of plasticity showed minimal support, Gravlee et al. (2003b) argued that, by and large, Boas got it right.

Further discussion by these two groups shows some commonality in results, although they reach somewhat different conclusions (Sparks and Jantz 2003, Gravlee et al. 2003a). It is clear from both studies that there is some impact of developmental plasticity. It is also clear that this plasticity is not the only factor influencing cranial variation. In particular, the differences between ethnic groups (which could reflect genetic and/or environmental factors) were roughly the same in both the U.S.-born and foreign-born samples. Figure 10.12 shows the mean CI for U.S.-born and foreign-born immigrants. The relative similarity of each ethnic group to all others is slightly different when comparing the U.S.-born and foreign-born children, and these differences are a reflection of developmental plasticity. However, the major pattern of difference between ethnic groups is roughly the same. In both U.S.-born and foreign-born groups, the Scottish and Sicilian samples are different from the others by having lower average CIs. This graph shows that cranial measures are affected by developmental plasticity, but this change does not erase or obscure differences between ethnic groups. Again, the variation in complex traits is often a mixture of genetic and environmental influences.

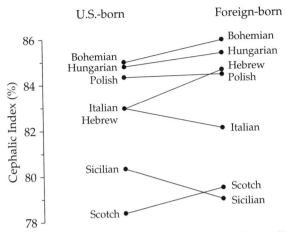

Figure 10.12. Comparison of average cephalic index between children of European immigrants born in the United States (U.S.-born) and those born in Europe (foreign-born). These averages are age-adjusted. (Data from Gravlee et al. 2003b.)

Aging

The effects of aging on anthropometric measures, particularly weight and stature, have been studied in both cross-sectional and longitudinal studies. Longitudinal studies provide a better estimate of the actual physical impact of aging because cross-sectional studies are often affected by secular changes as different generations are being compared. Longitudinal studies, however, measure changes in aging associated with specific individuals as measured over time. In general, weight increases during middle age and then declines among the most elderly, a pattern seen in some skinfold measures as well (Rossman 1977). A number of studies have shown that stature decreases with aging due to bone loss, compression of intervertebral discs, and flattening of feet.

An example of an aging study is the study of veterans in Boston described by Friedlaender et al. (1977). A group of 1,813 healthy white male veterans ranging in age from 30 to 70 years was measured twice over a 5-year period to obtain an idea of aging effects in different age groups. For example, men between 40 and 44 years of age lost an average of 4 mm of height over the next 5 years, while men over 70 years of age shrank 6 mm over a 5-year period. This study found significant shrinkage in both stature and sitting height. Weight increased with age for middle-aged men and stabilized during old age. The same pattern was found for calf and arm circumference. Head measures in general showed little change, except for an increase in head circumference with age.

Sexual Dimorphism

Sexual dimorphism refers to the difference in size between males and females. Some primate species, such as the gorilla, show very high levels of sexual dimorphism, while other species, such as the gibbon, show virtually none (Leutenegger 1982). Humans are moderately dimorphic.

Table 10.1 Sexual dimorphism among Irish adults, aged 35–49

Variable	Male Mean	Female Mean	Male/Female Ratio
Weight (kg)	74.3	65.5	1.13
Stature (mm)	1,725.2	1,592.5	1.08
Sitting height (mm)	917.9	844.2	1.09
Biacromial breadth (mm)	388.0	368.3	1.05
Head length (mm)	196.5	188.3	1.04
Head breadth (mm)	154.5	148.6	1.04
Head height (mm)	124.6	119.9	1.04
Minimum frontal diameter (mm)	109.6	104.2	1.05
Bizygomatic breadth (mm)	141.8	133.1	1.07
Bigonial breadth (mm)	109.7	102.3	1.07
Nose height (mm)	56.7	51.6	1.10
Nose breadth (mm)	36.2	32.4	1.12

Unpublished data (J. H. R.). See Relethford and Crawford (1995) for a description of the larger data set from which these data were taken.

Patterns of Sexual Dimorphism

For most anthropometric measurements, adult males are generally larger than adult females on average, even though the ranges of variation overlap and some females will be larger than some males. Table 10.1 provides an example of sexual dimorphism based on 2,145 adult Irish males and 320 adult Irish females between the ages of 35 and 49 years who were measured in the 1930s. This table reports the male and female averages for a number of anthropometric traits as well as the ratio of the male average to the female average, a simple measure of sexual dimorphism. A ratio of 1 indicates equal size of males and females, a ratio greater than 1 indicates that males are larger, and a ratio less than 1 shows that females are larger.

Males are larger than females for each anthropometric trait listed in Table 10.1, although the level of sexual dimorphism varies across traits. The greatest amount of dimorphism is for weight, where males are 13% heavier than females. Nose breadth shows a similar level of dimorphism, perhaps reflecting the fact that this measure includes a component of body fat, which is correlated with weight. Males in this sample are about 9% taller than females. Head measures show lower levels of dimorphism, 4%–7%.

As another example, Table 10.2 reports the male/female ratio for anthropometric traits collected on the population of Saltillo, Mexico, by Lees and Crawford (1976). These values show the typical larger body size in adult males but also show two traits (upper arm circumference and triceps skinfold) where females are larger than males. This reversal of pattern reflects the fact that adult females in general have a higher proportion of body fat, a difference that shows up more clearly in measures such as limb circumference and, in particular, skinfold measurements.

What Causes Sexual Dimorphism?

There is considerable debate over the reasons for differing levels of sexual dimorphism among primate species, as well as the implications for human evolution and variation. There is a general correlation between a species' mating system and the degree of sexual

Table 10.2 Sexual dimorphism among adults from Saltillo, Mexico

Variable	Male Mean	Female Mean	Male/Female Ratio
Weight (kg)	59.9	56.2	1.07
Stature (mm)	1,619.2	1,497.6	1.08
Sitting height (mm)	849.1	796.1	1.07
Arm circumference (mm)	264.6	274.1	0.97
Triceps skinfold (mm)	16.2	20.8	0.78
Head length (mm)	184.8	179.5	1.03
Head breadth (mm)	147.7	144.0	1.03
Head height (mm)	130.5	126.1	1.03
Head circumference (mm)	541.0	533.7	1.01
Minimum frontal diameter (mm)	115.7	111.2	1.04
Bizygomatic breadth (mm)	136.5	130.9	1.04
Nose height (mm)	52.7	46.9	1.12

Data from Lees and Crawford 1976.

dimorphism in body weight. While many primate species are **polygynous** (males mate with more than one female), some are **monogamous** (males mate with only one female). Leutenegger (1982) compiled male/female ratios of weight for 42 polygynous and 11 monogamous nonhuman primate species and found a significant difference. The average male/female ratio in monogamous species was 1.04 (not significantly different from 1.0). The average male/female ratio in the polygynous species was 1.40, indicating that males are, on average, 40% heavier than females.

The correlation between mating system and sexual dimorphism is not absolute; some polygynous species had values as low as 0.94 and some as high as 2.36. However, the average difference by mating system suggests that male–female differences are related in some way to a species' typical pattern of mating. Higher levels of sexual dimorphism in polygynous species are frequently considered to have resulted from competition among males to mate with females. Since, in many cases, the larger male would be more likely to win such a battle and mate with the female, he would then pass on any genes for larger body size to the next generation. Over time, the average difference between males and females would increase. While this seems to fit the evidence at first glance, studies of dominance disputes and mating practices among nonhuman primates show considerable variation and the impact of many factors. For one thing, mating is not always determined by a battle among males; female choice is also found in many primate societies. In addition, dominance disputes among males are not determined by size and strength alone but are also affected by a variety of factors ranging from the ability to cooperate to the rank of one's mother (e.g., Eaton 1976, Fedigan 1983). While male dominance and competition may be one factor affecting mating patterns, it is clearly not the only one.

What about differences in sexual dimorphism within a species? Studies of sexual dimorphism in human stature show variation among populations; some groups show little dimorphism whereas others show much more. Two different explanations have been offered to explain this variation: genetic and nutritional (Wolfe and Gray 1982). The genetic hypothesis extrapolates from studies of nonhuman primates and other mammals and suggests that there has been selection for larger body size among males in polygynous societies. Tests of this hypothesis offer no support (Wolfe and Gray 1982). The nutritional hypothesis proposes that female body size is, in general, less affected by nutritional fluctuations because of hormonal mechanisms that have evolved to meet the nutritional demands of pregnancy and breast-feeding. This model views female body size as being more constrained (or, to use the specific term, *canalized*). Males, however, do not have these extra nutritional demands and are more likely to vary in body size. According to this hypothesis, populations undergoing nutritional stress will show small body size for both males and females and little sexual dimorphism. When nutritional status is better, both males and females will show larger body size but males will show more gain, thus leading to increased sexual dimorphism. Several studies have found this pattern, which tends to confirm the nutritional hypothesis (e.g., Tobias 1975, Wolfe and Gray 1982).

ENVIRONMENTAL VARIATIONS

The discussion so far has focused for the most part on anthropometric variation within populations. This section examines anthropometric variation among populations, focusing on the correlation of anthropometric measurements with climate and habitat in order to illustrate long-term natural selection and short-term physiological adaptation. The use of anthropometric variation among populations in unraveling questions of population history will be discussed in Chapter 12.

Correlations with Climate

Some anthropometric measures show a correlation with specific climatic variables, including average annual temperature and humidity levels. These correlations suggest the action of natural selection, where anthropometric averages reflect past genetic adaptation to specific climates. The influence of natural selection does not explain all of anthropometric variation among populations but is a significant factor, among others, that affects mean values for a number of traits.

The Bergmann-Allen Rules

Humans are not the only species that show distinct patterns of environmental variation in physique. In the nineteenth century, a German zoologist named Carl Bergmann noted a relationship between body size and average annual temperature within a number of mammalian species. Specifically, he noted a tendency for populations within a species to have, on average, smaller body size and more linear body shape in hot climates and larger body size and less linear body shape in cold climates. Bergmann explained this relationship in terms of mammalian physiology. Mammals maintain a constant body temperature by converting food into energy and releasing it to the environment through a variety of mechanisms, promoting heat loss. In order to maintain a constant body temperature, mammals must lose heat at an appropriate rate. Too much or too little heat loss can induce physiological stress and even death (e.g., freezing to death). Mammals have a number of adaptations to regulate heat loss, including fur or hair and profuse sweating in some species (e.g., humans).

Bergmann reasoned that body size and shape is one of many factors that affect heat loss. His basic arguments can be seen by thinking about the geometry of an organism, contrasting an organism's body volume (which reflects absolute heat production) with its surface area (which reflects heat loss). The size of an organism affects volume and surface

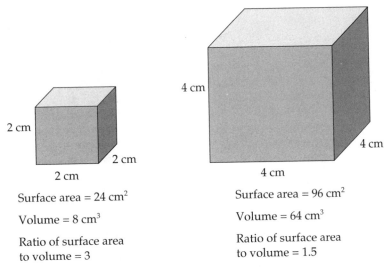

2 cm

2 cm

2 cm

2 cm

Surface area = 24 cm^2

Volume = 8 cm^3

Ratio of surface area
to volume = 3

4 cm

4 cm

4 cm

4 cm

Surface area = 96 cm^2

Volume = 64 cm^3

Ratio of surface area
to volume = 1.5

Figure 10.13. Geometric representation of Bergmann's rule relating body size to heat loss, showing how the larger cube has a smaller surface area/volume ratio and therefore loses heat less quickly than the smaller cube, which has a higher surface area/volume ratio.

area differently because volume increases as the cube of a measure, whereas surface area increases as a square of a measure. Figure 10.13 provides a simple example of this principle by comparing the volume and surface area of two cubes, one 2 cm in each dimension and the other 4 cm in each dimension. The volume of the smaller cube is $2^3 = 8$ cm^3 and the surface area is $6(2^2) = 24$ cm^2 (calculated as the surface area for each of the six sides and then multiplied by 6 to get the total surface area). The volume of the larger cube is $4^3 = 64$ cm^3 and the surface area is $6(4^2) = 96$ cm^2. The larger the cube, the greater the volume and the surface area will be. However, volume and surface area do not change at the same rate: volume increases as a cubic function and surface area as a square function. The more relevant quantity is the ratio of surface area to volume since this provides a rough index of an organism's ability to radiate the heat it produces. In this case, the smaller cube has a surface area/volume ratio of $24/8 = 3.0$ and the larger cube has a surface area/volume ratio of $96/64 = 1.5$. Even though the two cubes have identical shapes, the larger cube would radiate less heat relative to volume and would be more adaptive in cold climates, where heat loss would be disadvantageous. Based on this, Bergmann suggested that overall body size is greater in cold climates and smaller in hot climates.

Bergmann also showed that body shape affects heat loss. The reasoning behind this principle is shown in Figure 10.14, which contrasts the surface area/volume ratio in two shapes, a cube and a rectangular block, each with the same volume. The rectangular block has a higher surface area/volume ratio (4.25) than the cube (3.00), which means that it will lose heat more rapidly. Extending this principle to living organisms means that a slender physique loses more heat than a stocky physique, making it better adapted for hotter climates.

Bergmann's observations were extended by another zoologist, Joel Allen, to include heat loss from an organism's limbs. Allen predicted that organisms would have short, stocky limbs relative to body size in cold climates and long, slender limbs relative to body

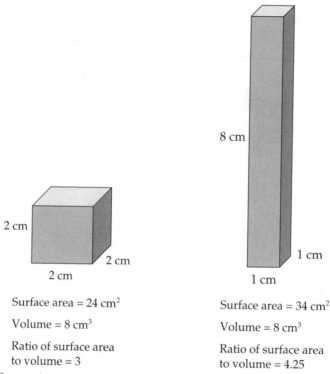

2 cm

2 cm

2 cm

Surface area = 24 cm²

Volume = 8 cm³

Ratio of surface area
to volume = 3

8 cm

1 cm

1 cm

Surface area = 34 cm²

Volume = 8 cm³

Ratio of surface area
to volume = 4.25

Figure 10.14. Geometric representation of Bergmann's rule relating body shape to heat loss, showing how the cube has a smaller surface area/volume ratio and therefore loses heat less quickly than the rectangular block, which has a higher surface area/volume ratio.

size in hot climates. Collectively, the observations made by Bergmann and Allen have become known as the **Bergmann-Allen rules**.

Body Size and Shape

A number of studies have confirmed that the Bergmann-Allen rules apply to human anthropometric variation. While the correlation between climate and anthropometric variation is not perfect and other factors also influence population averages, there is ample evidence that the Bergmann and Allen rules in general apply to human beings. Populations in cold climates tend, on average, to have greater body mass, stockier physique, and relatively shorter limbs. Populations in hot climates tend, on average, to have lower body mass, more linear physique, and relatively longer limbs. While variation exists within each population, the averages generally correlate with temperature (or latitude, which often serves as a proxy measure for temperature since temperature varies with distance from the equator). Ruff (1994) provides a history of these studies; several will be mentioned here.

Derek Roberts has compiled statistical evidence supporting the Bergmann-Allen predictions since the early 1950s. In his 1978 review, Roberts summarizes several significant correlations between anthropometrics and average annual temperature. As expected, average weight declines with increasing temperature, even after controlling for variation

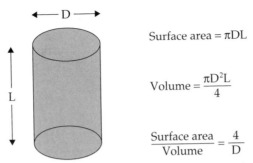

Figure 10.15. Ruff's modification of the Bergmann-Allen rules based on a cylinder. Simple geometry shows that the diameter of the cylinder controls the surface area/volume ratio and, therefore, the degree of heat loss. See Ruff (1994) for additional details.

in stature. Roberts also found a negative relationship between the cormic index (relative sitting height) and average annual temperature. A high cormic index shows a greater contribution of sitting height to total stature and, hence, a lower contribution of lower leg length. Thus, high values of the cormic index reflect relatively short legs and a low values reflect relatively long legs. The negative correlation between cormic indices and temperature shows that the lower limbs are relatively shorter in cold climates and relatively longer in hot climates, as expected from the Bergmann-Allen rules. The same pattern is also apparent in measurements of the upper limbs: arm length relative to stature increases with annual average temperature (Roberts 1978).

Christopher Ruff (1994) has investigated climate and anthropometric variation and demonstrated how the key factor affecting heat loss is the width of the body. Ruff starts with a cylinder as a rough approximation to body shape excluding the limbs. As shown in Figure 10.15, the surface area/volume ratio of a cylinder is dependent entirely on its width and not on its height. In order to test this hypothesis, Ruff collected data on bi-iliac breadth, a commonly used measure of body width (and one that is easily available on skeletal data as well), for a number of populations across the world. He used distance from the equator as a proxy measure of average temperature. Figure 10.16 shows the relationship of bi-iliac breadth to distance to the equator for 31 male and 25 female samples. As predicted from Ruff's model, bi-iliac breadth is larger as populations are located farther from the equator since an increase in bi-iliac breadth results in a lower surface area/volume ratio and thereby reduces heat loss. Populations in colder climates (farther away from the equator) have wider bodies than those living close to the equator. Ruff further found that this relationship holds even after statistically adjusting for differences in stature or weight. Also, he found significant correlations of both stature and weight with latitude, but these disappeared when controlling for variation in bi-iliac breadth. As predicted by Ruff, the critical variable was body width, and other anthropometric variables (e.g., stature, weight) showed significant correlations with climate only because of their covariation with bi-iliac breadth.

There is evidence suggesting that in recent generations the relationship between climate and body size and shape has been affected by nutritional differences. Katzmarzyk and Leonard (1998) looked at data on body size and shape and average annual temperature and noted that studies that collected data within the past 40 years show a somewhat different pattern from those that collected data in earlier times.

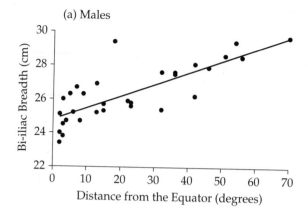

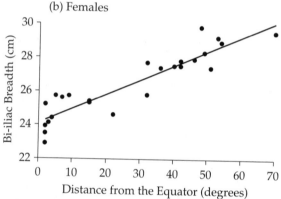

Figure 10.16. Relationship between body width (bi-iliac breadth) and distance from the equator for a global sample of (a) males and (b) females. The solid lines indicate the best-fitting linear regression equation. These data support Ruff's (1994) prediction that body width is smallest in hotter climates (near the equator) and larger in colder climates (farther from the equator). Referring to Figure 10.15, a wider body has a lower surface area/volume ratio and will lose heat less rapidly, which is adaptive in colder climates. (Data from Ruff 1994.)

Although the more recent samples still show the correlations expected under the Bergmann-Allen rules, the strength of these correlations has decreased somewhat. They interpret this change as reflecting recent changes in nutrition and health care, which have lessened the climatic influence to some extent. We are reminded that because complex traits reflect both genetic and environmental influences, the relative impact of these influences is likely to change over time, particularly in cases where there is considerable change in environment and culture.

Head Size and Shape

Several studies have found that worldwide variation in the size and shape of the human head is correlated with climatic variation. Beals (1972) and Beals et al. (1984) looked at variation in overall head measurements and CI and found that populations native to cold climates tend to have larger and relatively broader skulls than those in hot climates.

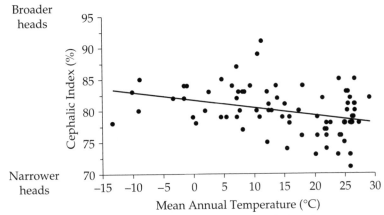

Figure 10.17. Relationship between cranial shape (cephalic index) and mean annual temperature for a global sample. The cephalic indices represent the average for males and females in each population. The *solid line* indicates the best-fitting linear regression equation. The cephalic index is greater (broader head) in colder climates because a round shape has a lower surface area/volume ratio, which minimizes heat loss (unpublished data courtesy of Kenneth Beals).

These results are expected by applying the Bergmann-Allen rules to the size and shape of the human head. Larger and relatively broader skulls lose less heat and are adaptive in cold climates; small and relatively narrower skulls lose more heat and are adaptive in hot climates (Fig. 10.17).

Nasal Size and Shape

A number of studies have found that nasal size and shape vary with climate. Several investigators have found a strong correlation between NI and temperature. Populations native to cold climates tend to have a low NI, reflecting relatively high and narrow nasal openings, while native populations in hot climates tend to have a high NI, reflecting relatively broad noses (Fig. 10.18). The geographic distribution of the NI is complicated by the fact that worldwide variation in the NI also reflects humidity levels, where relatively narrow noses are often associated with dry climates and relatively broad noses are associated with wet climates (Roberts 1978, Franciscus and Long 1991). Beals et al. (1984) devised an index of climate, taking both temperature and humidity into account, and found that worldwide variation in the NI correlated strongly. These correlations likely reflect the function of the internal nasal cavity, lined with mucous membranes, in warming and moistening inhaled air. High, narrow noses (with a low NI) maximize the amount of internal surface area relative to the volume of inspired air and are best adapted to cold, dry climates. Since the nose also functions to remove heat, broader noses (with a high NI) remove heat more quickly in a hot and humid environment (Franciscus and Long 1991).

The NI deals only with the shape of the nasal opening and does not take into account the size of the external nose. Carey and Steegmann (1981) investigated the relationship between the extent to which the nose protrudes from the face and a number of climatic variables. They found that there was a strong negative relationship between nasal protrusion and both temperature and humidity, where noses project more in cold and dry

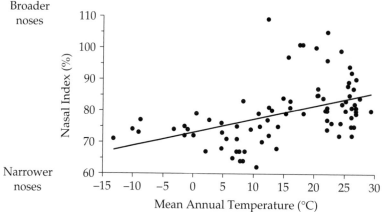

Figure 10.18. Relationship between nasal shape (nasal index) and mean annual temperature for a global sample. The nasal indices represent the average for males and females in each population. The *solid line* indicates the best-fitting linear regression equation. The nasal index is lower (narrower nose) in colder climates because this shape maximizes internal surface area, allowing a greater volume of air to be warmed and moistened (unpublished data courtesy of Kenneth Beals).

climates. While the actual warming and moistening of inspired air occurs in the internal nasal cavity, the projection of the external nose affects this function by causing more turbulence of inspired air, which increases contact of the air with the mucous membranes, thus allowing more efficient warming and humidification (Franciscus and Long 1991, Beall and Steegmann 2000).

We stress that all of the above studies of climatic correlations with anthropometrics are just that—correlations. As the old adage goes, correlation does not imply causation. Although the observed correlations are consistent with hypotheses of natural selection, direct correspondence between physical variation and differences in mortality and/or fertility remain to be demonstrated.

High–Altitude Adaptation

Some human populations live in **high-altitude environments**, defined as altitudes in excess of 2,500 m (8,250 feet) above sea level (Beall and Steegmann 2000). Harrison et al. (1988) estimate that there are roughly 25 million humans living in high-altitude environments. Over the past several decades, a number of studies have focused on anthropometric and physiological adaptations to high-altitude environments. Pioneered by Paul Baker and his students, the study of high-altitude adaptation has often focused on a contrast between the human biology of long-term natives and recent migrants to high altitudes. Such contrasts allow us to address the basic question of whether high-altitude adaptation is due to long-term natural selection, developmental acclimatization, or a combination of both.

A potential problem in living at high altitude is hypoxia, a relative reduction in available oxygen. Barometric pressure decreases as altitude increases. For example, the barometric pressure at 4,000 m is 60% that of sea level. While the oxygen content of the air remains the same at all altitudes (21%), reduced barometric pressure results in less available oxygen per inhalation (Beall and Steegmann 2000). At rest, hypoxia tends to occur in low-altitude natives above 3,000 m and somewhat lower (2,000 m) if the person is physically active and therefore using more oxygen (Frisancho 1993), although as always there is variation from one person to the next.

When a person moves from low to high altitude, there are several physiological changes that can occur to alleviate hypoxia, including an initial increase in respiration and increased red blood cell production. Other changes occur over longer periods of time, such as increased development of the right ventricle of the heart. Studies of human biology at high altitudes have also demonstrated changes in anthropometric measurements.

Chest Measures

Initial studies of high-altitude populations in Peru found two noticeable differences in the growth of high-altitude children compared to those from low altitudes. Chest dimensions of the high-altitude children were larger at all ages compared to those of children growing up at sea level (Fig. 10.19) (Frisancho and Baker 1970). This difference appears to be a direct response of the growth process to hypoxic stress at high altitudes. Similar patterns of chest growth have been observed in most high-altitude environments across the world (Frisancho 1993). Further, children migrating to high altitudes also show an increase in chest dimensions during growth, illustrating a developmental response to hypoxia. In general, the younger the age at migration, the greater the response (e.g., Greksa 1988).

Developmental acclimatization refers to the process by which a physiological change in the body due to an environmental stress occurs during the physical growth of an organism. Developmental changes of low-altitude native children to high altitudes suggest that increased chest dimensions represent developmental acclimatization rather than long-term natural selection (Frisancho 1993), although it is possible that the degree of developmental acclimatization itself might have a genetic basis (Beall and Steegmann 2000). Some studies have suggested populational differences, perhaps related to genetic factors, which affect the degree of developmental changes among migrants to high

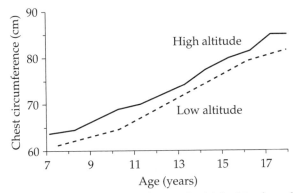

Figure 10.19. Distance curves for chest circumference for high-altitude and low-altitude populations in Peru. (High-altitude data from Table 3 and low-altitude data extrapolated from Figure 8, both in Frisancho and Baker 1970.)

altitudes. For example, Greksa (1988) found evidence of increased chest growth among Bolivians of European ancestry that had migrated to higher altitude, while Weitz et al. (2000) found no such effect among Han Chinese who had moved to high-altitude environments. Variation in chest dimensions is likely to involve both genetic and developmental factors, which interact in different ways in different populations (Greksa 1996).

Body Size

Studies of high-altitude populations have also compared different measures of body size (generally weight and stature) with low-altitude natives and immigrants from low altitude. In some (but not all) studies, it has been found that children who grow up at high altitude are typically shorter and lighter than their low-altitude counterparts. Frisancho (1993) suggests that smaller body size results from a slowing of the growth process, which functions as an adaptation to hypoxic stress and increased cold stress typical of many high-altitude environments. His hypothesis is that these stresses increase overall energy requirements, which, when coupled with the increased growth of oxygen transport systems, results in a decreased energy balance and less energy available for growth of overall body size. Thus, reduced growth in overall body size is a consequence of the specific energy demands in a high-altitude environment. This hypothesis has been critiqued, with others suggesting that reduced body size is not an adaptation to a reduced energy budget but instead reflects other stresses present at high altitude. For example, Leonard et al. (1990) suggest that three stresses are often present in high-altitude populations—hypoxia, cold, and nutritional stress—and the relative influence of each on overall body size is not always clear. In their study of Peruvian high-altitude natives, nutritional variation was found to be a primary factor, which in turn was affected by availability of land and income levels. Further research is needed to more clearly delineate the contributions of genetics and various environmental factors on the process of high-altitude adaptation.

CHAPTER SUMMARY

Anthropometry is the measurement of the human body, including measures of overall body size, lengths and breadths of different parts of the body, fat measures, and craniofacial measurements. Anthropometric data have been used for a wide range of studies of human variation, including analysis of growth and development, genetic and developmental adaptation, and population history. Most anthropometric measures show both genetic and environmental influences, often interacting in different ways dependent upon the specific population and environment.

The study of human growth focuses on changes in various anthropometric measures as a function of age, revealing characteristic patterns of human growth that are found throughout the species. Growth is often delayed because of inadequate nutrition and/or infectious disease. Various sociocultural influences, such as social class and family size, often result in differential patterns of nutrition and disease, causing sociocultural variations in growth patterns. Studies of secular change provide a further window on the relative roles of genetics and environment by comparing growth patterns over time within a population. Anthropometric measures often show a moderate amount of sexual dimorphism, usually showing males to be larger. An exception to this rule is body fat, which is frequently larger in females. Evolutionary explanations for sexual dimorphism have typically focused on variation in mating patterns and the presence of male competition for mates. Studies of sexual dimorphism in human populations suggest that nutritional factors are a more likely explanation.

A number of anthropometric measurements show strong patterns of environmental correspondence, suggesting past natural selection as an adaptation to a specific environment. Several anthropometric measures show significant correlations with average annual temperature, following the Bergmann-Allen rules, which predict smaller body mass, more linear body shape, and linear limbs in populations native to hotter climates and larger body mass and less linear body and limb shape in colder climates. Cranial shape is also associated with climate, apparently reflecting the same type of relationship between shape, temperature, and heat loss, the human head tending to be broader in colder climates and narrower in hotter climates. Nasal shape also shows a correlation with climate, reflecting the joint influence of temperature and humidity, with narrower noses found in climates that are cold and/or dry. This nose shape appears to be adaptive in allowing cold and dry air to be warmed and moistened. The external projection of the nose from the face may also affect adaptation. Studies of anthropometric variation have also noted differences between peoples native to high-altitude environments and those living at lower altitudes. High-altitude natives typically have enlarged chest size, which reflects adaptation to lower levels of oxygen at higher altitudes. Differences in chest measurements in many cases reflect developmental acclimatization, such that even migrants born at lower altitudes will develop larger chests during growth.

SUPPLEMENTAL RESOURCES

Beall C M and Steegmann A T Jr (2000) Human adaptation to climate: Temperature, ultraviolet radiation, and altitude. In: *Human Biology: An Evolutionary and Biocultural Perspective*, edited by Stinson S, Bogin B, Huss-Ashmore R, and O'Rourke D. New York: John Wiley & Sons, pp 163–224. An excellent review of studies on climatic and high-altitude adaptation.

Bogin B (1999) *Patterns of Human Growth*, second edition. Cambridge: Cambridge University Press. A solid review of the process of human physical growth and genetic, hormonal, and environmental influences on the growth process.

Frisancho A R (1993) *Human Adaptation and Accommodation*. Ann Arbor: University of Michigan Press. This text and the review paper listed above (Beall and Steegmann 2000) provide excellent surveys of the relationship of anthropometric variation to climate and adaptation to high-altitude environments.

Ruff C B (1994) Morphological adaptation to climate in modern and fossil hominids. *Yearbook of Physical Anthropology* 37:65–107.

11

Pigmentation

When we think about human variation or race, we probably think first about color. Of all human diversity, variation in pigmentation, particularly skin color and hair color, is the most visible and most commonly used to sort and classify human beings. Skin color (and to a lesser extent hair and eye color) has consistently been the primary trait used in racial classifications over the centuries. For example, Linnaeus's preliminary division of humanity into four varieties was based primarily on geography and skin color (Klass and Hellman 1971, Kennedy 1976). The reliance on skin color as a racial trait continued through the centuries in the racial classifications of Cuvier, Blumenbach, and others.

Anthropological studies of human skin color have shown that it is a continuous trait, where people across the world range from very dark to very light. When we put all these data together, there are no gaps in the distribution, making racial classification an arbitrary exercise when considering the total range of variation across the human species. However, skin color is not randomly distributed across the planet, and there is a strong geographic relationship. While, in general, Europeans are lighter than Africans, this geographic correspondence does not replicate perfectly geographic definitions of race. Instead, the relevant correlation is with the distance from the equator; as shown below, human skin color is darkest at or near the equator and becomes increasingly lighter with greater distance from the equator, north or south. This strong correlation with latitude suggests past natural selection in response to environmental factors that also vary by latitude, although there is continued debate over the relative impact of such factors. Compared to genetic and DNA markers and craniometric traits, variation among geographic regions is greater for skin color (Relethford 2002), a pattern consistent with natural selection that varies across environments.

This chapter reviews what is known about human pigmentation, focusing primarily on skin color but also considering variation in hair and eye color.

SKIN COLOR

What factors account for the enormous range of variation in human skin color? This section examines briefly the underlying physiology of skin color and its relationship to genetic variation and environmental factors.

The Biology of Skin Color

Physiology

The brownish black **melanin** pigment affects the majority of variation in human skin color. Melanin is produced by **melanocytes**, which are cells in the epidermis (the outermost layer of the skin). The melanin pigment is synthesized by the **melanosomes**, which are organelles inside the melanocytes. Packets of melanin are then distributed to the epidermal skin cells. While early studies suggested that all human populations have the same number of melanocytes, a recent review suggests that individuals with light skin have, on average, fewer melanocytes than those with dark skin (Beall and Steegmann 2000). Differences in degree of pigmentation (lighter versus darker) also depend on how many cells produce melanin, how they cluster together, the rate of melanin formation, and the rate of melanin transfer to the epidermal cells (Szabo 1967, Robins 1991).

Although most of the variation in human skin color is due to differences in melanin distribution, some color differences are also due to three other pigments: carotene, oxyhemoglobin, and reduced hemoglobin. Carotene is a yellowish pigment that normally contributes very little to variation in skin color except in light-skinned individuals suffering from a disease known as carotenemia, which results in a yellowish hue to the skin (Robins 1991). Hemoglobin pigments have a more noticeable effect, particularly in light-skinned people. Hemoglobin gives oxygenated blood its reddish color. Since light-skinned people have less melanin, the reddish color shows through the skin, giving a pinkish hue.

Measurement of Skin Color

When we look at someone's skin color and attempt to classify it into discrete categories such as "white" and "black," the result is necessarily crude and subjective, and many individuals do not fit neatly into one category or the other. Adding more categories, such as "brown," does not help because it becomes increasingly arbitrary where to draw the line between one color and the next. A more objective, standardized method of assessing skin color is clearly needed.

Early attempts to quantify human skin color also used crude standards, most frequently matching of the subject's color with a series of paper or tile standards (Byard 1981). Such measurements are not only inherently subjective but also affected by the source of external lighting (much like trying to match paint samples by eye). Additionally, the use of such standards imposes discrete groupings on a trait that varies continuously.

A more precise means of quantifying skin color variation based on **reflectance spectrophotometry** was developed by Edwards and Duntley (1939). Reflectance spectrophotometers provide an accurate means of objectively measuring skin color that is not prone to *interobserver error*, the error introduced when different observers measure different groups of people (Lees et al. 1978). Light in the visible spectrum at different wavelengths, ranging from blue to red, is bounced off of human skin. Lighter skin will reflect more light back than darker skin, thus allowing the percentage of reflected light to be used as an objective measure of skin color. By varying the wavelength of visible light, different measures of skin reflectance can be produced for different wavelengths. This method was cumbersome and expensive, requiring extensive laboratory equipment that is not feasible to transport around the world to populations of anthropological interest. By the 1950s, the necessary technology had developed in the form of portable abridged reflectance spectrophotometers. These machines, often used in the textile industry and other places where color matching is needed, are portable enough to transport to remote sites. The machines are referred to as "abridged" because they do not sample the entire visible spectrum but only a small number of wavelengths. Weiner (1951) and Lasker (1954) were the first to use these machines in anthropological studies.

Table 11.1 Wavelengths of filters used in E.E.L. and Photovolt reflectometers

Reflectometer	Filter	Wavelength (nm)
E.E.L.	601	425
	602	465
	603	485
	604	515
	605	545
	606	575
	607	595
	608	655
	609	685
Photovolt	Blue	420
	Triblue	450
	Green	525
	Trigreen	550
	Triamber	600
	Red	670

From Byard 1981.

These machines shine light through a glass filter of a given color, which then allows only that color to pass through to the subject's skin. The light is reflected back from the skin (with lighter skin reflecting more) and measured by a photocell. Two machines have been used most often in anthropological studies. The most commonly used machine is known as the **E.E.L.** machine, where the initials stand for the manufacturer, Evans Electroselenium Limited (today known as "D.S.L." for the company Diffusion Systems Limited). The E.E.L. machine uses nine colored filters, known as filters 601–609, to sample the visible spectrum. The filters range from blue to green to red. At one end of the spectrum, the number 601 filter only allows light at a frequency of 425 nm to pass, producing a deep blue light. At the other end of the spectrum the number 609 filter samples the visible spectrum at 685 nm (a red light). Another machine, the **Photovolt** reflectometer, uses six filters: two each for the blue, green, and red portions of the visible spectrum. Table 11.1 lists the filters and wavelengths for both machines. While the majority of anthropological studies have used the E.E.L. machine, some, particularly a number of studies on New World populations, have used the Photovolt machine. Unfortunately, the wavelengths sampled are not the same, and the two machines also differ in other ways, meaning that the results cannot be directly compared. Several formulae have been developed to convert readings from one machine to the other (Garrard et al. 1967, Lees and Byard 1978, Lees et al. 1979). Recently, two new handheld reflectometers have been described for their anthropological utility (Shriver and Parra 2000), although conversion formulae have not yet been developed to allow direct comparison with the older E.E.L. and Photovolt machines.

Skin reflectance is most often measured at the inner surface of the upper arm, slightly above the elbow (Fig. 11.1). This measurement site is easy to reach, typically socially acceptable to reveal, and a good index of skin color that has not been affected by environmental exposure, thus providing a measure of innate skin color. Some studies have also taken measurements of skin reflectance on the forehead and the back of the hand in order to also measure skin color at parts of the body that are likely to tan. The rationale

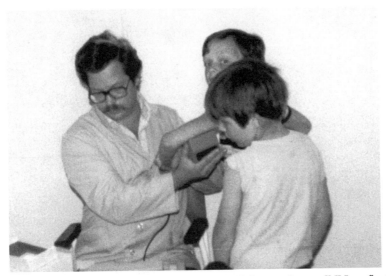

Figure 11.1. Skin color being measured by Dr. Francis C. Lees using an E.E.L. reflectance spectrophotometer. Skin color is quantified as the percentage of light reflected off of the skin at different wavelengths. (Photo by J. H. R.)

was that by comparing unexposed and exposed parts to the body, one could get an idea of a person's tanning capacity. This method has been dropped for the most part since Post et al. (1977) discovered that newborn children, who had not been previously exposed to sunlight, had differences in skin reflectance at different body sites. The foreheads of these children were 10%–12% darker than at the upper inner arm, showing an innate difference in pigmentation across the body such that any further comparisons would not necessarily indicate tanning potential.

The measurement of skin reflectance at different wavelengths allows a graphic illustration of skin color differences between samples. Figure 11.2 presents the average skin reflectance values for the nine E.E.L. wavelengths for four different populations. This graph reveals several features. First, the light-skinned population from the Netherlands shows greater skin reflectance at all wavelengths, followed by the sample from India, the Bantu, and the Fali Tinguelin population from Africa. Second, the differences between populations are most apparent in the red part of the visible spectrum; indeed, the E.E.L. filter 609 (685 nm) is often used as the best single measurement of skin color differences. Third, while reflectance generally increases with wavelength, the sample from the Netherlands shows a reduction in reflectance at 545 nm (filter 605). This reduction, known as the "hemoglobin dip," is due to the fact that hemoglobin has an absorption frequency in the green part of the visible spectrum, near filter 605. At this wavelength, the green light is absorbed by hemoglobin and therefore does not reflect back as much, making the skin reflectance lower than at other, nearby wavelengths. This dip is apparent only in light-skinned individuals, whose relative lack of melanin does not block the green light. Thus, readings at this wavelength are picking up on redness of the skin in light-skinned people due to oxygenated hemoglobin. Individuals with darker skin do not show this effect. In general, the readings at different wavelengths are highly correlated.

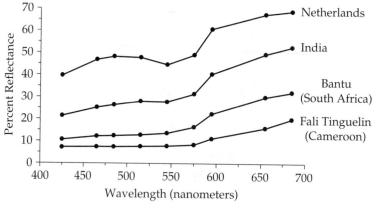

Figure 11.2. Average skin reflectance for four human populations of differing degrees of pigmentation. All four samples are males, measured at the nine wavelengths of the E.E.L. machine (see Table 11.1) at the upper inner arm. Note that the lighter the skin, the higher percentage of light reflected back at each wavelength. (Data From Rigters-Aris 1973a,b; Tiwari 1963; Wassermann and Heyl 1968.)

Korey (1980) presents a calculus-based method for deriving average skin reflectance that can suffice in many analyses of skin color variation.

Differences in skin reflectance between populations that are widely different in pigmentation reflect underlying ancestry. The children born to one light-skinned parent and one dark-skinned parent will tend to have, on average, medium-pigmented skin. This effect is shown in Figure 11.3, based on a study of English children born to light-skinned English women and dark-skinned African men (Harrison and Owen 1964). The skin

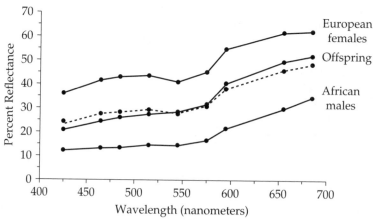

Figure 11.3. Average skin reflectance for a group of offspring resulting from the mating of European females and African males residing in Liverpool. Skin reflectance was measured at the nine wavelengths of the E.E.L. machine at the upper inner arm. The dashed line indicates the expected average between adult male and female skin color. The observed curve for the offspring (solid line) is very similar to this mid-parent average. (Data From Harrison and Owen 1964.)

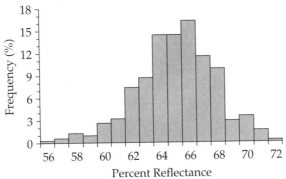

Figure 11.4. Histogram of skin reflectance as measured at E.E.L. filter 609 (685 nm) for 312 Irish male children. (Unpublished data as described in Relethford et al. 1985.)

reflectance curve of the offspring is very close to the expected curve based on a 50:50 contribution from each population. Methods have been developed using this observation to estimate degree of ancestry in admixed populations (Lees and Relethford 1978, Korey 1980, Relethford et al. 1983). Of course, these curves only indicate the average of the sample; the children do not all have the same skin color, and some are lighter and some are darker (Harrison and Owen 1964).

Within any population there is considerable variation in skin color, as shown in the example of Irish male children in Figure 11.4. Contrary to superficial visual inspection, not all people in any population have the same skin color; some are lighter, and some are darker. Some of this variation is genetic in nature, and some is environmental. The fact that some Irish boys are lighter than average and some are darker than average refutes the common misconception of human variation that "they all look alike." This is clearly not the case. Also, note that the distribution in Figure 11.4 is a normal ("bell-shaped") one, as expected in a complex trait that reflects the summation of a number of different influences. In the case of skin color, these influences are due to the action of different genes and the environment, the subject of the next section.

The Genetics of Skin Color

A number of studies have shown that skin reflectance has a moderate to high heritability (e.g., Post and Rao 1977, Clark et al. 1981, Frisancho et al. 1981), typically ranging from about 0.5 to 0.8. As always, differences in choice of population and method of estimating heritability produce a range of estimates. The higher heritabilities were derived from twin studies, which tend to inflate heritability estimates (Williams-Blangero and Blangero 1992). The most elaborate analysis to date simultaneously analyzed skin reflectance in Nepalese populations at three E.E.L. filters (601, 605, 609) in a variety of pedigrees of different sizes and found a heritability of 0.66 for filter 609 and somewhat lower heritabilities for the other filters (Williams-Blangero and Blangero 1992). These studies show that skin color variation is strongly affected by genetic variation in human populations.

Early studies of the quantitative genetics of skin color suggested that it was best explained by a polygenic model with three or four loci, each having approximately the same effect on skin color (e.g., Harrison and Owen 1964, Harrison et al. 1967). Later work by Byard (1981) and Byard and Lees (1981) showed problems with the underlying method, and Williams-Blangero and Blangero (1992) suggested that there are likely more loci affecting skin color variation. More recent work has suggested that a major gene model might be more appropriate than a polygenic equal and additive effects model. Recent work suggests that there may be three major genes that influence human skin color through their effects on melanin production (Sturm et al. 1998, Blangero et al. 1999). The greatest success to date in unraveling the genetics of human skin pigmentation is the identification that the melanocortin 1 receptor gene (*MC1R*) located on chromosome 16 is associated with variation in skin color (Rees and Flanagan 1999, Harding et al. 2000).

Variation in Human Skin Color

Human skin color varies both within and between populations. Variation within populations is affected somewhat by sex and age, both of which are related to hormonal influences. Variation between populations is strongly correlated with latitude, suggesting differing patterns of natural selection in the past.

Sex Differences

Among adults, males are generally darker than females, possibly reflecting differences in melanin (Edwards and Duntley 1939) and sex hormones (Byard 1981). While most studies have found adult males to be darker, some have found the reverse or no effect, suggesting that in some cases there are cultural differences in exposure to sunlight, even when skin reflectance is measured at the upper inner arm site (Byard 1981). In general, the differences between adult males and females are relatively small. The situation is a bit more complicated in children and adolescents, where the darker sex varies by both wavelength and the specific age of measurement (Relethford et al. 1985). The general tendency for women to be lighter than men (usually around 3%–4%) may reflect a need in women for more calcium, particularly during pregnancy and breast-feeding. Jablonski and Chaplin (2002) suggest that women are lighter to allow slightly more ultraviolet (UV) radiation to penetrate the skin and increase production of vitamin D.

Aging

In general, females tend to show a noticeable darkening during adolescence and adulthood. While some studies of males show some aging effect, others do not; and age-related trends are usually more noticeable in females. Kalla (1974) and Byard (1981) suggest that age-related changes in skin color, where they exist, are likely to reflect the effects of changing hormone levels on pigmentation. Relethford et al. (1985) note that Irish female children were significantly darker than Irish male children for filters in the green range of the visible spectrum and suggest that this difference relates to higher subcutaneous blood flow in light-skinned females. It is difficult, however, to generalize too much since both Byard (1981) and Williams-Blangero and Blangero (1991) note that the relationship of skin color and age can vary considerably even among closely related populations.

The Geographic Distribution of Skin Color

Throughout recorded history, people have observed a definite geographic pattern to human skin color. These casual observations have since been confirmed by detailed analysis. Human skin color varies with latitude such that indigenous populations at or near the equator tend to be dark-skinned and populations that live farther away from

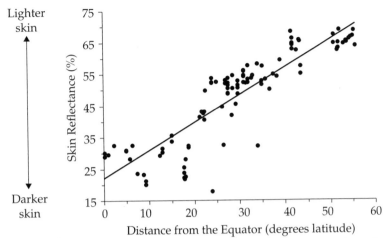

Figure 11.5. Relationship of skin reflectance (E.E.L. filter 609) and distance from the equator for 102 male samples across the Old World. (Data from Relethford 1997.)

the equator are lighter. These studies show significant positive correlations of skin reflectance with distance from the equator (Roberts and Kahlon 1976, Tasa et al. 1985, Relethford 1997, Jablonski and Chaplin 2000). Figure 11.5 provides an example of this relationship based on 102 male samples using E.E.L. filter 609, often taken to be the best single measure of melanin variation among human populations. While the relationship is not a perfect one, it clearly describes the average geographic distribution of human skin color, with latitude accounting for 77% of the total variation in skin color.

What reason could there be for this strong geographic distribution? Correlations of this magnitude suggest natural selection and some sort of environmental factor that also varies by latitude. Most often, the correlation between skin color and latitude is interpreted as a reflection of the geographic distribution of UV radiation, which is strongest at the equator and diminishes with increasing distance away from the equator. In this sense, latitude is considered a proxy for the average intensity of UV radiation. Jablonski and Chaplin (2000) note that skin reflectance is also strongly correlated with direct estimates of UV as obtained from spectrophotometric analysis of ozone mapping from space satellites.

The relationship of skin color, UV radiation, and latitude turns out to be more complicated. UV radiation tends to be higher in the Southern Hemisphere (below the equator) than at the same latitude in the Northern Hemisphere because of a variety of factors, including hemispheric differences in the ozone layer and the eccentricity of the earth's orbit, where the Southern Hemisphere is closer to the sun during its summer months (McKenzie and Elwood 1990, Relethford 1997, Relethford and McKenzie 1998). Given this hemispheric difference in UV radiation and the suggestion that worldwide variation in skin color is linked to UV radiation, we should expect to see a hemispheric difference in skin color at equivalent latitudes. Figure 11.6 presents the same graph as in Figure 11.5 but with each data point labeled as belonging to a Northern Hemisphere population or a Southern Hemisphere population. The Southern Hemisphere populations clearly tend to have lower average skin reflectance, and therefore darker skin color, than Northern Hemisphere populations at the same distance from the equator, which is

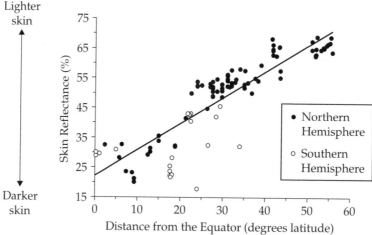

Figure 11.6. Relationship of skin reflectance (E.E.L. filter 609) and distance from the equator for 102 male samples across the Old World showing differences between populations located in the Southern Hemisphere (*open circles*) and the Northern Hemisphere (*filled circles*). (Data from Relethford 1997.)

expected if UV radiation levels are higher in the Southern Hemisphere. Relethford (1997) extended this observation by developing a mathematical model that allows the relationship between skin reflectance and distance from the equator to be different in each hemisphere. As shown in Figure 11.7, skin reflectance is lowest (darkest) near the equator and

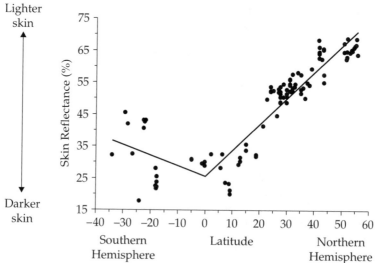

Figure 11.7. Relationship of skin reflectance (E.E.L. filter 609) and distance from the equator for 102 male samples across the Old World using a nonlinear model that allows for different slopes in the Southern and Northern Hemispheres. The change in skin reflectance with distance from the equator is less in the Southern Hemisphere than in the Northern Hemisphere. (Data from Relethford 1997.)

increases (lighter) farther away from the equator in both hemispheres but not at the same rate. Skin reflectance increases 8.2% for every 10 degrees of latitude in the Northern Hemisphere, but the rate of increase is less in the Southern Hemisphere, roughly 3.3% for every 10 degrees of latitude, as expected given higher UV radiation levels in the Southern Hemisphere.

THE EVOLUTION OF HUMAN SKIN COLOR

What events took place in the evolution of our species that led to such dramatic differences in skin color across the world? Comparing humans at different latitudes and with other primates suggests that there have been several major shifts in skin pigmentation during the past 6 million years or so of human evolution. Most of the skin of living primates is unpigmented except for exposed areas not covered with hair, suggesting a primitive feature in primates from which our ancestors changed. The best model for our own ape-like ancestors is probably the condition found in living chimpanzees, which have lightly pigmented skin covered with dark hair. As hominids adapted to the tropical savanna, the density of body hair most likely decreased and the density of sweat glands increased, both leading to a relatively naked skin best adapted for heat stress. As such, the hair no longer provided protection against the potentially harmful effects of UV radiation, and skin color most likely darkened as a selective response (Jablonski and Chaplin 2000).

Until almost 2 million years ago, all hominids lived in Africa and presumably had dark skin. As early humans expanded out of Africa into parts of Asia and Europe, they moved into northern latitudes with less UV radiation. The present-day distribution of human skin color suggests that, over time, populations living farther away from the equator became increasingly lighter. This distribution raises two basic questions: (1) Why specifically is dark skin more adaptive at or near the equator? (2) Why did skin color become lighter farther away from the equator? Can the same factor be responsible for both, or did different selective factors operate at different latitudes?

Selection for Dark Skin

Several models have been proposed to explain the evolution of dark skin in human populations at or near the equator, all linked to the greater amount of UV radiation found in that part of the world. The common element of these models is that dark skin acts as a natural "built-in" sunscreen (Jablonski and Chaplin 2002, 75).

Sunburn

UV radiation can cause severe damage to unprotected skin. The most immediate effect is sunburn. While we have a tendency to think of sunburn as a temporary, although painful, nuisance, severe sunburn can have major consequences. Severe sunburn can damage the sweat glands, which in turn can have a negative impact on a person's ability to handle heat stress (Byard 1981, Jablonski and Chaplin 2000). Given the fact that hominids first evolved in a tropical environment, the problem of heat stress may have been particularly acute if, as we expect, the transition to savanna life included a reduction in hair density and the increased production of sweat glands. Under these conditions, light-skinned individuals prone to sunburn and damaged sweat glands would be at a disadvantage and would be selected against. According to this scenario, dark skin appeared in the first hominids as an adaptive response to protect against UV radiation damage to sweat glands. Another consequence of severe sunburn is an increase in infection to damaged skin cells (Byard 1981, Rees and Flanagan 1999, Jablonski and Chaplin 2000), which could also influence relative survival and render light-skinned individuals at a disadvantage. It seems reasonable to suggest that darker skin evolved near the equator at least partially as a means of protecting against UV radiation damage.

Skin Cancer

Damage to the skin due to UV radiation over extended periods of time can lead to skin cancer, brought about by UV radiation damage to a gene that normally inhibits cancerous growths. The risk of developing skin cancer varies among human populations by as much as 100-fold, a difference that appears to be strongly related to skin color (Rees and Flanagan 1999). There are two lines of evidence that link skin color, UV radiation, and skin cancer rates. First, there is a strong correlation of skin cancer incidence and mortality with latitude, with both decreasing with greater distance from the equator. Second, skin cancer rates are lower in darker-skinned populations than lighter-skinned populations at the same latitude. In Texas, for example, the incidence of nonmelanoma skin cancer is 5 per 100,000 among nonwhite residents compared to 113 per 100,000 among white residents (Beall and Steegmann 2000).

Given these data, one suggestion for the evolution of dark skin in humans living at or near the equator is that light-skinned individuals would be selected against due to increased susceptibility to skin cancer. It has been argued, however, that skin cancer is unlikely to have been a significant factor in the evolution of skin color because cancer tends to occur later in life, often after the reproductive years (Jablonski and Chaplin 2000). Since natural selection operates through differential survival and reproduction, any disease influencing survival after the reproductive years would not be subject to natural selection. Others argue that while the incidence of skin cancer among younger individuals may be lower, it is still sufficient for the action of natural selection. Robins (1991) compared skin cancer rates among albinos and nonalbinos in Africa. Since albinos lack any pigment, they are more subject to skin cancer. All of the albinos showed skin cancer or premalignant lesions by the age of 20, suggesting early onset, at least in this sample. Skin cancer might have been a factor in the evolution of dark skin among our ancestors, although it was probably not the only factor. Rees and Flanagan (1999) suggest that although skin cancer was a likely factor in the evolution of dark skin in the tropics, risks related to severe sunburn were probably more important.

Vitamin D Toxicity

Vitamin D is an essential nutrient for humans. With the exception of fatty fish, such as salmon, most foods are generally low in vitamin D. In the United States, we are used to associating vitamin D with milk, but in this case the vitamin had been added to the milk during processing; cow's milk is generally low in vitamin D. Most vitamin D in humans is obtained from a biochemical synthesis in the skin brought about by the action of UV-B (wavelengths between 280 and 320 nm), which converts 7-dehydrocholesterol into a substance known as **previtamin D**, which then converts to vitamin D (Robins 1991).

Too little vitamin D can cause a variety of problems, as discussed below. In terms of the adaptive value of dark skin under high UV radiation levels, Loomis (1967) suggest that dark skin evolved to prevent toxic effects of producing too much vitamin D. High dosages of vitamin D are sometimes administered to people suffering from vitamin D deficiency, and in some cases an overdose can result in toxic levels of vitamin D, which can lead to excess calcium absorption, kidney failure, and death (Robins 1991). Loomis (1967) argues that dark skin evolved to protect against the possibility of vitamin D toxicity since dark skin prevents up to 95% of the UV radiation from reaching the deep layers of the skin, where vitamin D is synthesized.

While logical, this hypothesis must be rejected given measurements that show that humans do not synthesize toxic levels of vitamin D. Holick et al. (1981) examined the effect of artificial UV light on skin samples from light- and dark-skinned humans in order to determine the rate of previtamin D synthesis. They found that synthesis of previtamin D increases only up to a certain level, after which it remains at a plateau. After this point, 7-dehydrocholesterol converts into two inert substances rather than previtamin D. Therefore, vitamin D toxicity due to UV radiation is not possible, and the hypothesis is rejected.

Folate Photolysis

Several nutrients are subject to **photolysis**, chemical decomposition brought about by visible light. One such nutrient, **folate** (a compound belonging to the vitamin B complex group), has been implicated in the evolution of dark skin. Branda and Eaton (1978) found that folate concentrations in human blood plasma decreased significantly after brief exposure to UV light. In order to show that the same effect occurs in living humans, they further compared serum folate levels in 10 light-skinned patients who were undergoing therapeutic exposure to UV radiation with 64 healthy light-skinned people. The patients undergoing UV exposure had significantly lower serum folate levels than the controls, suggesting photolysis.

Branda and Eaton (1978) suggest that light-skinned individuals are likely to show folate deficiency in environments with high levels of UV radiation. Jablonski and Chaplin (2000, 2002) note that folate deficiency is clearly linked to differential survival and fertility. There is a link between folate deficiency and the incidence of neural tube defects, congenital malformations that affect the development of the nervous system. Two of these malformations, anencephalus and spina bifida, are more common in light-skinned populations and account for a notable percentage of infant deaths. It is also worth noting that Africans and African Americans, with darker skin, show lower levels of folate deficiency and neural tube defects. Folic acid prevents 70% of neural tube defects, and it seems reasonable to assume that low levels of folate can predispose individuals to neural tube defects during development of the embryo. Jablonski and Chaplin (2000, 2002) also note that there is a link between folate deficiency and differential fertility. Several studies have shown that induced folate deficiency leads to problems in sperm production and increased incidence of male infertility.

According to Jablonski and Chaplin (2000, 2002), folate photolysis was the major factor in the evolution of dark skin among early hominids in equatorial Africa. Those with lighter skin would be more likely to suffer from the photodestruction of folate, thus leading to increased mortality due to increased incidence of neural tube defects and decreased fertility due to the increased incidence of male infertility. Light-skinned individuals would thus be selected against because of differences in both survival and reproduction, while those with darker skin would be selected for because of the protective effect of dark skin on folate photolysis.

The folate photolysis hypothesis is supported by epidemiological and physiological data, and it seems reasonable to assume that folate deficiency was the major force shaping the evolution of dark skin in equatorial hominids. This does not mean, however, that other factors, such as sunburn and skin cancer, did not also influence overall fitness. Dark skin protects against UV radiation damage, which in turn can include both nutrient photolysis and damage to the skin and sweat glands.

Selection for Light Skin

The available evidence suggests a protective role of dark skin against the damaging effects of UV radiation in populations at or near the equator. The current distribution of human skin color shows that human populations that moved away from the equator, particularly those that moved far to the north, evolved lighter skin. The potential damage from UV radiation decreases as one moves away from the equator, but this means only that lighter skin *could* evolve; it does not explain why it did. To fully explain the current distribution of human skin color, we need a way to explain why light skin is found at distances farther from the equator. We can rule out random genetic drift since it is extremely unlikely that random variation in skin color in the absence of selection would lead to the observed high correlation of skin color and latitude.

Vitamin D Deficiency

The most widely accepted model for the evolution of light skin focuses on the role of insufficient UV radiation. UV radiation declines with increasing distance from the equator. This change means that the potential damage due to UV radiation also decreases, as shown by studies of latitudinal differences in skin cancer rates, among other findings reviewed above. For light skin to be favored in such environments, however, it is necessary to consider problems in survival and/or reproduction that would result from receiving too little UV radiation. Specifically, the idea here is that darker skin would be selected against (and hence lighter skin selected for) because reduced UV radiation leads to a deficiency in the production of vitamin D.

Loomis (1967) noted that nontropical regions receive low amounts of UV radiation during the year and suggested that individuals with dark skin would be at a selective disadvantage because they would suffer from a deficiency in vitamin D synthesis. The main effect of vitamin D deficiency is the childhood disease rickets, the defective growth of bone tissue. Severe rickets can obviously lead to reduced fitness, particularly given the active hunting and gathering life of our ancestors. Further, rickets can directly impact on differences in reproduction when pelvic bones are deformed during growth, thus leading to later complications and possible death during childbirth. The area of the pelvic inlet in women with rickets is only 56% that of healthy women, suggesting a greater incidence of difficulties during childbirth. Studies in the 1950s, performed prior to the increased use of vitamin supplements, showed that African American women had a much higher incidence of a deformed pelvis (15%) compared to European American women (2%) (Frisancho 1993). This suggests, but does not prove, a relationship with skin color as we must consider other genetic and/or environmental differences between the two ethnic groups. Nonetheless, this ethnic difference is consistent with the vitamin D deficiency hypothesis.

Dark skin reduces the amount of vitamin D synthesis, and it takes longer for previtamin D synthesis in dark skin. Holick et al. (1981) measured the rate of previtamin D synthesis in skin samples exposed to the same amount of UV radiation as found at the equator and found that it took 30–45 minutes for lightly pigmented skin to reach a maximal level and 3–3.5 hours for dark skin to reach a maximal level. These findings support the vitamin D deficiency hypothesis, which argues that dark-skinned individuals would be at a selective disadvantage outside of the tropics.

Not everyone agrees. Robins (1991) argues against the vitamin D deficiency hypothesis. He notes that rickets is a disease that came about in more recent times, with the spread of civilization and cities, and tends to be found in polluted cities but to be rare in rural areas. Robins (1991) also notes that vitamin D can be stored in body tissues, allowing people to synthesize enough vitamin D during part of the year (summer) to last through times of lower UV radiation (winter). Also, while dark-skinned individuals are less effective at synthesizing vitamin D, they are still capable of producing enough even at higher latitudes to allow survival. Robins's main argument here is that differences in vitamin D synthesis between light- and dark-skinned people are relative, and the reduced rate of synthesis in the latter would not pose a selective disadvantage. He cites several studies suggesting that, outside of polluted urban areas, dark-skinned people would be able to produce enough vitamin D. As further evidence, Robins also points out that there is little evidence of rickets in skeletons from preindustrial Europe, raising the question of whether rickets was present in sufficient degree to constitute a selective factor. Others have contested Robins's arguments, noting that the evidence for high levels of vitamin D storage is lacking (Frisancho 1993). Beall and Steegmann (2000) note that skeletal analysis of rickets focuses on the most severe manifestations of low vitamin D levels, and lower values might also have biomedical significance. Finally, Jablonski and Chaplin (2000) analyzed UV radiation levels across the planet to estimate the expected levels of previtamin D at different latitudes. Their analyses support the vitamin D deficiency hypothesis by showing that medium and dark skin would be at a disadvantage at higher latitudes, being less likely to be able to synthesize adequate levels of vitamin D.

Cold Injury

While many anthropologists have accepted the vitamin D deficiency hypothesis as the most likely explanation for the evolution of light skin, others have suggested that variation in temperature might also have an impact (Post et al. 1975). The cold injury hypothesis proposes that heavily pigmented skin is more susceptible to damage, such as frostbite, and would therefore be selected against in cold environments. Since average annual temperature tends to vary with latitude (hottest near the equator), this model further suggests that the observed correlation of skin color and latitude actually reflects, to some extent, an underlying correlation with temperature (recall from Chapter 10 that latitude is often used as a proxy for temperature in studies of human adaptation).

There is some statistical evidence for the influence of temperature on skin color variation. Roberts and Kahlon (1976) found that annual temperature was significantly related to skin reflectance even after variation in latitude was taken into account. More direct support for the cold injury hypothesis comes from medical records of soldiers, particularly during the Korean War, where winter conditions were severe. For example, Ethiopian troops, primarily dark-skinned, had the highest rates of frostbite of all nationalities, with rates almost three times that of the U.S. forces. This comparison is crude because the U.S. forces included both white and black soldiers. However, comparisons within the U.S. forces showed that frostbite among blacks was between four and six times that among whites, even after controlling for other factors (Post et al. 1975).

Physiological studies also suggest a relationship between pigmentation and susceptibility to cold injury. Post et al. (1975) cite a number of examples, including studies of piebald guinea pigs (having black and white–spotted skin). They froze black and white areas of the skin of anesthetized guinea pigs and found that cold damage was more severe in the darker skin. Cold injury was always more severe in the pigmented skin of the guinea pigs.

While some scientists have accepted the plausibility of the cold injury hypothesis (e.g., Robins 1991), others have not. Beall and Steegmann (2000) argue that black–white differences in cold injury might instead be due to known differences in the vascular response to cold and have nothing directly to do with variation in pigmentation. In other words, the cold injury hypothesis may be based on a spurious correlation. They also suggest that the guinea pig experiments are not directly relevant to human cold injury, primarily because the temperatures used on the guinea pigs were more extreme than typically found in likely human environments. Instead, they make the case, as do Jablonski and Chaplin (2000), that light skin evolved in response to vitamin D deficiency. Of course, these two models are not mutually exclusive. It is possible that both vitamin D deficiency and cold injury contributed to the evolution of light skin farther away from the equator. Although the evidence supporting the vitamin D hypothesis is strong, this does not preclude the possibility that differential cold injury was also a contributing factor. As is often the case in evolution, one's overall fitness reflects the net effect of phenotypic response in a given environment.

Sexual Selection

Another model also considers the evolution of light skin to be a consequence of past natural selection, but instead of relating light skin to environmental variation, this model rests upon the concept of sexual selection. First developed by Charles Darwin, sexual selection occurs when there is competition for mates or where there are preferences for mating with members of the opposite sex that have certain physical characteristics (Aoki 2002). A classic example of the latter is the preference of females in a number of species of birds and fish to choose brightly colored males as mates. Given this preference, there would be an advantage to having bright coloration, and any genes underlying this coloration would be selected for over time.

The sexual selection hypothesis as applied to human skin color rests on the assumption that human males prefer females with lighter skin color. Aoki (2002) suggests that if this assumption is correct, then the present-day geographic distribution of human skin color could be explained by a balance between natural selection and sexual selection. His model suggests that natural selection tends to favor darker skin but that (as discussed above) the intensity of this selection diminishes with increasing distance from the equator. Consequently, any preference for lighter-skinned mates would have its greatest impact when the counteracting natural selection for dark skin is lowest; that is, farther away from the equator. Aoki's model represents a balance between selection for dark skin and (sexual) selection for light skin.

What is the link between a preference for light skin and selection? Jones (1996) suggests that given a tendency for female skin color to darken with age, lighter skin might have acted as a signal for greater fecundity. However, this connection has not been demonstrated, and as noted earlier, the relationship between skin color and age varies even among closely related groups (Williams-Blangero and Blangero 1991).

Although the sexual selection model seems reasonable, it rests upon the critical assumption that there was widespread preference for light skin in past human populations. Van den Berghe and Frost (1986) performed a cross-cultural analysis of ethnographic accounts and found preferences for light skin color in mates in 47 out of 51 societies, with males expressing this preference more often. There remains the question of whether these preferences, recorded for recent human societies, apply to our past as well. In other words, is a preference for light skin part of our species' evolutionary history, or does it represent the widespread adoption of culturally based preferences, particularly standards spread via colonialism? Although some authors argue that a preference for light skin existed historically throughout much of the world prior to European colonialism (Jones 1996, Aoki 2002), we of course have no evidence regarding mating preferences from earlier historic or prehistoric times. Perhaps the major problem with the sexual selection hypothesis is showing a link between mating preference and mating behavior. The hypothesis would be strengthened with a demonstration that mating preferences have a significant impact on the actual choice of a mate. Although the sexual selection hypothesis has its proponents (e.g., Aoki 2002), other anthropologists feel that although sexual selection is perhaps a contributing factor, the evolution of light skin has been affected more by the link between UV radiation and vitamin D production (e.g., Jablonksi and Chaplin 2002).

No Selection?

Another view on the evolution of light skin is that there is no selective advantage for light skin, either through natural selection or sexual selection. Brace (2000) and Brace and Montagu (1977) have argued that when humans moved farther away from the equator, selection for dark skin decreased. Over time, mutations for light skin arose; and in the absence of selection against them, these neutral mutations became more common. The underlying assumption in this model is that any mutations would interfere with the production of melanin and therefore lead to lighter skin.

Although this neutral mutation model has not received much support, the idea that light skin may not reflect a selective advantage has resurfaced with Harding et al.'s (2000) analysis of the *MC1R* gene. Using statistical methods to detect past natural selection from genetic diversity measures, they found evidence for strong selection in Africa but a lack of evidence for selection in European populations. Instead, they argue that patterns of *MC1R* diversity in Europe are consistent with neutral expectations, suggesting a lack of significant selection for light skin. However, other analyses of *MC1R* diversity do suggest a selective interpretation (Rana et al. 1999). In addition, it is not clear to what extent variation (and selection) for the *MC1R* gene is related to variation and selection in skin color. Variation in *MC1R* explains only part of the variation in skin color, and it is possible that other genes affecting skin color are influenced by adaptation (Harding et al. 2000).

OTHER ASPECTS OF HUMAN PIGMENTATION

In addition to skin color, anthropologists have studied variation in hair and eye color, although to a lesser extent.

Hair Color

Although hair color is an immediately noticeable phenotypic trait, less is known about the underlying causes of hair color variation in the human species.

Biology of Hair Color

Hair color, like skin color, is related to variation in melanin, which in this case is extruded into the hair from the hair follicles. Although the term *melanin* has been used previously, there are actually two different types of melanin: (1) **eumelanin**, the black–brown pigment found in all melanocyte-producing tissues, and (2) **pheomelanin**, a pigment that is most often reddish brown and found in mammalian hair (and chicken feathers) (Robins 1991). While skin color is influenced only by eumelanin, hair color is due to the relative amounts of eumelanin and pheomelanin. In people with black hair, eumelanin dominates and masks the pheomelanin. Other combinations of eumelanin and pheomelanin result in a range of hair color from dark brown to blonde. Red hair occurs when pheomelanin is more frequent, but the exact shade depends on the contribution of eumelanin (Brues 1977).

Like skin color, hair color can be measured in several different ways. Some studies have used standardized color scales (e.g., Braüer and Chopra 1980), while others have used reflectance spectrophotometry, measuring hair color at the crown of the head, side of the head at ear level, and (given sufficiently long hair) the hair ends (e.g., Little and Wolff 1981). When the latter is used, reflectance tends to increase with wavelength, with red hair showing a noticeable increase starting in the green range of the visual spectrum (see Fig. 11.8).

The genetics of hair color is not clearly understood, although it appears that it is a polygenic trait, where darker shades are in general dominant over lighter shades. One study observed hair color in 100 German families and estimated a heritability of 0.61, showing moderate genetic influence but also a fair amount of environmental effect (Braüer and Chopra 1980). Hair and skin color may be related to some extent, particularly in individuals with red hair, who are generally paler and more prone to freckles (Little and Wolff 1981).

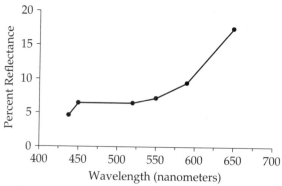

Figure 11.8. Percent reflectance of hair measured at the crown for 33 redheaded women using the Photovolt machine. The sharp increase in reflectance in the green range of the visible spectrum (around 550 nm) is characteristic of people with red hair. (Data from Little and Wolff 1981.)

Recently, the genetics of red hair color has become better understood in terms of variants of the *MC1R* gene and their effect on a switch in production from eumelanin to pheomelanin within melanocytes (Rees and Flanagan 1999, Harding et al. 2000). Three different mutations (Arg151Cys, Arg160Trp, and Asp294His) are associated with red hair. In one study, every person who had at least two of these mutations had red hair. Other studies confirmed this, showing that most redheads had two of the three mutations or were homozygous for one of them (Rees and Flanagan 1999).

Variation in Hair Color

One of the more obvious characteristics of hair color is how it changes with age. In general, hair color darkens with age (Robins 1991), particularly around puberty, which may reflect some hormonal effect (Brues 1977). Interestingly, lighter shades are sometimes found in young children even when the typical adult color is very dark; the best example is the observation that a number of Australian Aborigine children have blondish hair, although most adults are dark. Hair color can also lighten with increased age, producing a gray color that results from the mixture of pigmented and nonpigmented (white) hairs.

Outside of Europe, most human populations tend to have darker shades of hair color. The frequency of blond hair tends to be highest in northwestern Europe. However, there is variation even within smaller geographic areas, such as the noticeable cline showing an increasing frequency of blond hair from southern to northern Italy (Harrison et al. 1988). Red hair is also confined primarily to Europe, having the highest frequency in northern Europe, particularly Scotland and Wales (Little and Wolff 1981), and appears to be associated with a series of specific mutations of the *MC1R* gene (Rees and Flanagan 1999).

The cause of these geographic distributions is not clear. Some have suggested an adaptive value of dark hair color in hot climates because of the ability of darker hair to absorb heat. Since hair tends to act as an insulator, dark hair can absorb and then radiate heat that otherwise might be harmful in a hot climate (Brues 1977). This hypothesis has not been confirmed, but even if correct, it still leaves unexplained the higher frequency of lighter hair colors in Europe. Perhaps we are seeing an example of relaxed selection, such that dark hair is not as necessary in colder climates, thus allowing other shades to increase due to random genetic drift.

Eye Color

Eye color is perhaps even more complicated in terms of biology and measurement.

Biology of Eye Color

Eye color can be assessed using visual inspection, comparison with color scales, or comparison with glass eyes of given colors. In any case, there is a great deal of subjectivity in resolving between different shades, such as light versus medium brown. Measurement is complicated further by the numerous factors that influence eye color, including distribution of melanin and optical scattering. Eye color is to a large extent determined by the amount of melanin present in the anterior layer and stroma of the iris. Brown and dark brown eyes result from heavy concentrations of melanin. Blue eyes are not due to blue pigment but instead result from a lack of pigment combined with optical scattering; in the absence of melanin (or very little), the shorter blue wavelengths of light are scattered, giving rise to a bluish color much in the same way that the sky appears blue. Other shades, such as hazel and green, reflect a balance between the amount of melanin and optical scattering (Robins 1991). Mixed colors (each eye has a different color) sometimes occur, and it appears to be more common among females, suggesting a sex-linked effect (Brues 1977).

There is some correlation of eye color with skin and hair color, suggesting that there is some pleiotropic effect related to pigmentation. Among redheaded women, for example, those with lighter eye color tend to have lighter skin color and lighter hair color (Little and Wolff 1981).

Eye color is typically considered to be inherited. However, a study of 100 German families (Bräuer and Chopra 1980) estimated a heritability of 0.8, indicating that 20% of the variation in eye color did not reflect inheritance in this sample. It is possible that some of this variation is due to problems in measurement and coding of eye color. The genetics of eye color is sometimes taught in school as a simple Mendelian trait affected by a single locus, where the "brown-eyed gene" is dominant over the "blue-eyed gene." Such statements are only rough approximations, and although the genetics of eye color are still poorly understood, it is likely that eye color is polygenic to some extent. It is true that darker eye color does tend to be dominant over lighter eye color but only in a general sense if only "brown" and "blue" categories are used and other variations are lumped together (Brues 1977, Robins 1991).

Variation in Eye Color

Newborns typically have blue or violet–blue eyes, which will often darken quickly with age. Infants with violet–blue eyes tend to develop brown eyes (Brues 1977). Some studies have shown some darkening later in life (Robins 1991).

Although light-colored eyes are found to a limited degree across the world, they tend to be very common only in some European populations (Brues 1977). The reason for this geographic distribution is not clear. Perhaps the distribution of blue and other light-colored eyes reflects, to some extent, variation in overall pigmentation since it is light-skinned populations in which light-colored eyes are most common. Some suggestions have been made that eye color has undergone a different pattern of selection. For example, one hypothesis was that light-colored eyes had enhanced ability to perceive shorter wavelengths, which would presumably be adaptive in early hunting and gathering populations living in glacial Europe with foggy weather. However, this hypothesis was tested, and no such differences in perception were found (Robins 1991). Other suggested evolutionary adaptations of light-colored eyes (see Brues 1977) have also not been supported or rely on unreasonable assumptions. To date, there is no accepted model of the evolution of eye color.

CHAPTER SUMMARY

One of the most visible signs of human biological variation is the difference in skin, hair, and eye color. Skin color is influenced primarily by variation in the melanin pigment and is best measured using reflectance spectrophotometry, which measures the percentage of light reflected off of the skin at different wavelengths. The genetics of skin color is still not completely known, but many studies show a high heritability. Recent studies have shown that variation in the gene MC1R is associated with human skin color variation. Females are generally lighter than males by a small amount, and there is some evidence for slight darkening with age, particularly among adolescent females.

Human skin color shows a strong environmental correlation. Skin color is darkest in populations at or near the equator and is progressively lighter with increasing distance from the equator. The geographic distribution of skin color matches that of UV radiation, which is strongest at the equator and diminishes farther away from the equator. Skin color is somewhat darker in the Southern Hemisphere than in the Northern Hemisphere at equivalent latitude, paralleling the known incidence of UV radiation.

A number of models have been proposed to explain the evolution of human skin color differences. Explaining the evolution of human skin color requires an explanation for dark skin at or near the equator and for light skin in populations farther away from the equator. Several explanations have been offered for the origin and maintenance of dark skin in equatorial populations, all focusing on dark skin protecting against injurious effects of UV radiation, including folate photolysis, sunburn, and skin cancer. The evolution of lighter skin farther away from the equator has usually been explained by lowered levels of UV radiation causing a reduction in vitamin D synthesis, which in turn can lead to a number of medical problems, including rickets. According to this model, light skin evolved as an adaptation to reduced levels of UV radiation. Some have suggested a relationship between latitude, temperature, and skin color, noting that light skin is less susceptible to frostbite than dark skin. Others have suggested that light skin evolved due to sexual selection and a preference for mates with lighter skin color. It is likely that the evolution of human skin color at all latitudes reflects a balance between several selective forces.

Hair color is influenced by both the blackish brown eumelanin and the reddish brown pheomelanin, leading to a range in hair color from dark to light and including various red shades. Most human populations have dark hair, and the greatest amount of variation tends to occur in European populations, although the evolutionary reasons are not known. Eye color is caused by variation in melanin in the iris combined with optical effects such that eyes with little pigment appear blue. As with hair color, European populations show the greatest degree of variation in eye color.

SUPPLEMENTAL RESOURCES

Jablonski N G and Chaplin G (2000) The evolution of human skin coloration. *Journal of Human Evolution* 39:57–106.

Jablonski N G and Chaplin G (2002) Skin deep. *Scientific American* 287:74–81. A short introduction to current views on the evolution of human skin color.

Relethford J H (1997) Hemispheric difference in human skin color. *American Journal of Physical Anthropology* 104:449–457.

Jablonski and Chaplin (2000) and Relethford (1997) review the evidence for the geographic distribution of human skin color. Jablonski and Chaplin (2000) includes a complete review of different models for the evolution of human skin color.

Robins A H (1991) *Biological Perspectives on Human Pigmentation*. Cambridge: Cambridge University Press. A thorough review of human pigmentation focusing on skin color.

Section 4

Population Studies and Human Behaviors

12

Population Structure and Population History

One of the major goals in studying human biological variation is to determine overall genetic similarity between populations. The two basic questions are (1) Which populations are more similar to each other genetically? and (2) Why? Are populations more closely related because of shared gene flow or a common historical origin? Are some populations less genetically similar because of differences in population size and, hence, genetic drift, or are populations less similar because of isolation? What is the influence of geography, demography, and cultural variation on the relative similarity among populations? These questions and others are often addressed using genetic distance and other related measures of genetic similarity. Several studies of population structure and history have been discussed in previous chapters, such as the example of the Irish Travelers in Chapter 5. The current chapter expands on these previous examples by considering in greater detail analytical methods and by providing some additional case studies. Whereas previous examples were used to illustrate the usefulness of one or more measures of human variation, the current chapter illustrates basic models and methods applicable to all measures of human variation.

GENETIC DISTANCE

The answers to the above questions typically start with some estimate of **genetic distance**, an average measurement of genetic dissimilarity between pairs of populations. We start by computing various measures of genetic distance to determine which populations are the most different. As a simple hypothetical example, consider three populations (A, B, and C) where we know the frequency of a given allele in each. Let us imagine that the allele frequencies are as follows:

Population A = 0.70
Population B = 0.75
Population C = 0.55

Which population is the most different genetically for this particular allele? In this simple example, the answer is fairly clear: population C is the most different. The difference in allele frequency between populations A and B is only 0.75 − 0.70 = 0.05, whereas the difference between populations A and C is much higher (0.70 − 0.55 = 0.15), as is the difference between populations B and C (0.75 − 0.55 = 0.20). In this case, the genetic distance (dissimilarity) is greatest relative to population C.

The next step would be to determine why population C is the most genetically different. Is population C isolated from the other populations, thus leading to a reduction in gene flow and genetic similarity? Is population C smaller, thus perhaps leading to increased genetic drift? Does population C have a different language or some other cultural uniqueness that might reduce gene flow and lead to greater isolation and genetic distance? Is population C descended from people who moved into the region at a different time or from a different source? These questions need to be answered after considering both the pattern of genetic distance as well as relevant information on geography, culture, demography, and history.

Computing Genetic Distance

There are a wide variety of genetic distance measures, many of which are closely related (Jorde 1980, 1985). Many of these are computed directly from allele frequencies, although methods also exist for estimating genetic distance from quantitative traits (e.g., Williams-Blangero and Blangero 1989, Relethford and Blangero 1990, Relethford et al. 1997), surnames (Relethford 1988), and migration data (Rogers and Harpending 1986).

Choice of variables (**loci** and traits) is critical to a successful analysis of genetic distance. Because the purpose of such studies is to make inferences regarding the genealogical relationships between populations, they should be confined to neutral (or presumably neutral) traits. For example, consider a comparison of the skin color of central African and Australian Aboriginal populations. The average skin color is similar (dark), but this correspondence should not lead us to conclude that the two groups share a closer historical relationship than other populations. In this case, the biological similarity is due to adaptation to a similar environment—living near the equator. Convergence due to natural selection can mimic a historical relationship; therefore, we should confine our analyses to traits that are neutral or at least neutral over the range of environments being studied.

The above example presents another problem: we should avoid relying on any single locus or trait. Gene flow is expected to affect all loci to the same extent, but genetic drift will lead to differences among loci that might give a distorted picture of population relationships if we rely only on a small number of loci or traits. As an example, consider the frequency of the O allele for the ABO blood group for three populations (data from Roychoudhury and Nei 1988):

England = 0.660
Japan = 0.559
Nigeria = 0.664

It is clear from inspecting these numbers that England and Nigeria are more similar to each other and Japan, more distant. This conclusion is in disagreement with more comprehensive studies that rely on a large number of loci that show European and Asian populations to be more similar genetically than either is to African populations (e.g., Cavalli-Sforza et al. 1994). This disagreement is due in part to the use here of single populations to represent entire regions, each of which will show some variation across local populations. A preferred approach for a regional analysis would be to sample several local populations from each geographic region. The disagreement also reflects the problem of relying on a single locus because the effect of genetic drift could obscure the underlying history of relationships among populations.

As a simple example of how to compute genetic distances, consider the hypothetical example presented in Table 12.1, which lists allele frequencies for five populations (A–E). These hypothetical data consist of the frequencies of seven alleles for three loci (two loci

Table 12.1 Allele frequencies for five hypothetical populations

Locus	Allele	Population				
		A	B	C	D	E
1	1–1	0.50	0.49	0.47	0.56	0.55
1	1–2	0.50	0.51	0.53	0.44	0.45
2	2–1	0.68	0.65	0.69	0.63	0.60
2	2–2	0.32	0.35	0.31	0.37	0.40
3	3–1	0.50	0.55	0.58	0.55	0.56
3	3–2	0.30	0.28	0.29	0.40	0.38
3	3–3	0.20	0.17	0.13	0.05	0.06

have two alleles each, and the third has three alleles). Close examination of these allele frequencies provides some clues as to the pattern of relationship among the five populations. For locus 1, populations A, B, and C have fairly similar values, while populations D and E are different. The situation is less clear for locus 2; here, populations A and C are very similar, but population B is closest to population D. Locus 3 also provides mixed results for the three different alleles. For example, populations B and D are identical for allele 3–1 but quite different for alleles 3–2 and 3–3. What can we say about the overall average pattern of genetic similarity?

In general, as we add more loci or traits, it becomes harder to arrive at an answer to this question simply by examining a table of numbers in an attempt to infer the average pattern of genetic similarity. Genetic distances provide a way of summarizing the average pattern over all loci or traits. An example of a genetic distance measure is the one developed by Harpending and Jenkins (1973) for allele frequency data. For any given allele, the Harpending-Jenkins distance (labeled D^2) between population i and population j is

$$D_{ij}^2 = \frac{(p_i - p_j)^2}{\bar{p}(1 - \bar{p})}$$

where p_i is the allele frequency for population i, p_j is the allele frequency for population j, and $\bar{p}$ is the average allele frequency computed from all populations in the analysis (not just the two being compared at any one time). The top part of this equation expresses the difference between the allele frequencies, which is then standardized by the lower part of the equation, allowing these values to be averaged over alleles with different ranges of variation. Incidentally, this genetic distance measure is related mathematically to the F_{ST} measure discussed in earlier chapters.

For the hypothetical data in Table 12.1, the average frequency for allele 1–1 is

$$\bar{p} = \frac{0.50 + 0.49 + 0.47 + 0.56 + 0.55}{5} = \frac{2.57}{5} = 0.514$$

(For technical reasons, in any actual analysis, we would weight the average by the individual population sizes; but in this simple example, we assume that all five populations have the same size.) The genetic distance between populations A and B for allele 1–1 is therefore

Table 12.2 Harpending–Jenkins genetic distances between the five hypothetical populations listed in Table 12.1

	A	B	C	D	E
A	0	0.0640	0.1068	0.2126	0.2078
B	0.0041	0	0.0721	0.1860	0.1729
C	0.0114	0.0052	0	0.1752	0.1718
D	0.0452	0.0346	0.0307	0	0.0412
E	0.0432	0.0299	0.0295	0.0017	0

The numbers below the diagonal of the matrix are the D^2 values, and the values above the diagonal are the square root of the D^2 values. For example, the squared distance between populations A and B is $D^2 = 0.0041$, whereas the square root of this distance is $D = \sqrt{D^2} = \sqrt{0.0041} = 0.0640$.

$$D_{ij}^2 = \frac{(p_i - p_j)^2}{\bar{p}(1 - \bar{p})} = \frac{(0.50 - 0.49)^2}{0.514(1 - 0.514)} = \frac{0.0001}{0.2498} = 0.0004$$

whereas the genetic distance between populations A and C is

$$D_{ij}^2 = \frac{(p_i - p_j)^2}{\bar{p}(1 - \bar{p})} = \frac{(0.50 - 0.47)^2}{0.514(1 - 0.514)} = \frac{0.0009}{0.2498} = 0.0036$$

It is clear that for this particular allele the genetic distance between populations A and B is less than that between populations A and C. Of course, we could already see that from Table 12.1, but the advantage of genetic distance measures is that we then compute an average distance by averaging the genetic distances over all loci.

Table 12.2 shows the final values of the Harpending-Jenkins distances after averaging over all seven alleles. The lower portion of the matrix (below the diagonal) shows the overall genetic distances between all pairs of populations. The larger the value, the greater the genetic distance, indicating a pair of populations that are less genetically related than other pairs. The smallest distances are between populations D and E (0.0017), A and B (0.0041), B and C (0.0052), and A and C (0.0114). However, the distances between population A and populations D and E as well as between population B and populations D and E are larger. Overall, these distances show a pattern of two clusters of populations, the first cluster consisting of populations A, B, and C and the second cluster consisting of populations D and E.

The Harpending-Jenkins distance measure (as well as many other genetic distance measures) is actually a squared distance measure (D^2). For some purposes, including drawing graphic representations of genetic distances, some researchers prefer to use the square root of this measure, $D = \sqrt{D^2}$. These D values are shown in the upper portion of the matrix (above the diagonal) in Table 12.2. Although the actual distances change, the overall pattern is still the same—a cluster made up of populations A, B, and C and a cluster made up of populations D and E.

The Representation of Genetic Distance

The pattern of genetic distance shown in Table 12.2 is fairly easy to interpret, in part because only a small number ($n = 5$) of populations are used. For more complex cases, particularly involving large numbers of comparisons, interpretation of a genetic distance matrix is less easy because the number of comparisons increases rapidly. If the number of groups in an analysis is g, there will be $g(g - 1)/2$ genetic distances to consider. With only 5 populations, the number of distances is $5(5 - 1)/2 = 10$. For example, for 6 populations, the number of distances increases to 15; for 10 populations, the number of distances increases to 45; and for 15 populations, the number increases to 105. In each case, there are too many distances for us to be able to make any sense out of a long table of numbers. We need some way of capturing this information in an easy-to-interpret manner.

A common method of analysis in genetic distance studies is to make a graphic representation of the pattern of genetic distances, based on the old adage that a picture is worth a thousand words (or, in this case, many columns of numbers). A variety of methods exist for constructing a picture of relationships based on a matrix of genetic distances. Two are discussed here: cluster analysis and principal coordinates analysis.

Cluster Analysis

Cluster analysis is a method which produces groupings (*clusters*) of populations based on their overall similarity to each other. The placement of a population in a particular cluster depends on its genetic similarity to other groups as well as the particular clustering method that is used. A wide range of clustering algorithms has been developed (see Romesburg 1984). One widely used method, illustrated here, is the unweighted pair group method using arithmetic averages (UPGMA).

The first step in UPGMA analysis is to determine the smallest distance between pairs of populations to form the first cluster. For the purposes of this example, we will use the square root of the squared distance measures in Table 12.2. The smallest distance is between populations D and E (0.0412), and these two populations are placed into the same cluster at a distance of 0.0412. The next step is to determine the genetic distance from the remaining populations to this new cluster by averaging their initial distances. For example, the genetic distance between population A and the new cluster D–E is

$$\frac{(\text{distance between A and D}) + (\text{distance between A and E})}{2}$$

$$= \frac{0.2126 + 0.2078}{2} = 0.2102$$

Likewise, the distance between population B and cluster D–E is computed by averaging the distances between B and D and between B and E, giving 0.1795. Finally, the distance between population C and cluster D–E is computed as 0.1735. These new distances are then considered along with the distances among the remaining populations of A, B, and C. This entire method results in a genetic distance matrix between four groups—populations A, B, and C as well as cluster D–E. These genetic distances are summarized here:

A and B   = 0.0640
A and C   = 0.1068
A and D–E = 0.2102
B and C   = 0.0721
B and D–E = 0.1795
C and D–E = 0.1735

The smallest genetic distance is now between A and B (0.0640); thus, they form a new cluster, joining together at a distance of 0.0640. The process is repeated by looking at the genetic distances between three groups: A–B, C, and D–E. The genetic distances are again determined by averaging. For example, the distance between cluster A–B and C is computed by averaging the distances between A and C and between B and C, giving 0.0895. The distance between clusters A–B and D–E is then computed by averaging the distances between the original groups; that is, averaging A and D, A and E, B and D, and B and E, giving 0.1948. Finally, the distance between C and cluster D–E is computed by averaging the initial distances between C and D and between C and E, giving 0.1735. There are now three groups in the analysis, and the genetic distances between these groups are as follows:

A–B and C   = 0.0895
A–B and D–E = 0.1948
C and D–E   = 0.1735

The smallest distance here is between A–B and C, and they are joined to make a new cluster A–B–C at a distance of 0.0895. The final step is to determine the genetic distance between the two remaining groups, cluster A–B–C and cluster D–E. This distance is computed by averaging the original distances between A and D, A and E, B and D, B and E, C and D, and C and E, giving a value of 0.1877.

We now summarize the results of the clustering process by noting which populations joined which clusters at given levels of genetic distance:

Populations D and E clustered at a distance value of 0.0412.
Populations A and B clustered at a distance value of 0.0640.
Cluster A–B and population C clustered at a distance value of 0.0895.
Cluster A–B–C and cluster D–E clustered at a distance value of 0.1877.

These results are then used to construct a **dendrogram**, which is a graphic representation of the cluster analysis in the form of a tree that shows overall levels of genetic similarity. The dendrogram is produced by joining populations into clusters at a given level of genetic distance along the vertical axis. These results are shown in Figure 12.1; populations that join together closer to the bottom of the tree (such as D and E) are considered more genetically similar compared to groups that cluster toward the top of the tree, indicating less genetic similarity. The dendrogram in Figure 12.1 provides a clear and succinct representation of the hypothetical genetic distance matrix from Table 12.2. Populations A, B, and C are clearly more genetically similar to each other than to either D or E, which in turn form their own cluster. The main message from Figure 12.1 is the existence of two genetically different groups, a cluster with populations A, B, and C and a cluster with populations D and E. Some researchers prefer to perform cluster analysis on the D^2 values rather than the D values. In this case, the distances between the two clusters in Figure 12.1 would be changed, but the overall pattern of two clusters would be the same.

Principal Coordinates Analysis

Although useful in many cases, the dendrograms produced by cluster analysis are somewhat limited because they do not always clearly represent the pattern of relationships between clusters and may not always be that easy to interpret. An alternative method of representing genetic distances is **principal coordinates analysis**, a method that produces

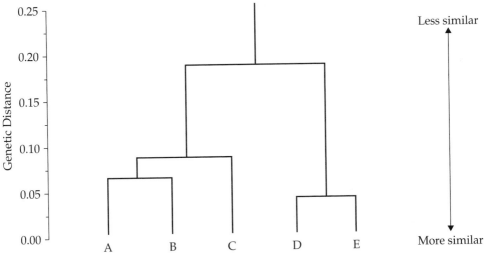

Figure 12.1. A dendrogram produced by cluster analysis that shows the pattern of genetic distances among the five hypothetical populations given in Table 12.2. Three populations (A, B, and C) are clearly genetically similar to each other and distinct from the other cluster (D and E).

two- or three-dimensional plots representing genetic distance. The resulting plot can be considered a genetic distance "map" because the results are interpreted the same way as those of a road map—the closer two points are on the map, the closer they are.

The method of computation used in principal coordinates analysis is complex and best performed on a computer (see Gower 1966 for computational details). The method produces a two-dimensional plot, where the distances between the points on the map are a useful approximation of the initial genetic distances. A two-dimensional principal co-ordinates plot based on the genetic distance matrix in Table 12.2 is shown in Figure 12.2.

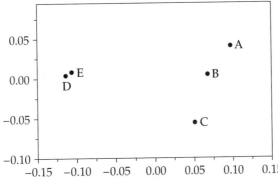

Figure 12.2. A principal coordinates plot of the genetic distances given in Table 12.2. Principal coordinates analysis is an alternative to cluster analysis for producing a picture of genetic relationships between populations. Note that A, B, and C are genetically similar and distinct from D and E. The numbers along the axis indicate coordinates in the reduced two-dimensional space.

This plot is immediately interpretable: populations A, B, and C plot next to each other, indicating close genetic relationships between them; the same applies to populations D and E, which also show close genetic similarity. Figure 12.2 also clearly shows the separation of these two groupings.

In this simple example, both cluster analysis and principal coordinates analysis show the same overall picture of genetic difference between two groups, the first consisting of populations A, B, and C and the second consisting of populations D and E. Cluster analysis and principal coordinates analysis (as well as other methods) show only the overall picture of genetic similarity. They do not tell us why these patterns exist or what they mean. In Figures 12.1 and 12.2, the pattern suggests some degree of genetic separation between the two major groupings. Is it geographic, with populations A, B, and C living close to one another but geographically isolated from populations D and E? If so, then we might be seeing the genetic impact of geographic distance limiting gene flow, thus allowing the two groupings to genetically diverge. Perhaps the separation reflects some cultural barrier to gene flow, such as a difference in language or ethnicity. A number of other scenarios are also possible. The point here is that we would need to have additional supporting information on these populations to answer these questions. The genetic distance maps by themselves do not provide the critical answer.

POPULATION STRUCTURE

Studies of genetic distance can be used to understand **population structure** and **population history**. Population structure deals with the factors that affect mate choice and genetic relationships between individuals within populations or subdivisions of a population. Population history is related but generally focuses on the genetic impact of historical factors—such as invasions, migrations, and other events that affect genetic exchange between populations—on genetic distances between a set of populations.

Factors Affecting Population Structure

Many cultural, demographic, and ecological factors influence mate choice, which in turn affects genetic distances between populations. We are interested here in examining those factors that limit or enhance genetic exchange (gene flow) between individuals or subdivisions of a population.

Isolation by Distance

One of the most common influences on mate choice and gene flow is geographic distance. Quite simply, people (and most other organisms) who are geographically separated are less likely to mate than people who are geographically close. This pattern persists among most human populations, both past and present. An example of the impact of geographic distance is shown in Figure 12.3, which graphs the frequency of marriages as a function of geographic distance for the town of Barre, Massachusetts. These data were obtained by examining 1,251 marriage records for the town for the time period from 1800 through 1849 as part of research conducted by one of the authors (J. H. R.). For each marriage, the geographic distance was calculated between the premarital residences of both bride and groom. The graph shows very clearly that the majority of all marriages occurred between individuals who lived less than 5 km apart. The frequency of

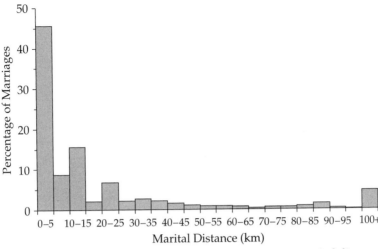

Figure 12.3. The percentage of marriages that took place at various marital distances (the distance between the premarital residences of bride and groom) for the town of Barre, Massachusetts, 1800–1849. (From unpublished data of J. H. R.)

marriages decreases sharply with geographic distance; fewer and fewer marriages took place over longer distances, although there were always a few marriages that occurred between individuals quite distant from one another, including a number where one spouse came from several states away.

The limiting effect of geographic distance on migration (marital migration in the above example) has a direct genetic effect: the proportion of gene flow decreases with geographic distance. As a consequence, populations located farther apart geographically will be less genetically similar, a phenomenon known as **isolation by distance**. As such, genetic distances and geographic distances will be highly correlated when isolation by distance occurs because the farther two populations are from each other, the more genetically different they will be. Specifically, genetic distances between pairs of populations will increase exponentially with geographic distance, leveling off at large geographic distances.

An example of this relationship is shown in Figure 12.4 for genetic distances between 10 Native American Papago populations from the U.S. Southwest based on a number of red blood cell **polymorphisms**. The points in Figure 12.4 indicate the genetic and geographic distances between each pair of populations (for 10 populations, there are 45 pairs of distances). The solid line indicates the best-fitting line according to a population genetics model relating geographic and genetic distances. The fit is not perfect, but the general pattern of isolation by distance is quite clear.

Many studies of human populations have found that isolation by distance explains at least some of the observed genetic distances between populations. However, there is variation in the actual impact of isolation by distance. In some cases, the effect of geographic distance is less than in others, often as a consequence of cultural factors affecting the likelihood of gene flow (e.g., changes in transportation technology). Comparison of isolation by distance studies across many human populations shows differences in rate of increase in genetic distance over space (Jorde 1980; Relethford and Brennan 1982; Relethford 1985, 1988). In addition, several studies have shown how the pattern of

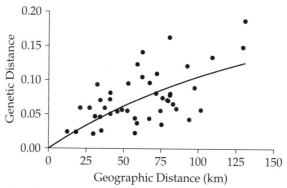

Figure 12.4. Isolation by distance. This is a plot of the relationship between genetic distance and geographic distance between 10 Native American Papago populations in the U.S. Southwest. Harpending-Jenkins (1973) distances were computed based on 22 alleles for nine genetic marker loci as reported in Workman et al. (1973). Geographic distances were taken from Workman and Niswander (1970). The solid line indicates the best fit of the isolation by distance model: $D^2 = 0.209(1 - e^{-0.007d})$, where D^2 is the Harpending-Jenkins genetic distance and d is the geographic distance in kilometers.

isolation by distance has changed over time, resulting in lower rates of spatial change in genetic distances, most likely reflecting cultural changes in the pattern and intensity of gene flow (e.g., Relethford and Brennan 1982).

Cultural Factors and Population Subdivision

Geographic distance is an important factor affecting mate choice in a population, but it is not the only one. A number of cultural factors also influence mate choice and lead to subdivision of a population into smaller and more endogamous subdivisions. Some of these factors include ethnicity, religion, language, and social class, to name but a few. Any factor that has an influence on mate choice could potentially have an impact on the genetic structure of a population. Of course, such factors will vary from one culture to the next, as well as over time.

An example of the genetic influence of language differences comes from the analysis of genetic structure on Bougainville Island in the Solomon Islands in Melanesia. Bougainville Island is relatively small (roughly 202 km long and 64 km wide) but contains a great deal of cultural variation, particularly linguistic: there are 17 major languages spoken on the island. Jonathan Friedlaender (1975) conducted an extensive survey of 18 villages on Bougainville Island, collecting demographic data, blood groups and proteins, anthropometrics, dental measures, and dermatoglyphics in an attempt to link patterns of biological variation to geographic, linguistic, and historical factors. He found that genetic distances were correlated with both geographic isolation and language differences (measured by computing a linguistic distance based on the frequency of shared words among the different languages). Populations farther apart geographically and/or speaking a different language tended to be the most genetically dissimilar.

Friedlaender also found, not unexpectedly, that geographic distance and linguistic distance were correlated; that is, villages located farther apart geographically tended to speak more dissimilar languages. This correlation raises an interesting question about the relative roles of geographic and linguistic distance. Is the correlation between genetic distance and linguistic distance simply a reflection of the fact that both genetics and linguistics are affected by geography, or do language differences act to increase genetic distances beyond the level expected under isolation by distance? Since all of the measures here are interrelated, clear separation of their relative impact is not always easy to determine. An additional analysis by Relethford (1985) found that both geographic and linguistic isolation have acted to increase genetic differences between Bougainville villages. In another study, including data from both Bougainville and other islands in the Solomon Islands, Dow et al. (1987) found that the correlations with linguistic distance remained after geographic distance was controlled for but not the reverse, suggesting that language differences had a greater impact on genetic differences than geographic isolation, at least for some of the traits analyzed.

Another example of a cultural factor that can influence the genetic structure of populations is religious affiliation. Differences in religious affiliation within a population can affect genetic variation in two ways. First, religious differences might form a partial barrier to gene flow throughout a population when people in one religion tend to marry and mate primarily within their own cultural group (a phenomenon known as *endogamy*). Second, religious differences might reflect different geographic origins of ancestors such that members of one religion came from one place and members of another came from somewhere else.

Several studies have focused on isolated religious groups moving to new locations, such as the immigration of Hutterite and Mennonite groups to North America. In general, these groups maintain cultural and genetic isolation, although this isolation has begun to break down in some groups, such as the Mennonites of Kansas and Nebraska (Crawford et al. 1989). Other studies have investigated the effect of religious affiliation in less isolated groups, such as Crawford et al.'s (1995) study of genetic structure among members of Catholic, Anglican, and Evangelical churches on Fogo Island in Newfoundland. In this study, the influence of religion was relatively minor relative to geographic distance. In another study, Relethford and Crawford (1998) examined the influence of religious affiliation (Catholic, Church of Ireland, Presbyterian) on the genetic structure of Northern Ireland. They found that there was a significant effect of religious affiliation, although it was relatively minor. Further analysis suggested that the effect of religion was in part due to differences in the geographic origin of immigrants of different religions.

Case Study: The Åland Islands

The Åland Islands, Finland, are an archipelago made up of 6,600 islands and skerries located between the eastern border of Sweden and the western border of Finland (Fig. 12.5). There are extensive historical demographic records from the Åland Islands over several centuries, allowing the opportunity to measure patterns of migration and how they have changed over time. Demographic data, including marriage records, provide inferences on past gene flow. Combined with estimates of population size, these data can be used to estimate genetic distances based on the likely effects of gene flow and genetic drift (Mielke et al. 1976, 1982). In addition, a large quantity of genetic data (blood groups and protein polymorphisms) were collected on the Åland Islands between 1958 and 1970 from a wide range of age groups. By analyzing subsets of these genetic data based on the year of birth, Jorde et al. (1982) derived genetic distances and related

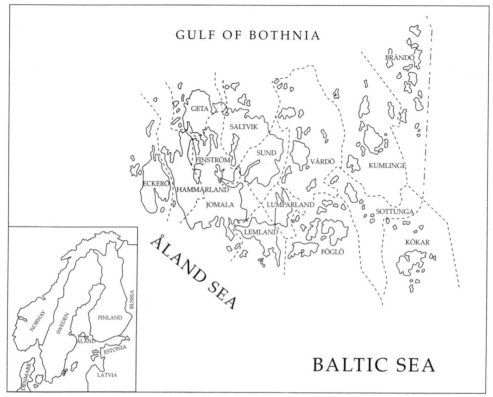

Figure 12.5. Map of the Åland Islands, Finland.

measures for three time periods: those born prior to 1900, those born between 1900 and 1929, and those born after 1929.

Many of the studies of the Åland Islands looked at the differences between 12 parishes. Marital migration data were used to estimate the genetic kinship, a measure of genetic similarity, between all pairs of parishes. As expected under isolation by distance, there was a moderate to strong correlation with geography. These correlations were stronger in early time periods (1750–1849) and diminished afterward, suggesting that, over time, geographic distance declined somewhat as a determinant of migration (Mielke et al. 1976, 1982).

The relationship between estimated kinship based on migration patterns and geographic distance can be summarized by fitting the data to the isolation by distance model, which predicts that kinship will decrease exponentially with geographic distance. Figure 12.6 compares the estimated isolation by distance curves for several time periods based on the Åland Island data. The vertical axis of the graph shows genetic kinship within populations; higher values indicate increased genetic similarity between individuals within a population, reflecting greater isolation. In general, an increase in gene flow into a population results in a decline in within-group kinship, a pattern shown clearly in Figure 12.6, where local kinship consistently decreases in more recent time periods. In addition, the slope of the curves flattens over time, another reflection of increased gene

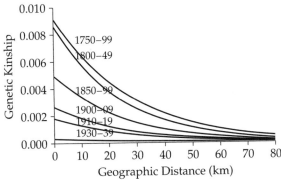

Figure 12.6. Changes in isolation by distance in the Åland Islands over time. The *curves* represent the fit of the isolation by distance model as applied to genetic kinship estimated from migration data: $\phi = ae^{-bd}$, where ϕ is the estimated kinship between pairs of populations and d is the geographic distance between populations in kilometers. a and b are parameters estimated from the fit of the data to the curve, where a is an estimate of local kinship and b is an estimate of systematic pressure, primarily gene flow. Only selected time periods are shown here, to illustrate the decline in isolation by distance over time. (Data from Mielke et al. 1976.)

flow leading to lower estimated genetic distance among populations separated in space. In other words, the limiting effect of geographic distance declined over time.

Closer examination of migration rates shows temporal changes more clearly. Relethford and Mielke (1994) examined exogamy rates for 15 Åland parishes in several time periods defined on the basis of the year of marriage. *Exogamy* is a measure of gene flow, defined here as the proportion of individuals that married into a parish from another parish. For example, if there were 100 marriages, there would be 200 individuals total (brides plus grooms). By examining his or her premarital residence, it is easy to determine if each married person came from the same parish (endogamy) or a different parish (exogamy). If there were three people who came from another parish, then the exogamy rate would be 3/200 = 0.015. Figure 12.7 shows the exogamy rate averaged over all 15 parishes for four time periods. There is only slight variation in the first three time periods (1750–1899), ranging in value from about 0.07 to 0.09. The average exogamy rate almost doubles (0.15) in the last time period (1900–1949). These results show greater gene flow in the twentieth century relative to earlier times.

Mielke et al. (1994) also looked at factors affecting the rate of migration between parishes and how they changed over time. They found a consistent effect of population size on migration over all four time periods between 1750 and 1949, where there was greater relative migration from larger to smaller parishes, typical of most human populations (Relethford 1992). As expected, they also found that geographic distance exerted a strong effect on migration between parishes; the farther apart, the lower the migration rate. Mielke et al. (1994) also found that the decay of migration with distance declined over time and was particularly noticeable after 1850. Combined with earlier studies, this analysis shows that Åland parishes became less isolated over time, particularly during the twentieth century. This change seems to be due to a number of different cultural and demographic changes during this time period, including improvement in transportation

Figure 12.7. Exogamy rate in Åland Island parishes for four time periods, defined by the year of marriage. Exogamy rate is the proportion of individuals in a parish who married into that parish from another parish. Note the increase in exogamy in the most recent time period (1900–1949). (Data from Relethford and Mielke 1994.)

technology—a steamboat fleet started operation after 1890, and there was an increase in private motorboats during the 1920s and 1930s.

Analysis of genetic distances based on blood groups and protein polymorphisms revealed several interesting trends (Jorde et al. 1982). The geographic nature of the Åland Islands affected patterns of genetic distance. Some parishes are located on the main island, whereas others are spread throughout the more isolated outer islands. F_{ST} values for the outer islands were greater than those for the main island, showing increased differentiation, reflecting greater isolation and reduced gene flow, in the outer islands. The increased isolation is due to two factors: (1) the outer island parishes are more geographically dispersed than those on the main island, leading to a greater influence of isolation by distance, and (2) travel between parishes was easier on the main island, whereas travel between outer island parishes necessitated sea travel, which in turn reduced the amount of gene flow. Genetic drift has also affected the magnitude of genetic distances. The outer island parishes were generally smaller and, hence, experienced more genetic drift compared to the larger main island parishes. Cultural and demographic changes appear to have had some impact on the genetic distances even over a short period of time. Jorde et al. (1982) found that the influence of genetic drift on genetic distances decreased over time, while the impact of gene flow increased over time.

CASE STUDIES IN POPULATION HISTORY

The history of any human population includes many events that affect genetic distances between populations. Humans colonize new regions. Migrations and invasions are accompanied by gene flow, often across large distances, as populations previously separated come into cultural and genetic contact with one another. Populations change in size, thus leading to changes in the magnitude of genetic drift. The history of how human populations interact genetically through mating leaves genetic signatures of past events, or what one of us calls "reflections of our past" (Relethford 2003a). The purpose of this section is to provide several case studies that illustrate this basic idea. The first example deals with the use of genetic data to determine the origin of the first colonists to the New

World, the first Native Americans. The second example discusses the genetic impact of the slave trade in North American history by examining the genetic impact of European admixture in African American populations. The third example describes the effect of repeated migrations on the genetic diversity of human populations in Ireland. These case studies rely on a wide range of measures of biological variation, including red blood cell polymorphisms, DNA markers and anthropometrics.

The Origin of Native Americans

The origin of Native Americans has been of interest to Western science ever since the voyages of Columbus over 500 years ago. Various ideas have been proposed to explain the origin of Native Americans, including the notions that they represented one of the "lost tribes of Israel" or refugees from the mythical continent of Atlantis. As scientific study replaced such wild speculations, it quickly became clear that the first human inhabitants in the "New World" were anatomically modern people who came from Asia, specifically northeast Asia (Crawford 1998). The Asian origin of Native Americans is supported by archaeological, linguistic, and biological evidence.

The biological evidence for a historical connection between Asia and the New World is very strong and includes analyses of genetic markers, cranial and dental remains, and mitochondrial DNA. Genetic analyses show several alleles that are found almost exclusively in Native American and Asian populations and are absent or at very low frequencies elsewhere in the world. One example is the Diego blood group, a locus that has two alleles—Di^a and Di^b. The Di^a allele is absent in most European, African, and Oceanic populations but is found in many Asian and Native American populations (Fig. 12.8). There is some variation within regions, such as South America, where the Makiritare have a very high frequency of the Di^a allele while the Yanomama lack the allele; but the general pattern is quite striking. Other loci studied to date also show an Asian connection (Szathmary 1993).

The genetic similarity of Native American to Asian populations is quite clear when considering a wide number of genetic markers simultaneously. Figure 12.9 shows the genetic distances between Native Americans and eight other geographic regions based on 120 alleles for a variety of blood groups and protein and enzyme polymorphisms (Cavalli-Sforza et al. 1994). This graph shows clearly that Native Americans are most similar genetically to populations in northeast Asia.

Mitochondrial DNA has also been used to establish a connection between Native Americans and northeastern Asia. Analysis of restriction fragment length polymorphism (RFLP, see Chapter 8) reveals related groups, known as **haplogroups**, which share a set of mutations. The mitochondrial DNA found in most Native Americans clusters into five different haplogroups, known as A, B, C, D, and X. These haplogroups are also found in ancient DNA recovered from Native American skeletal remains. The striking feature of these five haplogroups is that they are rare or absent elsewhere in the world except for northeastern Asia, a finding consistent with an origin of Native Americans in this part of the world. These results provide further confirmation of an Asian origin for Native Americans, although the debate continues over the exact number and location of the original founding groups (Schurr 2000, Relethford 2003). Haplogroup X is particularly interesting because initial studies found it in Native American populations as well as in parts of Europe but not in northeast Asia. For a time, this pattern suggested that haplogroup X was somehow introduced into the New World from Europe, giving rise to

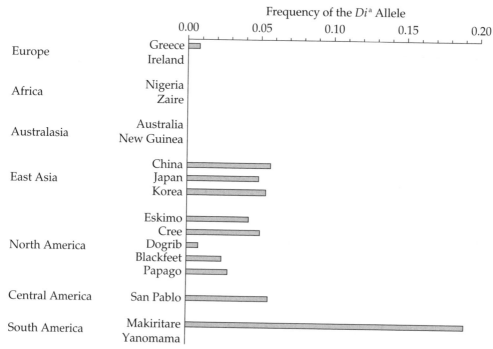

Figure 12.8. Frequency of the Di^a allele for the Diego blood group in selected human populations. Note that this allele is more common in both Native American and Asian populations and virtually absent in Europe, Africa, and Australasia. (Data from Roychoudhury and Nei 1988.)

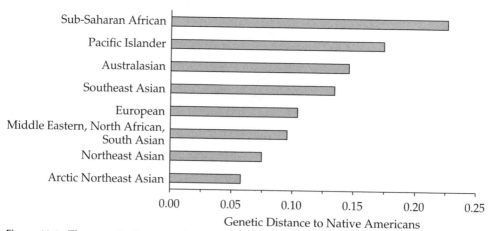

Figure 12.9. The genetic distance of eight geographic regions to Native Americans. Note that Native Americans are most similar to northeastern Asian populations. These distances are based on 120 alleles, most of which are for red blood cell loci. (Data from Cavalli-Sforza, Luca, *The History and Geography of Human Genes.* © 1994 Princeton University Press. Reprinted by permission of Princeton University Press.)

speculations regarding contact with Europe before the voyages of Columbus. However, a more recent study (Derenko et al. 2001) found haplogroup X in parts of northeast Asia. This finding means that we do not have to postulate any prehistoric European contact. The presence of haplogroup X in both Europeans and the New World seems instead due to migration from northeast Asia to both of these regions.

The close relationship between Native American and East Asian populations is also seen in a variety of physical characteristics, including straight black head hair, little body hair, and broad cheekbones, among others (Crawford 1998). An example of a physical characteristic shared by Native Americans and Asians is the frequency of shovel-shaped incisors, a dental variant formed when the front teeth have ridges on the outer margins. Although shovel-shaped incisors are found across the world, the frequency of this trait is rather low for people of African or European ancestry but tends to be very high in both Asian and Native American populations, providing further illustration of their genetic relationship (Fig. 12.10).

Although there is little doubt about an Asian origin of Native Americans, the dynamics of this origin continue to be widely debated. The specific debates concern the number and origin of initial migrations. Various authors have suggested anywhere from one to four major migrations and, in the case of multiple migrations, perhaps different founding populations (e.g., Szathmary 1993, Crawford 1998; Powell and Neves 1999, Schurr 2000). The key question is whether the observed genetic diversity in Native Americans can be explained by a single wave of migration, followed by drift and other historical events, or multiple waves. For example, Schurr (2000) argues that the present-day distribution of mitochondrial DNA haplogroups supports at least two distinct migration events, one from northeastern Asia (Siberia) and the other from East Asia. Karafet et al. (1999) come to the same conclusion based on their analysis of Y-chromosome polymorphisms.

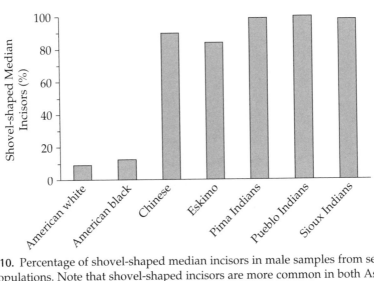

Figure 12.10. Percentage of shovel-shaped median incisors in male samples from selected human populations. Note that shovel-shaped incisors are more common in both Asian (Chinese) and Native American populations and relatively rare in African American and European American samples. (Data from Bass 1995.)

Another area of long-standing controversy involves the timing of the migration(s), with some arguing for a fairly early colonization (14,000–12,000 years ago) and others suggesting an older date, perhaps more than 25,000 years ago (Goebel 1999). Final resolution of this debate will require confirmation of archaeological sites older than 15,000 years; although such dates have been reported, they remain highly debated.

Admixture in African Americans

Human history contains many examples of migrants from geographically distant origins mating and thus experiencing **admixture**, a form of gene flow. Although the United States has experienced immigration from across the world, much of its early colonial history can be characterized by two main groups of immigrants: European settlers, primarily from northwestern Europe, and enslaved Africans, primarily from western Africa. Although ethnic differences are often formidable barriers to gene flow, it is clear that mating took place between those of European ancestry and those of African ancestry, often the result of forced intercourse between enslaved women and European slaveholders. In addition, peaceful unions have taken place, becoming more frequent in recent times following some relaxation of cultural barriers.

The net effect of history has been to introduce European admixture into the gene pool of some African American populations. From a population genetic standpoint, it is of interest to estimate the magnitude of this admixture. Ignoring selection and drift for the moment, the genetics of African Americans can be considered using a simple **dihybrid model** of gene flow, as illustrated in Figure 12.11. Here, a hybrid population (H) is formed through gene flow from two parental populations, 1 and 2. The amount

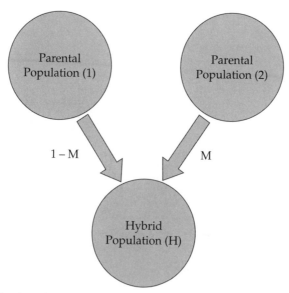

Figure 12.11. Dihybrid model of admixture, where the genes in a hybrid population come from two parental populations in relative proportions, $1 - M$ (from population 1) and M from population 2. Given data on existing allele frequencies in all three populations, the overall admixture proportion M can be estimated (see text).

of overall admixture from population 2 is labeled M, meaning that a proportion M of the genes in the hybrid population comes from parental population 2. Consequently, the proportion of gene flow from parental population 1 is $1 - M$ such that the total proportions add up to $1(1 - M + M = 1)$. For any allele, the frequency in the hybrid population, p_H, is a function of the allele frequencies in the parental populations (p_1 and p_2) and the degree of gene flow from each group. The dihybrid model is as follows:

$$p_H = (1 - M)p_1 + Mp_2$$

Thus, if gene flow were the only evolutionary force acting upon the hybrid population, then its allele frequency will reflect this proportional mixing. If this is the case and if we have estimates of the values of the allele frequencies, we can estimate the proportion of admixture from parental population 2 by taking the above equation and using some simple algebra:

$$p_H = (1 - M)p_1 + Mp_2$$

$$p_H = P_1 - Mp_1 + Mp_2$$

$$Mp_1 - Mp_2 = P_1 - P_H$$

$$M = \frac{p_1 - p_H}{p_1 - p_2}$$

As an example, consider this method as applied to the A allele for the ABO blood group using data given by Workman et al. (1963) in their analysis of the African American community of Claxton, Georgia. The frequency of A in the Claxton African American sample was 0.158. They used an average frequency for west Africa of 0.148 to approximate the African parental population and a frequency of 0.246 from the European American community in Claxton as an approximation of the European parental population. Using the above formula and letting H = African Americans, 1 = Africa, and 2 = Europe, gives the following:

$$M = \frac{p_1 - p_H}{p_1 - p_2} = \frac{0.148 - 0.158}{0.148 - 0.246} = 0.10$$

The estimate of $M = 0.10$ suggests that 10% of the current gene pool of Claxton African Americans is of European ancestry and, therefore, that the remainder (90%) is of African ancestry.

Several qualifications regarding this method are needed since the underlying model makes several assumptions (Reed 1969). First, we assume that the parental populations can accurately be identified. Second, we assume that the allele frequencies in the contemporary samples used to represent the parental populations have not changed significantly—in this case, that there has been little or no change in allele frequencies in the past several hundred years in African or European populations—so that contemporary allele frequencies can be used to estimate past allele frequencies in those regions. A third assumption is that gene flow is the only evolutionary force that has acted upon the hybrid population and that there has been little, if any, drift or selection. A fourth assumption is that all of our samples are unbiased so that we are not seeing random errors due to small sample size (Reed 1969). Some of these assumptions can be dealt with in part by choosing appropriate samples of sufficient size and basing admixture estimates on a large number of loci.

We also assume that the underlying model is correct and that there are only two sources of gene flow into the African American population. This might apply in some cases, but in situations where historical evidence suggests additional genetic contributions from Native Americans, the underlying model of two parental populations would be violated (in which case we could use a trihybrid model, see Pollitzer 1964).

We would never want to estimate admixture based only on a single allele. We expect variation across alleles due to genetic drift and sampling error and are best off using an average over many alleles and loci. A variety of methods exist for estimating admixture from multiple alleles and loci (Chakraborty 1986, Chakraborty et al. 1992). Comparison of estimates from different alleles and different sets of alleles also allows us to potentially differentiate between neutral loci and those affected strongly by natural selection. In their analysis of Claxton, Workman et al. (1963) estimated M for 14 different alleles, which fell into two clusters. The first group, consisting of 9 alleles, gave estimates of European admixture ranging 10%–22%, most clustering close to the average estimate of $M = 13\%$. Alleles in the second group gave much higher estimates, most in excess of 40%. This second group includes the hemoglobin S allele, known to be affected by natural selection (see Chapter 6). Workman et al. (1963) concluded that the first group provided the best estimate of European admixture, whereas the second group included alleles affected by both admixture and natural selection. In the case of the hemoglobin S allele, there is evidence suggesting selection against it among African Americans, thus leading to a lower allele frequency than expected due to admixture alone, which in turn will lead to a higher estimate of European admixture.

Admixture estimation also depends on how different the two parental populations are in the first place. If, for a given gene, the parental populations are quite similar, then it will be difficult to estimate admixture. The most precise admixture estimates are those derived from DNA markers known as **population-specific alleles (PSAs)**, alleles that are either absent in one of the parental populations or present in both but with large differences between them (Parra et al. 1998). Esteban Parra and colleagues (1998, 2001) applied this method to a number of samples of African Americans (Table 12.3). The average amount of European admixture across all of their samples is roughly 15% (which therefore means that the amount of African ancestry is roughly 85% on average). Note, however, that there is considerable variation in these rates, ranging from a low of 3.5% among the Gullah of South Carolina to a high of 22.5% in New Orleans. In addition, these are averages based on samples; there is also considerable variation in ancestry within each sample such that some individuals have a small amount of European ancestry while others have considerably more. In their analysis of admixture in Columbia, South Carolina, Parra et al. (2001) found that some individuals had less than 10% European ancestry while others had more than 50%. These studies show that there is genetic diversity both within and among African American populations. Any racial classification that places all African Americans into the same category is flawed biologically. Although all African Americans have a common cultural identity, they show considerable genetic diversity.

Parra and colleagues (1998, 2001) also examined sex differences in admixture by looking at estimates obtained from mitochondrial DNA, which estimates maternal ancestry, and Y-chromosome markers, which estimate paternal ancestry. In each sample of African Americans that they analyzed, the amount of European ancestry from the male line was more than that from the female line. In other words, European genes were introduced more from men than women. This finding is consistent with the history of enslavement in the United States, where male European slaveholders mated with enslaved African American women.

Table 12.3 Estimates of European admixture in African American populations based on allele frequencies of population-specific alleles

Population	European Ancestry (%)
Gullah, South Carolina	3.5
Jamaica	6.8
Charleston, South Carolina	11.6
Low Country, South Carolina	11.8
Philadelphia (sample 1)	12.7
Philadelphia (sample 2)	13.8
Baltimore	15.5
Detroit	16.3
Houston	16.9
Columbia, South Carolina	17.7
Maywood, Illinois	18.8
New York	19.8
Pittsburgh	20.2
New Orleans	22.5

Data from Parra et al. 1998, 2001.

The Population History of Ireland

The island of Ireland (presently made up of two nations, the Republic of Ireland and Northern Ireland) has a fascinating history consisting of numerous settlements and invasions. Ireland was initially inhabited over 4,000 years ago by perhaps as many as four different waves of migrants. Since that time, there have been a number of invasions and additional waves of immigrants, including Viking invasions starting in the eighth century, the Anglo-Norman invasion of 1169, and the introduction of English immigrants following the Articles of Plantation of James I (Hooton et al. 1955, Hackett et al. 1956, Sunderland et al. 1973). Ireland also experienced massive emigration and population decline following the Great Famine of 1846–1851, when there was repeated failure of the potato crop (Kennedy 1973).

Several studies have investigated genetic variation across Ireland in an attempt to link one or more of these historical events to the current patterns of biological diversity (e.g., Hooton et al. 1955, Hackett et al. 1956, Relethford and Crawford 1995, North et al. 2000). This case study provides a summary of one of these studies (Relethford and Crawford 1995), an analysis of genetic distances based on anthropometric measures originally collected throughout Ireland during the 1930s (Hooton et al. 1955). The data discussed here consist of 10 anthropometric measures of the head and face for 7,214 adult Irish men. Comparisons were made between men born in 31 different counties (there are 32 counties in Ireland, 26 in the Republic of Ireland and 6 in Northern Ireland; but there were insufficient data for one county, thus leaving 31).

Genetic distances between the 31 counties were estimated from these data using quantitative genetic measures developed by Williams-Blangero and Blangero (1989) and Relethford and Blangero (1990). Although anthropometrics are affected by both genetic and environmental factors, they can be used in many cases to provide insight into the genetic relationship between populations. The genetic distance map of the 31 counties is

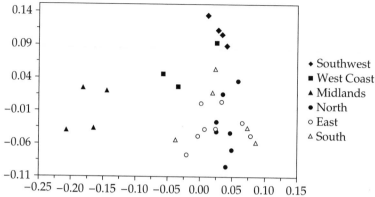

Figure 12.12. Principal coordinates plot of the genetic distances between 31 Irish counties based on 10 anthropometric measures of the head. Individual counties are labeled according to their geographic location. The horizontal axis separates the four counties in the Irish midlands from all other counties. The vertical axis tends to separate counties to the west of the midlands (southwest, west coast) from those to the east (north, east, south). (Adapted from Relethford and Crawford 1995.)

shown in Figure 12.12. Two major patterns were discovered by Relethford and Crawford (1995), both highlighted in Figure 12.12. First, the four counties making up the Irish mid-lands (located in the geographic center of the island) are clearly distinct from all other counties along the horizontal axis of the genetic distance map. Second, the vertical axis tends to separate those counties on the western coast of Ireland from those located more to the east.

The distinctiveness of the midland counties most likely reflects the genetic impact of Viking invasions. The Vikings first came into contact with Ireland in 794 on a small island outside of Dublin. Somewhat later, several Viking settlements sprang up in various coastal communities. In 832, some Norwegian Vikings sailed up the Shannon River into the midlands, forming a permanent settlement of over 10,000 men. During the tenth and eleventh centuries, Danish Vikings attacked, with some sailing into the midlands via the Shannon River. Viking influence in Ireland continued until roughly 1200, and historical accounts suggest frequent mating. According to Relethford and Crawford (1995), the distinctiveness of the midlands in the genetic distance map is due to this differential gene flow; the Vikings had more genetic impact in the midlands than along the coast, where their genes were swamped by later immigrations. If so, then the midland populations should be more similar to Denmark and Norway than any other region of Ireland. Relethford and Crawford (1995) tested this hypothesis by examining the estimated genet-ic distances between Denmark, Norway, and six geographic regions of Ireland. As shown in Figure 12.13, the results support the Viking hypothesis because the midlands are closest to Denmark and Norway on the genetic distance map. North et al. (2000) found similar results in their analysis of blood groups and protein and enzyme polymorphisms.

The west–east division shown in Figure 12.12 also has a historical explanation. Irish history included a number of major settlements over the past millennium, primarily from England, Scotland, and Wales. Tills et al. (1977) suggest that the eastern counties were settled heavily by immigrants from England and Wales and the northern counties by immigrants from England and Scotland. Thus, the northern, eastern, and southeastern

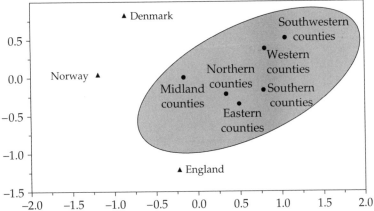

Figure 12.13. Principal coordinates plot of the genetic distances between the six geographic regions in Ireland (enclosed within the gray oval) and Denmark, Norway, and England based on anthropometric measures of the head. The Irish midlands plot closer to Norway and Denmark than any other Irish sample. The northern, eastern, and southern regions of Ireland plot closest to England. These patterns suggest differences in the history of settlement throughout Ireland (see text). (Adapted from Relethford and Crawford 1995.)

parts of Ireland were disproportionately settled by immigrants from elsewhere in northwestern Europe, which in turn led to some genetic distinctiveness between the west and east of Ireland. Figure 12.13 provides further support for this hypothesis; the northern, eastern, and southeastern parts of Ireland are more similar to England (a major source of many of the immigrants) than the western regions. It is clear that a variety of historical influences, coming from different sources and at different times, have helped shape genetic diversity in Ireland.

GLOBAL GENETIC DIVERSITY AND THE EVOLUTION OF THE HUMAN SPECIES

The case studies discussed thus far have centered on the specific population history of different human groups. Another area of extensive research by anthropologists and geneticists is the relationship between genetic variation and the history of our entire species. This section deals with examining patterns of genetic diversity across the entire planet, with an eye toward understanding the factors that could have shaped this diversity in our species' past.

Evolutionary Context

To understand the relevance of looking at global genetic diversity from an evolutionary perspective, it is necessary to consider what we know from the fossil record and comparative genetic studies of the primates. Molecular studies estimate that the evolutionary lines leading to living humans and living African apes diverged roughly 5–7 million years ago, a date which is consistent with the earliest fossils of human ancestors in Africa. One or more of these early ancestors gave rise in Africa to a number of different species of what are called "australopithecines," who walked upright but had small ape-sized brains. By 2.5 million years ago, there is some evidence for the first members of the genus *Homo*, characterized by larger brains and stone tool technology.

One species, *Homo erectus*, appeared in Africa roughly 2 million years ago and is our ancestor. *H. erectus* was similar to later humans in terms of its upright anatomy and brain size of roughly 70% of the average size for living humans. Some populations of *H. erectus* left Africa and dispersed into Southeast Asia and parts of Europe by 1.7 million years ago (Swisher et al. 1994, Gabunia et al. 2000). By 500,000 years ago, these early humans were spread out over parts of Africa, Asia, and Europe and brain size had increased to approach that of living humans. Anthropologists debate exactly what to call these "archaic" humans; some advocate placing them in several different species, while others suggest these are an early form of our own species, *Homo sapiens*.

The fossil record shows that "anatomically modern" humans appeared by 160,000 years ago in Africa (White et al. 2003). Modern human anatomy then spread over the Old World such that since 28,000 years ago all fossil humans can be classified as modern *H. sapiens*. The debate that concerns us here is the evolutionary origin of modern *H. sapiens*. During the time that anatomically modern humans appeared in Africa and then dispersed, there were other archaic humans living outside of Africa, including the Neandertals of Europe and the Middle East. Although it is clear that some archaic humans evolved into modern humans, there is considerable debate regarding the nature of this transition. Some anthropologists argue for a **replacement model**, where modern humans arose as a new species (*H. sapiens*) in Africa and then spread across the Old World, replacing existing populations of non-African archaics. Thus, the Neandertals and other archaic populations became extinct and did not contribute to our ancestry. Other anthropologists argue for a **continuity model**, where populations dispersing out of Africa mixed genetically with the non-African archaics. Under this model, Neandertals and other archaic humans belonged to the same species as modern humans.

The debate over modern human origins can be thought of in terms of personal ancestry (Relethford 2003a). Under the replacement model, all of our ancestors lived in Africa. Under the continuity model, some of our ancestors lived in Africa but some lived outside of Africa. Most anthropologists today agree that modern human anatomy appeared first in Africa. The debate is over whether there was any genetic mixture with populations outside of Africa or if these archaic groups became completely extinct.

Patterns of Genetic Diversity

What can we learn about the origin of our species by looking at genetic data? The basic assumption is that whatever happened in the past, those events left a signature on our genetic diversity. The task then is to examine genetic diversity in living humans and figure out what led to the patterns we observe. Several lines of evidence are explored here in terms of what they can tell us about the likelihood of replacement or continuity.

Gene Trees

The properties of coalescent models were discussed in Chapter 8. Here, we focus on global analysis of DNA that has provided clues to identifying common ancestors. Perhaps the most famous example of this type of study is Cann et al.'s (1987) analysis of mitochondrial DNA variation. They examined mitochondrial DNA sequences from 147 women having recent ancestry in Africa, Europe, Asia, and Australasia. They found 133 distinct mitochondrial DNA sequences that showed division into two groups, one consisting exclusively of women of African descent and the other including both African and non-African samples. Further, they estimated that the common ancestor was African and lived roughly 200,000 years ago. Further analyses of mitochondrial DNA, some using different methods, have arrived at the same general conclusion. At first glance, these results appear to confirm the replacement model because the pattern we see of recent African ancestry is compatible with the view that we all have common ancestry that can be traced back to Africa some 200,000 years ago. However, some have argued that the same pattern could occur if modern humans mixed with archaic populations outside of Africa (Templeton 1998). In other words, the mitochondrial DNA data may be tracking the movement of new DNA sequences within a species and not necessarily the movement of a new species.

Gene tree analysis has been applied to Y-chromosome polymorphisms as well as nonrecombining sections of nuclear DNA. While most gene trees show a recent African ancestry, not all do, and there is some genetic evidence for non-African ancestry in excess of 200,000 years ago. The genetic evidence is compatible with an African origin but perhaps not an exclusive African origin. In other words, there is some evidence for genetic continuity outside of Africa.

One of the major problems with gene trees is that they focus on the evolutionary patterns of a single gene or DNA sequence. What is needed is a way to combine these different analyses. Templeton (2002) has developed such a method and suggests that the overall patterns presented by gene tree analysis are best explained by a model of multiple dispersals out of Africa. Templeton's findings suggest that there were two major dispersals of human ancestors out of Africa following the initial dispersal of *H. erectus* 1.7 million years ago: the second dispersal dates between 400,000 and 800,000 years ago, and the third dates to roughly 150,000 years ago. Templeton's analysis suggests that in each case there was some genetic mixing with preexisting populations outside of Africa. If so, then our ancestry might be mostly, but not exclusively, African (Relethford 2001b; 2003a).

Levels of Genetic Diversity

There is a notable geographic distribution of genetic diversity in living humans: genetic variation is higher within sub-Saharan Africa than within other geographic regions. This pattern has been found for mitochondrial DNA and microsatellite DNA (see Fig. 12.14 for an example) as well as craniometric variation and skin color (Relethford 2001b). Why would sub-Saharan Africa show more genetic diversity than the rest of humanity?

One possibility has to do with the relationship between mutation and genetic diversity in neutral genes. Mutations accumulate over time, leading to increased genetic diversity. Under the replacement model, modern humans arose as a new species in Africa and then dispersed later outside of Africa, giving rise to populations in Europe, Asia, and elsewhere. If true, then human populations in Africa have been around longer and have therefore had more time to accumulate mutations, thus leading to higher levels of genetic diversity. Populations in Europe and Asia, however, have a more recent origin and, therefore, have had less time to accumulate mutations.

The relationship between mutation, time, and genetic diversity in modern human origins has been questioned (Rogers and Jorde 1995). More significantly, mutation is not the only factor that affects genetic diversity. Another important factor is genetic drift.

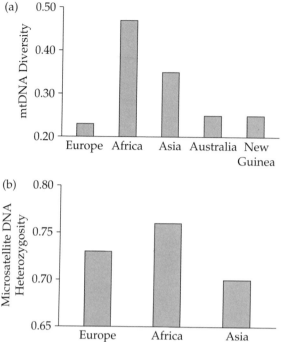

Figure 12.14. Regional variation in genetic diversity. *A*: Regional differences in mitochondrial DNA diversity. (From Cann et al. 1987.) *B*: Regional differences in microsatellite DNA heterozygosity. (Data from Relethford and Jorde 1999.)

Recall from Chapter 3 that genetic drift ultimately leads to one allele becoming either fixed (with an allele frequency of 1.0) or extinct (with an allele frequency of 0.0). Either way, genetic drift acts to remove variation from a population. The key factor affecting the magnitude of genetic drift is population size: the smaller the population, the more likely is genetic drift. Higher levels of genetic diversity might therefore reflect larger population size in Africa than elsewhere during much of human evolution, a suggestion supported by archaeological estimates of prehistoric population size (Hassan 1981). Several genetic studies have analyzed a model of genetic diversity that takes population size into account and have found that patterns of genetic diversity are best explained by a larger African population (Relethford and Harpending 1994, Relethford and Jorde 1999). These findings do not distinguish between replacement and continuity other than to show that while replacement is possible, it is not the only explanation.

Genetic Distance

Although the human species is rather genetically homogeneous and differences between regions are not that large, most studies of genetic distance between human populations across the world show sub-Saharan Africa to be the most distinct (Fig. 12.15). This pattern has been found in studies of red blood cell polymorphisms, DNA markers, and craniometrics (Jorde et al. 1998; Relethford 2001b).

Why would sub-Saharan Africa be more different? Again, one explanation is that modern humans arose as suggested by the replacement model. Here, modern humans

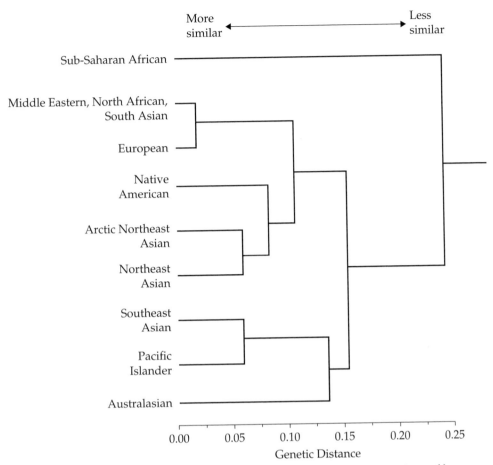

Figure 12.15. Dendrogram of genetic distances between nine regional groupings of human populations. These distances are based on 120 alleles, most of which were for red blood cell loci. Note the distinctiveness of sub-Saharan Africa relative to other regions. (Data from Cavalli-Sforza, Luca, *The History and Geography of Human Genes*. © 1994 Princeton University Press. Reprinted by permission of Princeton University Press.)

began in sub-Saharan Africa, and then some portion of the species split off to move out of Africa. This non-African population, probably first moving through the Middle East, later split into populations that colonized Europe, Asia, and Australasia. If so, then we would expect to see smaller genetic distances between the non-African populations because they split off later in time. An analogy helps illustrate this idea. People are more similar to their siblings than to their first cousins because siblings share more recent ancestors (parents) than first cousins (grandparents). However, we might simply be seeing a pattern caused by differences in the rate of gene flow between geographic regions. All that is needed to produce the same picture is to have somewhat less gene flow into Africa than out of Africa (Relethford and Harpending 1994, 1995). If so, then all we might be seeing is a reflection of past patterns of gene flow that could have occurred continually throughout time (continuity) or that took place after replacement. Again, the genetic evidence fits both models.

Insights from Ancient DNA

All of the above discussion is based on inferences that can be made about our past using genetic data from living populations. The discovery of polymerase chain reaction–based methods (see Chapter 8) now allows us to make additional inferences based on the extraction of ancient DNA from fossil specimens. Several studies of ancient humans and possible human ancestors have been performed, providing in each case a small amount of mitochondrial DNA to be sequenced. Several of these studies have been of European Neandertal specimens dating as recently as 29,000 years ago and perhaps as old as 70,000 years ago (Krings et al. 1997, 1999, 2000; Ovchinnikov et al. 2000). These Neandertal sequences are potentially quite valuable because of the debate over the evolutionary relationship of Neandertals to living humans. Under the replacement model, the Neandertals were a different species that became extinct, while under the continuity model, the Neandertals are likely to have contributed to some of our ancestry.

Comparison of the Neandertal mitochondrial DNA sequences with those of living humans has raised some interesting issues. The Neandertal sequences are quite a bit different from those of living humans. For example, one specimen differed from living humans on an average of 27 substitutions out of the 360 bp sequence that was analyzed. In comparison, living humans differ only an average of 8 substitutions for the same sequence. What does this difference mean? There is no doubt that Neandertal mitochondrial DNA is different, but the question is whether this difference is sufficient to classify them as a separate species (Relethford 2001b; Gutierrez et al. 2002).

Mitochondrial DNA sequences have also been extracted from ancient human fossils that are definitely modern, thus allowing some additional comparison. One of these sequences, extracted from the Lake Mungo 3 skeleton from Australia, is different from both Neandertals and living humans (Adcock et al. 2001), raising the possibility that mitochondrial DNA sequences have changed within modern humans. However, a recent analysis of European moderns dating back 24,000 years shows little change in mitochondrial DNA over time (Caramelli et al. 2003). Although the analysis of ancient DNA may provide greater insight into our species' ancestry, we still need a more comprehensive series of samples. At this point, the best we can say is that the demonstrated difference in Neandertal mitochondrial DNA might be a reflection of replacement in Europe or a reflection of a small Neandertal population mixing with a larger group of modern humans moving into Europe. Thus, we cannot clearly rule out either replacement or continuity in Europe. Of course, the issue of ancient DNA is not without other problems, most notably the potential problem of contamination. If an ancient fossil shows modern mitochondrial DNA, how can we be sure we are seeing the DNA sequence of that specimen or the DNA sequence of a living human who has handled the fossil?

Conclusion?

Given all the above evidence on genetic variation, can we see anything definite about the evolutionary history of our species? Although the issue of replacement versus continuity cannot be resolved at present, we can provide some generalizations that may be useful in considering future research. Genetic data from living human populations are consistent in showing a strong African influence. Combined with the fossil record to date, all lines of evidence point to an African origin of modern human anatomy some 200,000 years ago. What remains less clear, however, is whether there was replacement of or some mixing with archaic populations outside of Africa (Stringer 2003). Elsewhere, one of us has suggested that the most likely hypothesis at present is that most, but not all, of our ancestors came from Africa, thus combining a recent African origin with some mixture and genetic continuity outside of Africa (Relethford 2001b; 2003a).

CHAPTER SUMMARY

Studies of genetic variation within and among human populations allow inferences regarding population structure and population history. There are many different measures of genetic distance between populations, but all share the underlying method of using information from many different loci or traits in order to get the best estimate of underlying populational relationships. Population structure is concerned with those factors that affect mate choice and genetic relationships between individuals and subgroups within a population. Isolation by distance is a common factor affecting population structure; typically, populations farther apart geographically tend to be more genetically different. Other factors have also been shown to influence human population structure, including cultural factors such as language differences and religious affiliation.

Anthropological geneticists have conducted many studies relating genetic variation to population history. Three examples have been discussed here: *(1)* the origin of Native Americans, *(2)* admixture in African Americans, and *(3)* the population history of Ireland. All three case studies illustrate how historical patterns have affected the movement of human populations and the genetic impact of these movements. A major focus in anthropological genetics today is the use of genetic data to unravel the population history of our entire species. Fossil and genetic evidence point to a recent African origin of modern humans, but it is not clear whether earlier humans outside of Africa were replaced completely or mixed with moderns dispersing out of Africa. Further work is needed to resolve this debate.

SUPPLEMENTAL RESOURCES

Cavalli-Sforza L L and Cavalli-Sforza F (1995) *The Great Human Diasporas: The History of Diversity and Evolution*. Reading, MA: Addison-Wesley.

Cavalli-Sforza L L, Menozzi P, and Piazza A (1994) *The History and Geography of Human Genes*. Princeton: Princeton University Press. A comprehensive analysis of genetics and population history, including discussion of global history as well as the relationship of genetics and population history in each geographic region.

Crawford M H and Mielke J H, eds. (1982) *Current Developments in Anthropological Genetics, Ecology and Population Structure*, volume 2. New York: Plenum Press. A collection of papers focusing on the analysis of population structure and history in different ecological and cultural settings.

Olson S (2002) *Mapping Human History: Discovering the Past Through Our Genes*. Boston: Houghton Mifflin.

Relethford J H (2003) *Reflections of Our Past: How Human History Is Revealed in Our Genes*. Denver: Westview.

Cavalli-Sforza and Cavalli-Sforza (1995), Olson (2002), and Relethford (2003) are written for a general audience and summarize studies that use genetics to address questions of population history.

13

Genetics, Behavior, and Human Variation

A common theme throughout this book has been the complexity of factors affecting human biological variation. This complexity is apparent even when considering genetic markers that are completely inherited and not affected by environmental or developmental influences, such as blood groups (Chapter 4), blood proteins and enzymes (Chapter 5), and DNA markers (Chapter 8). Even without the complications of non-genetic influences on a person's genotype or phenotype, the distribution of allele frequencies can reflect a complex set of interactions among the four evolutionary forces, which in turn can be influenced by a variety of environmental, demographic, and cultural factors. The analysis of human variation is even more complex when dealing with quantitative traits (Chapters 9–11), where phenotypic variation is also affected by environmental and developmental factors.

The analysis of human behavioral characteristics becomes even more complex given our greater uncertainty about the relative influence of "nature" and "nurture." Further, behavioral traits, such as intelligence and sexual orientation, are often culturally and politically charged subjects, having greater impact on our lives and how we view people than ABO blood type or cormic index. Even the simplest discussions of the relationship of human biology and behavior, such as an analysis of intelligence quotient (IQ) scores, is likely to raise accusations of racism or political correctness, depending on the political views of the participants. This is not to suggest that scientific analysis of the relationship of genetics and human behavior is impossible but to point out that this analysis and discussion of this topic are frequently shaped by social and political ideologies (Marks 1995).

The primary purpose of this chapter is to examine briefly some of the ways in which human behavior is considered in terms of human biological variation. Several examples (dyslexia, sexual orientation, IQ scores) are considered in terms of some basic questions: (1) How are behaviors defined and measured? (2) Is there a genetic component of a behavior? (3) What are the relative roles of genetic and environmental variation on variation in a behavior? (4) What do we know about the distribution of a behavioral trait within our species, and does it reflect variation in genetics, environment, or both? (5) Can a behavioral trait be understood in terms of evolutionary principles?

HUMAN BEHAVIORAL GENETICS

The study of **human behavioral genetics** attempts to understand the interaction of genetic and environmental factors on behavioral traits. Much research in human behavioral genetics has traditionally concerned itself with attempts to partition phenotypic variation into genetic and environmental components and with the estimation of heritability. As such, much of human behavioral genetics is the application of quantitative genetics theory (Chapter 9) to complex behavioral traits. Gilger (2000) refers to such studies as the "classic approach" to human behavioral genetics and notes that a new approach has been developing that attempts to link molecular genetics with the study of behavior by identifying specific genes that influence behaviors. Note that the attempt to link specific genes with behaviors does not imply that there is a direct correspondence or that a particular gene for a given behavior exists.

Genetics, Environment, and Behavior

As described in Chapter 9, much of quantitative genetics is based on the fundamental equation partitioning total phenotypic variance (V_p) into two components, additive genetic variance (V_a) and environmental variance (V_e):

$$V_p = V_a + V_e$$

Standard quantitative genetic methods, including twin studies, family studies, and adoption studies are used to estimate the relative proportions of phenotypic variance due to genetic variance (heritability) and environmental variance (Segal and MacDonald 1998).

Although these methods have already been described in Chapter 9, it is important to review briefly what such studies can and cannot tell us. First, demonstration of significant heritability establishes only that a trait has some genetic component. Heritability is a relative measure, showing the proportion of phenotypic variation due to genetic variation; as such, changes in the degree of environmental variation will lead to changes in heritability. For example, reduction of the amount of environmental variation will result in an increase in heritability. Second, estimates of heritability tell us nothing about the relative role of genetics for individuals. For example, a heritability of 0.6 for IQ does not mean that a given individual's IQ score is 60% due to genetics and 40% due to environment. The numbers apply only to the overall partitioning of variation within a sample. Third, heritability estimates do not necessarily tell us anything about the relative role of genetics in differences between groups. Imagine an example where two groups are different for a given trait and that this trait also has a high heritability. Can we automatically assume therefore that the primary reason for the difference in the two groups is genetic? No, because we might also be seeing two groups that are genetically similar but exposed to different environments. Heritability is a measure of variation within groups and does not necessarily tell us anything about genetic differences between groups. As discussed below, misconceptions regarding heritability abound when considering behavioral traits.

Methods of Behavioral Genetics

Classic quantitative genetic models can provide us with a clue as to whether genetic factors influence a behavioral trait. Some genetic influence has been found for a wide variety of behavioral traits, including schizophrenia, autism, reading and language disabilities, eating disorders, and a variety of personality and cognitive traits, among others. Plomin et al. (1994) note that such studies merely establish the importance of genetic influences and that more detailed analyses are needed to understand better the interaction of genetic and environmental influences. In particular, they note the importance of studies that deal with developmental changes in the relative influence of genetics and environment, the need for multivariate analysis to consider interactions among different behavioral traits, and the need to consider whether outcomes often labeled as "normal" and "abnormal" are actually extremes along a continuum of phenotypes.

In recent years, the study of behavioral genetics has also gone beyond tests for an overall genetic influence to a more detailed attempt to understand the underlying genetics and physiology by identifying specific genes (and possibly their actual influence). The search for quantitative trait loci (QTL) is one example, attempting to locate genes that have varying effects on a given phenotype. Some preliminary analysis suggests association of certain genes/chromosomes with several behavioral traits, including mental retardation, Alzheimer's disease, hyperactivity, and sexual orientation, among others (Plomin et al. 1994).

All of these approaches can help inform us about the role of genetics in complex behavior but represent only a starting point in a much more complex analysis. Evidence of a genetic association should not be read as evidence for *genetic determinism*, the view that one's genes uniquely control one's destiny. Although there are some human traits that are completely controlled by genetics and fixed for life, such as blood type, most physical and behavioral traits represent a complex mixture of genetic and environmental effects (and interrelationships). Plomin et al. (1994) note that behavioral genetics actually provides some of the best evidence for nongenetic influences. Most of the behavioral traits reviewed by them show at least as much nonheritable as heritable influence on total phenotypic variation. What is often lacking is a clearer understanding of these environmental effects and how they covary with genetic factors. The long-standing debate over nature (genetics) versus nurture (environment) has often been phrased in terms of an either–or dichotomy. In truth, most complex traits (physical and behavioral) represent the combined influence of both genetics and environment. The proper answer to the question *Is trait X due to nature or nurture?* is often *Yes*.

CASE STUDIES IN BEHAVIORAL GENETICS

Both "classic" quantitative genetic approaches and molecular studies have been applied to a number of human behavioral traits. To illustrate some of these basic principles, this section provides information on three examples of studies of human behavioral genetics: (1) dyslexia, (2) sexual orientation, and (3) IQ.

Dyslexia

Dyslexia, also known as "specific reading disability," is a chronic neurological condition defined by the inability to recognize and comprehend symbols, especially written words, in subjects with normal intelligence (Smith et al. 1998). Symptoms include poor reading skills and a tendency to reverse letters and words when reading or writing. A standard definition focuses on word accuracy relative to IQ (Paulesu et al. 2001).

Prevalence of Dyslexia

Estimates of prevalence in the United States suggest that up to 10%–15% of school-aged children suffer from dyslexia (DeFries et al. 1987). Cross-cultural studies show that prevalence rates vary across different languages, suggesting a relationship with the specific type of language. The prevalence rate in Italy, for example, is roughly half that in the United States (Paulesu et al. 2001).

The Genetic Basis of Dyslexia

It has long been observed that dyslexia tends to run in families. The *familial risk rates* (a measure of the probability of a sibling having a trait if it is present in another sibling) are high, ranging 35%–45% (Pennington et al. 1991). Twin studies have shown that the concordance of dyslexia is higher in monozygotic twins than dizygotic twins, also suggesting a strong genetic component (DeFries et al. 1987). Heritability estimates are generally high (0.5–0.7) (DeFries et al. 1991).

Several studies have used family studies to test different models for the genetic basis of dyslexia. One study, by Pennington et al. (1991), examined the relatives of dyslexics for dyslexia in four samples of families from three different states (Colorado, Washington, and Iowa). Segregation analysis was used to test for a major gene model (compared with a polygenic model with numerous loci having an equal and additive effect). Three of the four samples showed evidence supporting a major gene. Further analysis showed that the best-fitting model was where this major gene was either dominant or semidominant, and there were further indications of some variation in expression between the sexes.

Recent molecular studies have examined the association between dyslexia and QTL. Several studies have suggested an association with genetic markers on chromosomes 6 and 15 (Smith et al. 1998). The linkage with markers on chromosome 6 is most apparent, linking dyslexia to specific genes in the human leukocyte antigen (HLA) region. Gilger (2000) notes that it is interesting that other behavioral traits, such as autism and schizophrenia, also show QTL on chromosome 6. Of course, a statistical association of dyslexia and a QTL does not tell us about the specific relationship between genetics, neurological development, and behavior, although it is a start.

Language Orthography and Dyslexia

Much of the genetic analysis of dyslexia has been conducted on families in the United States. There have been some doubts about the universality of dyslexia because, as noted above, prevalence rates vary across different languages, with lower rates found for languages with a "shallow" orthography, where more letters and letter combinations are uniquely mapped to specific sounds. Languages with "deep" orthography are those where the mapping between letters and letter combinations and sound is more ambiguous.

Paulesu et al. (2001) examined differences in behavioral tests and neurology in dyslexic subjects and a control group in three countries: France, the United Kingdom, and Italy. Both French and English are languages with deep orthography, whereas Italian is a language with shallow orthography. The goal of their study was to look for similarities that might suggest a universal phenomenon. They found that dyslexics performed significantly below controls on various reading tests in all three countries. However, Italian dyslexics scored better than either English or French dyslexics on reading tasks (faster reaction time and greater accuracy), even though their performance was still lower than that of Italian controls (see Fig. 13.1 for an example).

Paulesu et al. (2001) also examined neurological functioning using positron emission tomographic scans on dyslexics in all three countries. They found that dyslexics, regardless of country, showed the same reduction in activity in a region of the left hemisphere of the brain and concluded that dyslexia is a universal neurological phenomenon. However, the impact of this neurological difference (presumably linked to a genetic difference) on reading skills is less in certain languages such as Italian. They further suggest that languages with deep orthography, such as English and French, can "aggravate the

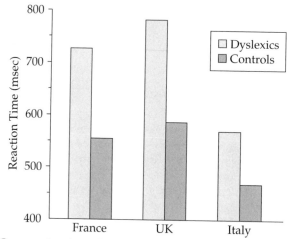

Figure 13.1. From Cross-national results for word reading reaction time in dyslexics and controls. Note that although reaction time is greater among dyslexics in all three countries, the actual magnitude is less in Italy, perhaps reflecting the different orthography of the written language. (Data from Paulesu et al. 2001.)

literacy impairments of otherwise mild cases of dyslexia" (p 2167). We think that this case study is an excellent example of how specific cultural factors (in this case, language) can affect the phenotype (reading ability). Although the evidence for neurological (and genetic) correlates of dyslexia is strong, the prevalence rates in a given population will also depend on environmental factors. Genetics is not destiny but instead increases the probability of an outcome in a specific environment.

Sexual Orientation

What are the factors, genetic and/or environmental, that affect a person's sexual orientation? This question is complex and further complicated by the problem of defining sexual orientation.

Measuring Sexual Orientation

Can individuals unambiguously be defined as homosexual or heterosexual, or are there gradations of sexual orientation? Marks (1995) notes that the usual use of the terms *homosexuality* and *heterosexuality* is to a large extent a cultural construct which attempts to dichotomize variation in human sexual orientation. For example, how should a man who has been involved only in sexual relationships with women but who has frequent homosexual fantasies be classified? What about a man who has had only occasional homosexual fantasies or experiences? What impact do cultural conditions have, particularly those affecting degree of acceptance, on nonheterosexuality, and how do these vary across time and society? For example, how is a person who has homosexual fantasies and self-identifies as a homosexual but maintains a heterosexual lifestyle because of cultural restrictions classified?

How is sexual orientation assessed? Most studies use the **Kinsey scale**, which considers four variables, each measured on a 7-point scale ranging from 0 to 6:

1. *Self-identification:* Individuals self-identify as exclusively homosexual (0), predominantly homosexual (1), bisexual with an orientation to the same sex (2), bisexual with an equal orientation to both sexes (3), bisexual with an orientation to the opposite sex (4), predominantly heterosexual (5), or exclusively heterosexual (6).
2. *Romantic/sexual attraction:* A measure of the degree to which an individual finds others attractive romantically and sexually. For males, this variable is measured by the degree of attraction to females (0–6) and for females, the degree of attraction to males.
3. *Romantic/sexual fantasy:* A measure of the degree to which in individual fantasizes about males and females. For males, this variable is measured by the degree to which females are involved in fantasies (0–6) and for females, the degree to which males are involved.
4. *Sexual behavior:* A measure of an individual's actual sexual experiences. For males, this uses a scale (0–6) based on female sexual partners, and for females, it is based on male partners.

If there is a high correlation between these four factors, then an average composite is often used. For each variable (or a composite), analysis is typically based on either the actual score or by placing individuals into one of three groups: individuals with a score of 0 and 1 are classified as homosexuals, individuals with a score of 2–4 are classified as bisexuals, and individuals with a score of 5 or 6 are classified as heterosexuals (Pattatucci 1998).

The Kinsey scale is not without problems. Although much work has gone into assessing the validity and reliability of these measures, there continue to be problems with volunteer bias as well as the pressure to give socially acceptable answers, among others (Pillard and Bailey 1998). Pattatucci (1998) further notes that it is almost impossible to obtain random samples of homosexuals, who continue to be a marginalized population. She notes that "the data reported in sexual orientation studies, regardless of how convincing, apply only to the cohort studied and cannot necessarily be extrapolated to lesbian, gay male, bisexual, and heterosexual populations at large" (p 373).

The Distribution of Sexual Orientation

Variation in study design and the degree to which a sample is or is not random probably contribute somewhat to the variation in prevalence rates. Some studies suggest the prevalence of male homosexuality ranges 4%–10% (Hamer et al. 1993), while others suggest only 3%–4% are exclusively homosexual (Pillard and Bailey 1998). There are differences between the distribution of Kinsey scores for males and females. Pillard and Bailey (1998) note that the distribution of scores is bimodal in men, who are more most likely to be either exclusively heterosexual or homosexual, with fewer cases of bisexuality, whereas bisexuality is more common in women. Pattatucci (1998) found similar results: few males were bisexual (Kinsey score of 2–4), but roughly one-third of women were bisexual. Again, these prevalence rates cannot easily be extended to the general population in most cases and further may reflect cultural differences regarding bisexuality (Pattatucci 1998).

Biological Correlates of Homosexuality

Several studies have reported evidence of biological associations with homosexuality. Although such studies do not demonstrate causality, they are suggestive of some

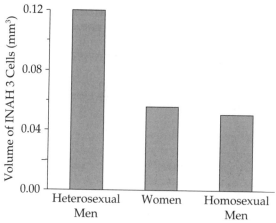

Figure 13.2. Comparison of the volume of INAH 3 cells in the anterior hypothalamus of the brain in male heterosexuals, women, and male homosexuals. (Data from LeVay 1991.)

underlying biological predisposition to sexual orientation. LeVay (1991) examined differences in the structure of the anterior hypothalamus, a region of the brain linked to male-typical sexual behavior. He compared results from three samples: women, heterosexual men, and homosexual men. He found that the size of one group of cells (interstitial nuclei of the anterior hypothalamus 3 [INAH 3]) was twice as large in heterosexual men as in either women or homosexual men (Fig. 13.2). Although potentially interesting, this study does not demonstrate a direct causal relationship (correlation does not imply causation), and it is possible that both observed sexual orientation and brain anatomy are linked to yet a third factor. Nonetheless, the study is suggestive of some biological basis for sexual orientation (although see Marks 1995 for a critique).

Another suggested biological correlate with sexual orientation is the finger-length ratio, which compares the length of the second digit (2D, or index finger) with the length of the fourth digit (4D, or ring finger). Finger-length ratios are affected by the level of prenatal exposure to androgens (male sex hormones). Women tend to have roughly equal lengths of 2D and 4D, whereas men tend to have a shorter 2D. Williams et al. (2000) examined a sample of 720 adults in San Francisco and found that the 2D:4D ratio was higher in women (as expected) and that this difference was greater for the right hand, suggesting greater sensitivity to fetal androgens. They also found that the right-hand 2D:4D ratio of homosexual women was significantly less than that of heterosexual women but not significantly different from that of heterosexual men. Their conclusion is that homosexual women have received greater amounts of fetal androgens, which presumably act to masculinize sexual orientation. Further, these results suggest that events before birth can influence sexual orientation. In another study of 2D:4D finger-length ratios, Robinson and Manning (2000) found a lower ratio in 88 homosexual men compared to 88 heterosexual controls and a relationship between the ratio and the degree of exclusive homosexuality.

Genetics and Sexual Orientation

Again, these studies must be viewed as suggestive rather than conclusive. They show biological correlates with sexual orientation and further suggest that there may be some

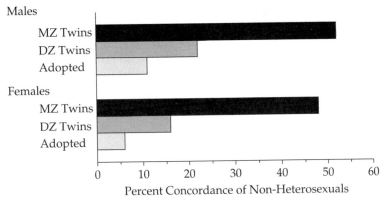

Figure 13.3. Percent concordance of nonheterosexuality among monozygotic (MZ) twins, dizygotic (DZ) twins, and adopted siblings of nonheterosexual individuals. (Data from Pillard and Bailey 1998.)

underlying genetic component. Behavioral genetic analyses provide somewhat firmer support for a genetic influence on sexual orientation. Pillard and Bailey (1998) examined rates of concordance for nonheterosexuality (including both homosexuality and bisexuality) among three different groups of siblings: monozygotic twins, dizygotic twins, and adopted siblings. For both males and females, the percentage of concordance (both siblings classified as nonheterosexuals) was highest in monozygotic twins, lower in dizygotic twins, and lowest in adopted siblings (see Fig. 13.3). This ranking is expected if there is some genetic predisposition to homosexuality since monozygotic twins are more closely related genetically than dizygotic twins, who in turn are more genetically related than adopted siblings.

Classic quantitative genetic studies have suggested a moderate to high heritability for sexual orientation, ranging 0.31–0.74 for males and 0.27–0.76 for females (Hamer et al. 1993). These values suggest some genetic influence but also a considerable amount of variation related to environmental factors. Such results do not support a model of genetic determinism (i.e., genetics causes sexual orientation) but instead show the wide range of nongenetic influences on complex behaviors. Further, Pattatucci (1998) notes that sexual orientation is multifactorial in nature and states that "it is highly improbable that any single genetic variation or allele will be present in all homosexual individuals and absent from all heterosexual individuals" (p 368).

Hamer et al. (1993) found an interesting pattern in rates of concordance among male relatives of self-identified male homosexuals. As was found in other studies (Pattatucci 1998, Pillard and Bailey 1998), most subjects self-identified as either strongly or exclusively heterosexual (Kinsey score of 5–6) or strongly or exclusively homosexual (Kinsey score of 0–1). Family histories were collected for all subjects, and pedigree analysis was conducted on 76 randomly chosen male homosexuals. Rates of concordance (the percentage of male relatives that were also homosexual) were compiled for different degrees of relationship, as shown in Table 13.1. Compared to the frequency of homosexuality in the general population (2% in this particular study), 13.5% of brothers were also homosexual. In general, the concordance rate is lower in uncles (who are less genetically related than brothers) and even less in male cousins. Perhaps more interesting is the pattern

Table 13.1 Rates of homosexual orientation in the male relatives of homosexual male subjects

Relationship to Homosexual Male Subject	Percentage Homosexuality	
	Significantly Greater than in General Population	Not Significantly Different from General Population
Brother	13.5	
Maternal uncle—mother's brother	7.3	
Paternal uncle—father's brother		1.7
Maternal cousin—mother's sister's son	7.7	
Maternal cousin—mother's brother's son		3.9
Paternal cousin—father's sister's son		3.6
Paternal cousin—father's brother's son		5.4

Data from Hamer et al. 1993.

of significant results. Significant rates of increased homosexuality were found only in those male relatives who were related through the maternal line: brothers, maternal uncles (mother's brothers), and mother's sister's sons. However, paternal uncles, mother's brother's sons, and paternal cousins showed rates of homosexuality that were not significantly different from the general population. The same general results were obtained when the authors confined their analysis to cases where one or more brothers were homosexual.

These results suggest some degree of maternal transmission. Hamer et al. (1993) further suggest that this is due to X-chromosome linkage since males (who have one X chromosome and one Y chromosome) receive their X chromosome from the mother. If there is a gene (or genes) that increases the probability of male homosexuality, then genetically related gay men should share DNA markers on the X chromosome that are located close to these hypothetical genes. The authors examined this possibility by performing linkage analysis on the families of 40 pairs of gay brothers, analyzing 22 markers on the X chromosome. They found a significant association with region Xq28 on the X chromosome and suggested that a gene affecting male sexual orientation lies within the 4 million base pairs of this region.

A subsequent study (Hu et al. 1995), using an independent sample of families with two gay brothers, found the same results—significant linkage between the probability of male homosexuality and the Xq28 region of the X chromosome. This study also looked at families with lesbian sisters and found no significant linkage with this marker, suggesting that "a locus at Xq28 influences sexual orientation in men but not in women" (p 253). According to Hu et al. (1995), sexual orientation develops along different pathways in men and women.

Evolutionary Considerations

If we accept the findings to date that there is some genetic component to sexual orientation, then we might ask how such genes might be perpetuated over many generations, given the fact that exclusive homosexuality does not lead to any offspring and, thus, no passing on of the genes. Robinson and Manning (2000) note that the prevalence of male homosexuality is too high to explain by recurrent mutation and that therefore there is some evolutionary advantage maintaining a higher allele frequency.

Pillard and Bailey (1998) phrase this question in terms of the relative costs and benefits of such genes: "If a genetic predisposition for a homosexual orientation exists, what advantage could it confer to pay the cost of lost reproduction to the individual?" (p 360). They note that three different explanations have been proposed. The first is the hypothesis that there is a reproductive advantage to heterozygotes such that certain alleles would be maintained at higher frequencies. The second explanation is that there is a relationship between genetic susceptibility to male homosexuality and altruism so that closely related kin benefit from the altruistic actions of gay family members, thus acting to increase the likelihood of passing on certain genes (shared by close kin). The third explanation is that if male homosexuality is linked to the X chromosome, then a gene providing a reproductive advantage in females would persist over time. It is important to note that Pillard and Bailey (1998) find that there is, at present, no evidence to support any of these explanations. Also, all evolutionary explanations assume that an orientation toward homosexuality results in an actual increase in homosexual behavior and a subsequent decrease in heterosexual behavior. In different societies and at different times, it is likely that there may have been greater discrepancies between orientation and actual behavior, depending on cultural constraints.

Summary

The evidence is suggestive that some part of male homosexual orientation is influenced by genetic factors, although the evidence is less clear for females and the relationship between genetic and environmental factors may differ between the sexes. However, the relationship, if confirmed, does not appear to be deterministic; and as is the case for many complex behavioral traits, environmental influences are very significant. We cannot reduce sexual orientation to a simple model of genetic determinism, and the relative roles of genetic and environmental factors are likely to operate in different ways in different cultural contexts. Future research in this area would be enhanced by a wider cross-cultural approach, taking into consideration cultural differences in definitions and perceptions of sexual orientation.

IQ Test Scores

Few behavioral traits are as controversial as IQ test scores. The past century has seen numerous debates over IQ, ranging from its definition and relationship to intelligence to the role of genetic and environmental influences to debates over "race" and intelligence. The debates over IQ have not been conducted in academic isolation; these debates are related to, and often influence, public policy ranging from immigration quotas to sterilization (Gould 1981).

IQ Tests and Intelligence

What is *intelligence*? It is a term we all use frequently but, as is the case with terms such as *love* or *race*, it can mean many things to many people. Are we referring to accumulated knowledge, problem-solving ability, verbal skills, quantitative skills, or many other components? A continuing debate in psychology is the definition of intelligence and how it can be measured.

There have been many attempts to measure intelligence using various biometric variables, such as cranial size (Gould 1981). The English geneticist Francis Galton was interested in the inheritance of intelligence and analyzed both anecdotal evidence (e.g., historical prominence) as well as some anthropometric measures he though might reflect intelligence, including sensory acuity and reaction time. The first intelligence test was developed at the beginning of the twentieth century in France by Alfred Binet, who was assigned by the French Ministry of Public Education to develop a test that could identify any child who "because of the state of his intelligence, was unable to profit, in a normal manner, from the instruction given in ordinary schools" (Macintosh 1998, 12). The purpose here was not to rank intelligence, as in many later variants of the test, but rather to identify children who required additional help. Binet sought to develop a test that would measure intellectual ability rather than academic knowledge, although it is important to note that he did not feel that such tests necessarily reflected innate intelligence. The outcome of his test was to assign each child a "mental age" based on his or her performance relative to peers at different ages. Thus, a child who successfully passed a test that was successfully completed by the majority of 6-year-olds but did not pass the equivalent test for 7-year-olds would be assigned a "mental age" of 6 years (Macintosh 1998).

Psychologist Lewis Terman devised the Stanford revision of Binet's test, leading to the modern Stanford-Binet test. Terman further modified the test to express the individual's mental age relative to chronological age. This ratio, known as the IQ, is computed as follows:

$$IQ = \frac{\text{Mental age}}{\text{Chronological age}} \times 100$$

If, for example, a person has a mental age of 14 and a chronological age of 12, his or her IQ would be $\frac{14}{12} \times 100 = 117$. Modern IQ tests are designed so that the average for a reference population is 100, with a standard deviation of 15. An average of 100 indicates that the average person has a mental age equal to his or her chronological age. Given a normal distribution, a standard deviation of 15 means that approximately 68% of all people will score within one standard deviation of the mean (or between 85 and 115) or roughly 95% of all people will score within two standard deviations of the mean (or between 70 and 130). Although Binet's objective was to devise a test for diagnostic purposes such that the truly important finding was whether a child passed or did not pass a given test, IQ tests soon became devices for ranking intelligence, with the inevitable conclusion that higher IQ scores necessarily equated to higher intelligence and, more unfortunately, greater "worth" to society (see Gould 1981).

Many variants of the IQ test have been used since its development, including versions developed during the First World War for testing of literate (alpha test) and illiterate (beta test) recruits. The Stanford-Binet test was later replaced by the Wechsler tests for adults (Wechsler Adult Intelligence Scale) and children (Wechsler Intelligence Scale for Children). The Wechsler tests are made up of 11 subtests, six of which test verbal abilities and five of which test performance (see Table 13.2). Another commonly used test is Raven's Progressive Matrices, which is a nonverbal test designed to measure general abstract ability by completing sequences of abstract designs (Macintosh 1998).

What is the relationship of IQ tests to intelligence? There is considerable disagreement. Some scholars argue that intelligence is multifaceted and cannot be measured by any single test (e.g., Bodmer and Cavalli-Sforza 1976, Hunt 1995). Others argue that IQ measures general intelligence, usually referred to as g (Herrnstein and Murray 1994, Bouchard 1998, Jensen 1998). The concept of g is based in large part on the observation

Table 13.2 Subtests of the Wechsler intelligence test

Test	Subtest
Verbal	Information (general knowledge)
	Vocabulary (word definitions)
	Comprehension (meaning)
	Arithmetic (solving word problems)
	Similarities (how are two things alike)
	Digit span (repeating a series of digits)
Performance	Picture completion (completion of familiar figure)
	Picture arrangement (arrange pictures to tell a story)
	Block design (forming certain patterns using colored cubes)
	Object assembly (simple jigsaw puzzles)
	Digit symbol (timed test to fill in symbols corresponding to digits 1–9)

Adapted from Mackintosh 1998.

that there are often high correlations between the scores on the different components of IQ tests, thus suggesting some common ability across tests. A statistical method known as factor analysis is often used to quantify this general factor of intelligence, and consequently it is often treated as a single measure of overall intelligence. There is no doubt about the correlation between test scores, but there is debate over whether the g factor represents an underlying component of intelligence or a statistical artifact (Gould 1981, but see Jensen 1998). Gould (1981) further argues that our search for a single thing labeled "intelligence" that can be measured by a single test distorts the nature of intelligence and variability.

Although this section focuses on the specific use of IQ tests, it is worth mentioning that some studies focusing on variation in human intelligence use brain size (or cranial capacity) as a proxy measure for intelligence (e.g., Rushton 1995). During the past decade, a number of studies have shown significant moderate correlations between brain size, as measured by magnetic resonance imaging, and IQ scores (e.g., Willerman et al. 1991). These correlations average about 0.4, implying a moderate association between brain size and IQ (and, presumably, intelligence). However, as noted by Schoenemann et al. (2000), these studies failed to control for environmental differences between families that might elevate the correlation. To deal with this potential problem, Schoenemann et al. examined the relationship between IQ scores and brain size within families, thus controlling to some extent for environmental differences between families, and found essentially zero correlations. Given the problems noted by some psychologists even working directly with IQ scores, it seems inadvisable to use proxy measures such as brain size.

Heritability

If we assume for the moment that intelligence can be defined in such a way as to measure it with a standard IQ test, there is the question of genetic and environmental influences on IQ test scores. Is one's IQ score a reflection of genetics, environment, or both? Although few argue that intelligence is completely a reflection of genetics or environment, there is considerable difference of opinion regarding the relative influence of nature and nurture on IQ test scores, particularly when the debate is extended to questions of "racial" differences (Flynn 1980).

Twin and adoption studies provide support for some genetic influence on IQ scores. In general, monozygotic twins show higher correlations in IQ scores than dizygotic twins or adopted siblings (Macintosh 1998). Numerous studies have been conducted to estimate the heritability of IQ. Although there is general agreement that some portion of IQ is heritable, there is a wide range in the specific estimates of heritability. Bouchard (1998) reviewed a number of twin studies using different research designs and concluded that most of these studies show that heritability of IQ is high, ranging 0.6–0.8. Bouchard et al. (1990) reviewed their long-term study of twin pairs in monozygotic and dizygotic twins who were separated at birth and raised apart. They conclude that heritability of IQ is roughly 0.7.

Devlin et al. (1997) note a problem with uncritical acceptance of high heritabilities for IQ. Although similarity between relatives is due in part to genetic factors, it can also be due to shared environments. Specifically, they argue that many estimates of the heritability of IQ based on twins raised apart are biased upward because of a failure to take into account the shared prenatal maternal environment of twins. Earlier studies often assumed that maternal environmental effects (sharing the same womb) were negligible such that any similarity of monozygotic twins raised apart in different environments was due entirely to genetic factors. Devlin et al. (1997) developed a statistical model that incorporated maternal environment and applied it to data from 212 earlier studies. They found that a shared maternal environment explained 20% of the similarity in IQ of twins. Once this effect is taken into consideration, the narrow sense heritability (due to additive genetic factors) drops to 0.34. Their study also points to a strong effect early in life of the maternal environment, which could have major effects on brain development (McGue 1997). It appears that there is some genetic influence on IQ scores, although perhaps not as high (in terms of relative heritability) as once thought. Some recent research is now attempting to locate QTLs that are related to IQ scores (Daniels et al. 1998).

Environmental Influences

In this context, the term *environment* is broad, referring to all nongenetic factors that could affect IQ scores, including both sociocultural and physical environmental factors, and operating in both prenatal and postnatal life. Numerous studies have shown significant correlations of environmental factors with IQ test scores, including education, social class, family income, neighborhood, family size, birth order, nutrition, disease, environmental pollution, and mother's health during pregnancy, among others (e.g., Loehlin et al. 1975, Gould 1981, Macintosh 1998). Although ample evidence exists for environmental correlations, these do not necessarily prove causality. Some have argued, for example, that some correlations with environment might reflect correlations with genetics, as would be the case if one's educational attainment reflected innate intelligence. It is also not always clear how the various environmental effects interact in terms of a relationship with IQ scores. Macintosh (1998, 139) concludes the following:

> the most plausible conclusion suggested by these data is that there is no single environmental factor that has a magical and permanent effect on children's IQ scores. What the data imply is that there are a very large number of factors, each of which has no more than a small effect on IQ. In most modern societies, many of these factors happen to go together, such that a child benefiting (or losing out) from one will probably be fortunate (or unlucky) enough to benefit (or lose out) from another. But in terms of psychological mode of operation, they are probably quite independent: there are no grounds for believing that lead poisoning and loss of schooling depress a child's IQ for the same reason.

One of the clearest indications that environmental factors in general do affect IQ scores comes from comparison of test scores across several generations. Changes over a few generations are too quick to reflect genetic change, which operates at a much slower rate. These changes instead reflect phenotypic responses to a set of changed environmental conditions.

Direct comparison of IQ scores across generations is complicated by the fact that revisions to IQ tests are standardized to produce a mean of 100 in the reference population. Political scientist James Flynn (1984) found a way around this problem when he became aware of a large number of studies that had given different versions of an IQ test to the same group of subjects, such as a group completing both the 1932 and 1971 versions of the test. Flynn collated American IQ data from studies where two or more versions of an IQ test were given to the same group of subjects with at least a 6-year gap between testing sessions. He found 73 studies including almost 7,500 subjects that spanned a period of 46 years. Flynn's results showed that test scores were higher on the later exam. Since these tests are standardized to different reference groups, a higher score on the later exam shows that the average IQ of the American population had increased. Flynn's results suggested a total gain of 13.8 points over 46 years, or 3 points per decade. The greater the interval of time between test versions, the higher the gain in average IQ (see Fig. 13.4).

Flynn (1987) later tried a different approach by comparing average test scores over time in 14 nations and standardizing each test against the earliest test in the sample (assigned a mean IQ of 100). For example, a study of IQ in the Netherlands in 1962 gave an average IQ, relative to the earlier test in 1952, of 106.2, showing a 6.2-point increase. Flynn's analysis showed consistent gains in average IQ ranging from 5 to 25 points in a single generation (see Fig. 13.5).

This recent increase in IQ scores, dubbed the "Flynn effect," is clear evidence that some environmental factor(s) operates on IQ scores since these changes are too rapid to reflect genetic change. What is less clear is what these results imply. Flynn (1987) argued

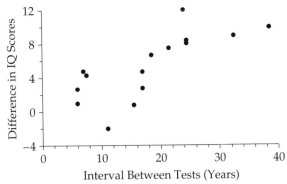

Figure 13.4. Comparison of the difference in IQ scores for individuals who have taken two or more versions of an IQ test with at least a 6-year gap between test versions. The horizontal axis refers to the interval (in years) between the test versions. In general, the greater the interval of time between tests, the greater the gain in mean IQ score. Since the tests at each time period are standardized according to the average IQ during the year the test was first used, a difference in test scores for those who took two or more of these tests at a later time indicates an increase in average IQ. (Data from Flynn 1984.)

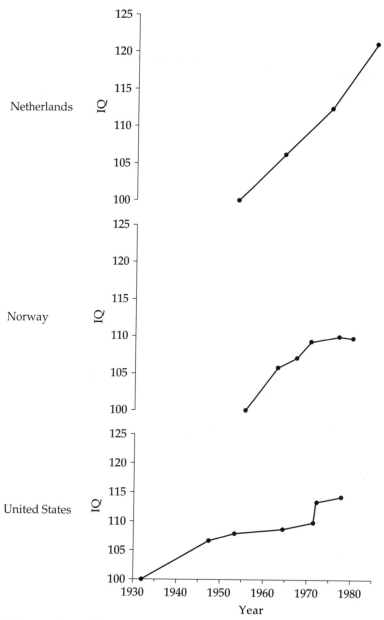

Figure 13.5. A comparison of average IQ scores over time in three nations, with each score standardized relative to the first average in the series. In all three cases, average IQ has increased, a phenomenon known as the "Flynn effect." (Data from Flynn 1987.)

that these changes show that IQ tests are poor assessments of basic intelligence and that "IQ tests do not measure intelligence but rather a weak causal link with intelligence" (p 171). Neisser (1997) reviews several explanations for this increase, including an increase in test-taking ability, improvements in nutrition, years of schooling, changes in child-rearing, changes in early childhood experiences, and increased exposure to visual media. No conclusive results are yet available. Although we still lack an adequate explanation of the Flynn effect, the existence of this generational change does show that IQ scores are affected by changing environmental conditions.

Group Differences in IQ

The available evidence suggests that IQ, like many complex traits, is affected by both genetics and environment and that a simple debate over nature versus nurture is useless. However, the existence of both genetic and environmental influences often raises the question of which one is more influential. Discussion of the relative impact of genetic and environmental influences on IQ is perhaps nowhere more heated than in the continuing debate over group differences in IQ. The greatest debate has arisen over the implications of average IQ differences in blacks and whites in the United States. Many studies show that the average IQ of American blacks is 15 points lower than that of whites, although the distributions for the two groups overlap (see Fig. 13.6). Although much attention has been given to these "racial" differences, it must be kept in mind that there are other sources of variation that account for more of the variation in IQ than "race." Jensen (1998) reports results from one study that looked at several sources of variation in IQ scores among black and white schoolchildren. This analysis examined variation in IQ scores simultaneously between "races" (black/white), social classes, families, and individuals within families. The results show that the greatest source of total variation explained (39%) was between siblings within families and the second greatest source of variation explained (26%) was between families within the same race and social class.

There is little debate over the average 15-point difference between American blacks and whites. What is less clear, and vigorously debated, is the meaning of this difference. Is the black–white difference genetic, environmental, or both? It might seem intuitively obvious to some that at least some portion of the black–white difference in IQ must be genetic. The chain of reasoning usually goes like this: if IQ scores at least partially reflect

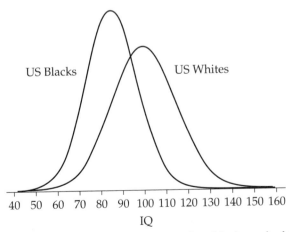

Figure 13.6. Normal distributions of IQ scores in American blacks and whites. Mean IQ: blacks = 85, whites = 100. Standard deviation in IQ: blacks = 12, whites = 15. Note the extensive overlap in test scores. (Data from Jensen 1998, 353.)

genetic influences and if races are, by definition, genetically different, then it stands to reason that at least part of the black–white difference in IQ reflects a genetic difference. Ignoring the problems with many assumptions (including the use of "races" as defined in the United States as biological populations), the real problem is extrapolating from one trait to another. The fact that two groups differ in some genetically influenced traits does not imply that they will differ in all genetically influenced traits. Some genetic (or genetically influenced) traits do vary between groups, and some do not. We cannot extrapolate from one trait to another and must test each for the hypothesis that group differences reflect genetic differences.

Arthur Jensen's (1969) paper "How Much Can We Boost IQ and Scholastic Achievement?" argued that a large amount of the black–white difference in IQ scores reflected genetic differences (with the additional implication that educational programs based on a large environmental component might therefore be less effective than once proposed). The basic idea here is that evidence of high heritability shows that the major component of variation in IQ is genetic and, therefore, that the major difference between average black and white IQ scores is genetic. The problem with this line of reasoning, as noted earlier, is that heritability is a relative measure of genetic variation *within* groups and does not necessarily tell us anything about the relative genetic component of variation *between* groups (Macintosh 1998, see also Block 1996).

Is it possible for a trait with a high heritability to show group differences that are not genetic in nature? Yes. As an example, consider a hypothetical case where we take two samples of mice, both of which are genetically diverse but which have the same set of allele frequencies such that there is genetic variation between individuals within groups but not between the two groups. Now, imagine that one group of mice grows up in a favorable environment, with ample food, water, space, and other factors conducive to growth, whereas the other group grows up in a less favorable environment, with minimal food and water, higher rates of disease, and other factors that inhibit growth. Let us now consider body size, a trait that is affected by both genetic and environmental factors. Within each group, assuming that the environment is the same for all mice, any variation in body size will be due solely to genetic variation; thus, heritability will be 100%. However, the only difference between groups in this example is environmental because the genetic composition of the two groups is the same. Thus, the fact that a trait has a high heritability does not necessarily imply that any difference between groups must be genetic. It depends on the distribution of genes and environmental factors within each group.

The fact that IQ is suggested to have a moderate to high heritability (although see Devlin et al. 1997) tells us nothing about the cause of group differences. Lacking any other data, the most we can say is that group differences might reflect genetic differences, environmental differences, or both. It is necessary to conduct tests to determine which of these hypotheses is more likely to be correct (another factor might be test bias). Macintosh (1998) describes a hypothetical experiment to untangle genetic and environmental influences on group differences: take random samples of black and white children at birth and raise them in matched environments.

Although he argues that a perfect experiment is unattainable, Mackintosh (1998) notes that several studies have approximated these conditions. One example is a study conducted by Eyferth on the illegitimate children of African American soldiers and German women following the Second World War. These children were raised by the mothers or foster parents in Germany and not by the fathers. Eyferth then compared the IQs of these children with those of white German children matched on a number of characteristics. This study provides a comparison of two groups of children raised in similar environments. If there is an innate genetic difference between groups, then their IQ scores should be different. However, the results showed that there was no significant difference between the two groups of children, suggesting a lack of "racial" difference in IQ due to genetic factors (Mackintosh 1998). By extension, this type of study suggests that any observed difference in black and white IQ scores must therefore reflect environmental differences.

A number of studies have examined IQ scores in black children who are adopted into white families, often with mixed results (Mackintosh 1998). Weinberg et al. (1992) note that black children raised by white families score higher than black children raised by black families, suggesting that environmental differences in family characteristics are responsible for the observed difference in average IQ scores between blacks and whites. However, these adopted children still tended to score lower than the birth children of the adoptive parents, a result somewhat consistent with a genetic hypothesis. Further, a 10-year follow-up found that after a decade the scores of black children had declined to the same level as their biological parents, evidence sometimes used to argue that changes in the environment (brought about through adoption) have little lasting effect on IQ scores. However, it is important to note that the largest drop in IQ scores occurred within the white adopted children. Mackintosh (1998) notes that although adoption studies have provided mixed results, there is little evidence supporting a genetic hypothesis for group differences in IQ because "it will never be possible to bring up black and white children in truly comparable environments" (p 155).

Another approach to group differences in IQ scores is to examine variation in the proportion of black and white ancestry. Such studies also help to get around the glaring problem in many studies of IQ and race of using samples of "blacks" and "whites" as homogeneous genetic populations. Such use ignores variation in ancestry (see Chapter 12), lumping everyone with some degree of African or European ancestry into the same group. Examination of individual ancestry, however, can aid in testing the hypothesis that black–white differences in IQ are due to genetic differences. If this hypothesis is correct, then there should be a correlation between IQ score and the degree of European ancestry. That is, black children with more European ancestry should have higher IQ scores than those with less, or little, European ancestry. Such studies have generally found no correlation between European admixture and IQ scores (see studies reviewed by Flynn 1980, Loehlin et al. 1975, Macintosh 1998).

Group Differences and the Evolution of Intelligence

The debate over whether genetic differences between groups are responsible for the observed difference between groups in IQ (and, presumably, intelligence) has also been approached in terms of evolutionary models. If group differences in IQ reflect, at least to some extent, genetic differences, then what evolutionary model can be used to explain these genetic differences? Then again, if there are no underlying genetic differences between groups, how can that be explained in evolutionary terms?

J. Philippe Rushton (1995) has proposed a controversial model where racial differences in IQ reflect genetic differences between groups resulting from differential selection for different traits in differing environments. Rushton applies an evolutionary model focusing on the difference between r-selection and K-selection. These concepts are from the field of evolutionary ecology and used to describe contrasting patterns of adaptation in differing species. r-selected species maximize fertility by having more offspring, whereas K-selected species maximize survival by having fewer offspring but providing more parental care. Rushton argues that there is a difference among human populations in the relative degree of r-versus K-selection: Africans are the least K-selected, followed by Europeans and then by Asians. According to Rushton, since Asians are the most K-selected, they have the largest brains and the highest IQ scores but the smallest genitalia and the lowest frequency of sexual intercourse. Africans, however, are considered to be the most r-selected and, thus, maximize fertility over intelligence. Central to Rushton's model is the idea that Asians evolved higher intelligence because of the demands of "more challenging environments" (p 7). Rushton's views have been widely criticized (e.g., Relethford 1995, Lieberman 2001). A major problem is that there is no evidence for any geographic region being more cognitively demanding than another. Further, Rushton's use of geographic aggregates (e.g., "Asian") does not deal with the fact that such broad geographic areas include a variety of environments; in the case of Asia, local environments range from tropical to arctic. Another problem is that it is not clear if r- and K-selection models, developed to explain differences between species, can be successfully used when explaining variations within a single species united by gene flow.

Brace (2000) takes a different approach, arguing that intelligence (in all facets) is a broad adaptive strategy crucial to all human populations, regardless of specific environments. Brace maintains that culture is the primary human adaptation and that problem solving, language, and other cultural traits are needed in any and all human environments. Under this view, no specific environment is more challenging. What matters is the ability to solve problems and to pass information on from generation to generation via language. According to Brace, all human groups possess the same basic levels of ability that have been selected for throughout our species over time. Although genetic variation exists between individuals and within groups, genetic differences underlying intelligence do not vary across populations because the same selective pressures are applied everywhere. As such, any IQ differences between groups today must reflect environmental differences. It is likely that these debates will continue, but at present there is no compelling evidence that genetic differences underlie group differences in IQ.

CHAPTER SUMMARY

The field of human behavioral genetics seeks to understand the influence of genes on complex behavioral traits. Rather than argue from the perspective that human behavior is primarily a function of "nature" or "nurture," much of behavioral genetics seeks to understand the complex interaction of genetic and environmental factors on human behavior. Heritability is a statistical measure of the relative degree of total variation due to genetic variation among individuals within a sample. Heritability tells us only that genetics plays a role in a given trait and not the exact contribution of genes to a behavior in any specific individual.

Dyslexia is a neurological condition affecting reading ability found in as many as 10%–15% of American children. Twin and family studies show a strong inherited risk for dyslexia, which appears to be related to a semidominant major gene. Although there is a strong genetic component, environmental factors also influence the prevalence of dyslexia within families. Recent research comparing France, Italy, and the United Kingdom suggests that the actual prevalence of dyslexia might also reflect differences in language structure.

Sexual orientation is a more complex behavior in terms of measurement definition and variation across the sexes. Although some physiological studies have demonstrated correlation of male sexual orientation with brain structure, it is less clear if this is evidence of cause, effect, or correlated response. Twin and family studies suggest moderate to high heritability, and more recent analysis suggests that male sexual orientation may be in part related to a region of the X chromosomes and that male homosexuality may be partially inherited maternally. Even given this recent evidence, however, there is considerable evidence that both genetic and environmental factors influence sexual orientation.

IQ is perhaps the most studied and controversial trait dealt with in behavioral research. There is still wide debate over the exact meaning of IQ scores. Are they good measurements of innate intelligence, measures reflecting one's ability to take an IQ test, or both? Twin and family studies have consistently demonstrated a heritable component; however, the magnitude varies considerably, and more recent work in behavioral genetics suggests previous estimates of the heritability of IQ may be biased upward. There is evidence for both genetic and environmental influences on IQ scores. Group differences in IQ test scores, such as found in American whites and blacks, appear to be due to environmental differences.

SUPPLEMENTAL RESOURCES

Gilger J W and Hershberger S L, eds. (1998) Special Issue: Human Behavioral Genetics: Synthesis of Quantitative and Molecular Approaches. *Human Biology* 70: 155–432. A special issue of the journal *Human Biology* containing 14 articles on state-of-the-art approaches to human behavioral genetics, including articles on language, reading disorders, cognitive ability, and sexual orientation.

Gould S J (1981) *The Mismeasure of Man.* New York: W W Norton. An excellent historical critique of the abuse of various measurements of intelligence as applied to group differences.

Hamer D and Copeland P (1996) *The Science of Desire: The Search for the Gay Gene and the Biology of Behavior.* New York: Touchstone Books. An interesting, popular book illustrating what we know and do not know about genetics and sexual orientation.

Macintosh N J (1998) *IQ and Human Intelligence.* Oxford: Oxford University Press. A comprehensive review of IQ tests, including measurement, genetic and environmental influences, and group differences.

Glossary

achondroplasia Nonproportional dwarfism that is inherited in dominant fashion and caused by mutations in the *FGFR3* gene on the fourth chromosome.

acrocentric State of having the centromere of a chromosome located very near the end of the chromosome.

additive genetic variance Average squared deviation (around the average) that is due to the additive effects of alleles.

adenine One of the four nitrogenous bases found in DNA and RNA. Pairs with thymine in DNA and uracil in RNA.

admixture Gene flow between populations, often specifically defined as gene flow between individuals from geographically distant populations, such as between European settlers and enslaved Africans in the United States.

adolescent growth spurt Rapid increase in the rate of growth in human body size that occurs during the period of sexual maturation.

agglutinate The act of red blood cells clumping together.

allele conversion *See also* **gene conversion** for comparison. Process that occurs when small segments of an amino acid sequence are replaced with the homologous region of another allele.

alloantigen Antigen of the HLA system. Term is used to distinguish antigens of the HLA system from other antigens. *See also* **human lenkocyte antigen (HLA)**.

allotype Any of a number of allelic variants characterized by antigenic differences. Usually used to designate allelic variants of the constant region of the immunoglobulin light and heavy chains (e.g., Am, Gm, Inv, and Km).

alu insert Mutation that involves the insertion of a particular DNA sequence (approximately 300 bases long) characterized by a number of "alu" restriction sites (AGCT).

analysis of variance Form of statistical analysis that partitions the **variance** in a variable into different factors. For example, the variance in stature could be partitioned into within-sibship and between-sibship components.

annealing Attachment of primers (by base complementation) to denatured DNA during polymerase chain reaction when the reaction mixture is cooled.

anthropometer Device used for the measurement of a number of anthropometric factors, consisting of a rod with one or two movable ends used to measure intervening distance.

anthropometrics Measurements of the human body, head, and face.

antibody Protein produced by the immune system in response to a specific substance (**antigen**) that is typically foreign (e.g., bacterium) to the organism. The antibody is capable of binding with that antigen.

anti-codon Three base sequence of a transfer RNA that matches (by complementation) a three base sequence of messenger RNA.

antigen Any substance, usually foreign to the body, capable of initiating the production of antibodies.

assortative mating Nonrandom mating on the basis of phenotype (which may or may not equal genotype).

autosomal First 22 pairs of chromosomes, excluding the twenty-third pair (the sex chromosomes).

average fitness Relative fitnesses weighted by their genotype frequencies.

balanced polymorphism The case where selection favors heterozygotes (called *heterozygote advantage*). The heterozygote is relatively more fit than either of the two homozygotes. When this happens, both alleles will be maintained. Stable equilibrium will eventually occur if selection pressures remain the same over time.

Bergmann-Allen rules Zoological rules describing the relationship between average temperature and average size and shape of populations of mammals. Bergmann's rule predicts that mammals in hot climates will be smaller and have a more linear body shape than those in cold climates. Allen's rule extends this prediction to the size and shape of the limbs.

biacromial breadth Measure of body width across the shoulders (more specifically, between the acromion process of each scapula).

bigonial breadth Measure of facial breadth, taken as the distance between the bottom back portions of the lower jaw.

bi-iliac breadth Measure of body width at the hips.

bizygomatic breadth Measure of facial breadth, taken as the distance between the outermost parts of the zygomatic processes (cheekbones).

blastula Early developmental stage in amphibians reached by the zygote dividing many times to form a hollow ball of cells. In humans, the comparable stage is called a *blastocyst*.

body mass index (BMI) Index of body fat computed as $weight/stature^2$, where weight is measured in kilograms and stature is measured in meters. BMI >25 is considered overweight and BMI >30 is considered obese.

broad sense heritability Proportion of the total genetic variance out of the phenotypic variance.

cellular defense system Defense system that is primarily responsible for protecting the body from viral infections. System recognizes and kills infected host cells.

central limit theorem Classical theorem from probability theory that states that with increasing amounts of data many distributions will tend toward a normal ("bell-shaped") form. For example, if enough loci influence a quantitative trait, then that trait will be expected to follow a normal distribution.

centromere The one unpaired section of an individual chromosome that has undergone replication to form sister chromatids.

cephalic index Measure of relative head width computed as (head breadth/head length) × 100. A high cephalic index indicates a relatively broad head.

chain termination reaction A form of the polymerase chain reaction which uses a single primer and labeled dideoxynucleotides in addition to the regular nucleotides. When the dideoxynucleotides are incorporated into the growing chain, the chain is terminated, which allows for the DNA sequence to be determined.

chest breadth Size of the chest measured from side to side.

chest depth Size of the chest measured from front to back.

chromosome DNA wrapped around proteins to form a discrete structure found in the nucleus.

classical satellite DNA Fairly long DNA sequences that are repeated in tandem in the genome. Classical satellites are much longer than mini satellites, which are in turn micro satellites.

cline (clinal distribution) Plot (or map) of the changes in allele, genotype, or phenotype frequencies over a geographic area.

cluster analysis Method of representing similarity between objects in a hierarchical fashion by noting which pairs of groups form clusters. When applied to genetic distances, cluster analysis produces a picture of overall genetic similarity between a set of populations.

coalescent model Model in which all DNA molecules are traced back in time to a single "most recent common ancestor." Thought of as a "family tree," the coalescent model traces the "coming together" of pairs of descendant DNA molecules into single ancestral molecules.

coding strand Another term used for the non-template strand of DNA because the coding strand has the same base sequence as the produced RNA strand (save for having thymine instead of RNA's uracil).

codominant In its most general sense, codominant alleles are ones where both homozygotes and the heterozygote are phenotypically distinguishable. For example, in the ABO system the genotypes *AA*, *AB*, and *BB* are all phenotypically distinguishable when tested with antibodies A and B (AA reacts with anti-A, AB reacts with anti-A and anti-B, BB reacts with anti-B).

coefficient of relationship Two times the inbreeding coefficient that would arise in the potential offspring of two individuals. It can also be defined as the expected proportion of alleles in common for two individuals or as the expected additive genetic correlation between two individuals.

collagen Structural protein found in many connective tissues.

continuity model Model of modern human origins where there is genetic continuity between archaic and modern samples across the Old World. Specifically, this model states that modern humans dispersing out of Africa mixed genetically with preexisting archaic groups outside of Africa. This model contrasts with the replacement model.

cormic index Measure of the relative height of the upper body computed as (sitting height/stature) × 100. A high value indicates a relatively long torso and relatively short legs.

correlation Measure of association between two variables, ranging between −1 and 1. A correlation of 1.0 means complete positive association, a correlation of −1.0 means complete negative association, and a correlation of 0.0 means no association.

crossing over Physical process by which two chromosomes exchange DNA.

cross-sectional study As applied to the study of human growth, a study where individuals of different ages are measured at a single point in time.

cultural diffusion Spread of ideas, technology, and behaviors from one region or group to another region or group (society).

cytosine One of the four nitrogenous bases found in DNA and RNA. Pairs with guanine.

demic diffusion Process of colonization of a region by a population. Term is often used in contrast with *cultural diffusion*.

denature Conversion of double-stranded DNA into single strands, which can be accomplished by, among other things, heating the DNA. The first part of a polymerase chain reaction cycle is the denaturation phase, which is achieved by heating the DNA to 94°C.

dendrogram Graphic representation of the results of a cluster analysis, where populations are arranged in a "tree" showing overall levels of genetic similarity.

deoxyribonucleic acid (DNA) Molecule formed of two chains of deoxyribose (a sugar) alternating with a phosphate, with each chain wrapped around the other in a double helix. Internal to this backbone are paired bases (composed of adenine, guanine, thymine, and cysteine). DNA is transcribed into RNA.

developmental plasticity Change in morphology that occurs during the physical growth of an organism following movement into a different environment.

differentiation Process by which cells become specialized during development.

dihybrid model Admixture model where the genetic composition of a hybrid population is due to the proportional amounts of gene flow from two parental populations.

diploid Presence of paired maternally and paternally derived chromosomes. For humans, the diploid number is 23 pairs of chromosomes, or 46 total.

distance curve Graph used in analyzing human growth that shows size as a function of age.

DNA polymerase Enzyme that connects DNA bases together to form a new DNA strand during replication.

dominance Relationship between two alleles at a locus, such that one allele "masks" the other in determination of the pheontype. For example, in the ABO blood group the A allele is dominant to the O allele because the AO and AA genotypes are phenotypically indistinguishable as far as reactivity to antibody A.

dominant allele Allele at a locus that "masks" a recessive allele in the production of the phenotype.

dominance variance Genetic variance at a locus due to dominance/recessive relationships of the two alleles at the locus.

dyslexia Neurological disorder where a person suffers from some inability to recognize and comprehend written words but has normal intelligence.

E.E.L. One of two portable abridged reflectance spectrophotometers used by anthropologists to measure human skin color, where the initials stand for the name of the initial manufacturer, Evans Electroselenium Limited. The other commonly used machine is the Photovolt reflectometer.

electrophoresis Method, using electricity, to separate proteins, allowing genotypes to be determined.

embryonic stem cell Cell that has differentiated and will give rise to a "population" of specialized cells during further development.

endosymbiont hypothesis Hypothesis that mitochondria represent ancient bacteria that have been incorporated into cells and now exist in a symbiotic relationship with the cells.

environmental correlation Relationship between two traits caused by environmental effects.

environmental variance Average squared deviation (around the average) that is due to environmental (nongenetic) effects.

epistasis Interaction between two or more gene loci that produces phenotypes different from those produced if loci acted individually.

Eppendorf tube Small plastic vial with a lid, used to contain the reaction mixture for a polymerase chain reaction.

erythroblastosis fetalis Hemolytic disease of the newborn caused by Rh incompatibility between the fetus and the mother—mother is Rh− (dd) and fetus is Rh+ (Dd).

erythrocyte Red blood cell.

eumelanin Black–brown form of the melanin pigment, found in all tissues that produce melanocytes, including skin, hair, and eyes.

exons Portions of a gene which are expressed (translated) as amino acid sequences within a protein. *See also* **introns**.

extension Third phase of a polymerase chain reaction cycle, in which DNA polymerase adds bases to growing chains of DNA.

favism Hemolytic condition that occurs upon ingestion of fava beans.

fixation When an allele frequency becomes 1.0.

fixed effect Any known factor that may influence a quantitative trait. Such effects can be genetic or environmental.

folate Compound of the vitamin D complex group.

frameshift mutation Mutation in a structural gene caused by deletion or addition of one or more nucleotide pairs, causing a "shift" in the reading frame of all codons following the site of mutation.

gamete Haploid reproductive cell (*see also* **ova** and **spermatids**) formed by meiotic division.

gel electrophoresis Molecular genetic technique used to size-separate DNA. *See* http://www.bergen.org/AAST/Projects/Gel/ for more detail. *See also* **electrophoresis**.

gene conversion Caused by mismatched repair of DNA when a direct copy of a gene is inserted in the middle of another gene.

gene (allele) frequency Proportion of a particular allele in a sample.

gene pool All the alleles at a locus represented in united gametes. For diploid loci, the gene pool is size 2N, where N is the number of individuals.

gene tree "Family tree" that traces extant genes to an ancestral gene.

genetic distance Average amount of genetic dissimilarity between a pair of populations. The greater the genetic distance between a pair of populations, the more genetically dissimilar they are. Genetic distances can be obtained from a wide range of data and typically are used to make inferences regarding the effect of mutation, gene flow, and genetic drift on neutral genes or traits.

genetic marker Mendelian genetic trait (usually a polymorphism) that can be used to characterize the genetic structure of a population (e.g., the ABO and MN systems). Term is also used now more frequently for describing DNA variants used in family studies to detect the presence of closely linked genes that may cause disease.

genotype Specific constitution of genes at one or more loci.

genotype by environment interaction Differential response of genotypes to environmental conditions. In quantitative genetics, more correctly referred to as "polygenotype by environment interaction."

germ cell Diploid cell that will give rise to the haploid gametes through meiotic division.

glycoprotein Protein with an attached carbohydrate.

guanine One of the four nitrogenous bases found in DNA and RNA. Pairs with cytosine.

haplogroup Group of related haplotypes defined by a set of shared restriction fragment length polymorphism mutations.

haploid Having half the number of chromosomes. In humans, the diploid number is 23 pairs of chromosomes (for a total of 46), while the haploid number is 23 (unpaired) chromosomes.

haplotype Sequence of genes, restriction sites, or nucleotides inherited as a linked unit from one parent. Haplotypes may be composed of a string of alleles since more than one locus may be involved. A combination of alleles from closely linked loci on a single chromosome, e.g., allelic combinations of the HLA system.

Hardy-Weinberg equilibrium Expected genotype frequencies given allele frequencies.

head breadth Measure of the greatest width on the back of the skull.

head height Vertical distance from the ear hole to the top of the head.

head length Measure of the greatest length of the skull.

heme Large, nonprotein molecule containing iron that can carry a molecule of oxygen.

hemizygous Present in the haploid rather than the diploid state. Any loci located on the non-recombining portion of the Y chromosome are hemizygous, since male have a single copy (while females have none).

hemolytic disease of the newborn (HDN) life span of an infant's red cells is shortened by the action of specific antibodies derived from the mother by placental transfer. Effect varies from little inconvenience to *in utero* death of the fetus. A common HDN is *erythroblastosis fetalis*, caused by Rh incompatibility.

heterozygous Condition in which both alleles at a locus are different.

high-altitude environment Altitude greater than 2,500 meters (8,250 feet), where people often show some degree of hypoxic stress.

HLA *See* **human leukocyte antigen**.

homozygous Condition in which both alleles at a locus are the same.

human behavioral genetics Interdisciplinary field that uses quantitative genetics theory and, more recently, molecular genetics to assess the relative impact of genetics and environment on behavioral traits.

human leukocyte antigen (HLA) Complex of genes found on the short arm of chromosome 6 that code for glycoproteins. These antigens "tag" nucleated cells as "self."

hypoxia Oxygen starvation.

identity by descent (IBD) Case where two alleles are the same because both came from a common ancestor.

immunoglobulin One of several classes of antibodies (e.g., IgG).

inbreeding Production of offspring by mates who are "related by blood."

inbreeding coefficient Probability of identity by descent for the two alleles at a locus.

indel Insertion or deletion at a specific DNA location. Because it is not generally possible to determine the ancestral state (i.e., whether the original mutation was for an insertion or a deletion), the two terms are contracted into *indel*. *See* http://research.marshfieldclinic.org/genetics/indels/ for an online database of indels.

intelligence quotient (IQ) Result of an intelligence test where a person's "mental age" is expressed as a ratio of chronological age and multiplied by 100. Thus, a person with an IQ of 100 has a mental age equal to his or her chronological age.

introns Intervening sequences within a gene which are transcribed but not translated into amino acids within a protein. *See also* **exons**.

isolation by distance Populations located geographically farther apart will tend to be less genetically similar. As applied to genetic distance measures, isolation by distance will create a pattern where genetic distance increases with geographic distance.

karyotype Standardized way of showing the chromosomes from longest to shortest and paired with their homologs.

Kinsey scale Standard measurement of sexual orientation consisting of four variables (self-identification, romantic/sexual attraction, romantic/sexual fantasy, and sexual behavior), each measured on a 7-point scale.

kinship As used in population genetics, it refers to the genetic similarity between pairs of populations, as opposed to genetic distance, which measures dissimilarity. Kinship can be estimated from genetic and demographic data.

kinship coefficient Measure of relationship between two individuals, defined as the inbreeding coefficient that would be produced in the offspring of the two individuals.

lagging strand One strand out of double-stranded DNA that is synthesized in pieces and ligated (connected) together into one strand during replication. *See also* **leading strand**.

leading strand One strand out of double-stranded DNA that is continuously synthesized during replication. *See also* **lagging strand**.

leukocytes White blood cells.

linear regression Type of statistical analysis that fits a straight line through paired data, such that the sum of the squared vertical deviations of the points off of the line are minimized. For example, in the linear regression of daughters' statures on midparent statures, a line is fit such that the sum of the squared deviations of the daughters' statures off of the line are minimized.

linkage (linked) Occurrence of two or more gene loci close together on the same chromosome.

linkage disequilibrium Nonrandom association of alleles at different loci.

locus (pl. loci) Physical location on a chromosome. Occasionally used more broadly to refer to a "hereditary unit," such as when we state that mtDNA and NRY act like loci.

long interspersed nuclear elements (LINES) DNA sequences of about 4,000 to 7,000 base pairs that are dispersed throughout the genome. There are about 100,000 LINE insertions in the human genome.

longitudinal study As applied to the study of human growth, a study where the same individuals are measured at different points during their lives.

lymphocyte Subgroup of white blood cells. T lymphocytes control antibody production and cell-mediated immune responses, while B lymphocytes are the main producers of antibodies.

lysis (lyse) Breakage of a cell caused by rupturing the cell membrane and cell wall.

major gene Gene that has a large effect on a quantitative trait.

major histocompatibility complex (MHC) Group of closely linked genes that code for antigens that play a major role in the immune system and in tissue incompatibility and recognition. *See also* **human leukocyte antigen**.

maximum likelihood estimation Form of statistical estimation in which the probability of obtaining the observed data is maximized across the one or more parameters in a model.

meiosis Cell division that occurs to produce spermatids or ova. The division begins with a diploid cell and ultimately produces one or more haploid cells.

melanin Primary pigment responsible for variation in skin color.

melanocyte Cell in the epidermis of the skin that produces melanin.

melanosome Organelle inside melanocyte that synthesizes melanin.

messenger RNA (mRNA) RNA molecules that are built from template DNA and code for particular proteins.

metacentric Condition of having a centromere placed in a chromosome near the center of each sister chromatid so that the long and short arms of the chromosome are approximately equal in length.

MHC *See* **major histocompatibility complex.**

minimum frontal diameter Measure of facial breadth, taken as the distance between points distal to the eye orbits.

mismatch distribution Device for summarizing variation in DNA which compares all pairs of individuals and counts the number of differences in the DNA of each pair. The mismatch distribution then shows how individuals differ from one another.

mitochondria Cell organelles found in the cytoplasm that are important in metabolism and which contain their own DNA in a circular structure.

mitosis Cell division which results in two identical daughter cells.

monogamous Mating system characterized by an exclusive sexual bond between an adult male and an adult female.

monogenesis Doctrine that held that all human races were the result of a single origin (creation) as described in the Scriptures. Subsequent changes in traits (skin color, hair form, etc.) were the result of such factors as different climates and modes of living.

most recent common ancestor (MRCA) Single "node" at the top of a coalescent tree, from which all further alleles (or chromosomes) are descended.

multivariate quantitative genetics Quantitative genetic model for two or more traits.

mutation rate Probability that a gene will mutate, usually given as the number of mutations per gamete per generation.

narrow sense heritability Proportion of additive genetic variance out of the phenotypic variance.

nasal index Measure of relative nose shape computed as (nose breadth/nose length) × 100. A high nasal index indicates a relatively broad nose.

9 bp deletion Specific deletion of 9 base pairs found between two coding sequences in mitochondrial DNA.

nondisjunction Lack of separation between homologous chromosomes during meiosis. When a gamete with non-disjunction combines with an ordinary gamete there will be three copies of a particular chromosome (a "trisomy") rather than the usual two copies.

nonrecombining Y (NRY) Major portion of the Y chromosome which cannot exchange material (cross-over) with the X chromosome. Consequently, the NRY is inherited as a unit in a son from his father.

non-template strand Single strand of DNA which does not serve as a template for producing RNA, but which does have the same sequence as the produced RNA (save for having thymine instead of uracil). *See also* **template strand.**

nose breadth Width of the nose at its base. In living humans, this measure includes a certain amount of fat contained in the fleshy part of the nose.

nose height Vertical length of the nose.

nuclear transplantation Also known as "cloning," in which the nucleus from a diploid cell is placed in an ovum which has had its haploid nucleus removed.

nucleotide Sugar, phosphoric acid, and nitrogenous base in DNA.

nucleus Cellular organelle that contains the chromosomes. Most cells contain a single nucleus, but some cell types are multinucleate (e.g., striated skeletal muscle cells) and others lack a nucleus (e.g., red blood cells).

osteometrics Measurements of bones.

ovum (pl. **ova**) Haploid gamete (an "egg") formed by females in their ovaries.

phenotype Some observable or measurable aspect of an organism.

phenotypic variance Average squared deviation of a trait around the average.

pheomelanin Reddish brown form of the melanin pigment found in mammalian hair. Combined with eumelanin, it determines hair color.

photolysis Chemical decomposition caused by light.

Photovolt One of two portable abridged reflectance spectrophotometers used by anthropologists to measure human skin color. The other commonly used machine is the E.E.L. reflectometer.

pinta Mild skin disease caused by the bacterium *Treponema carateum*, occurring in humid, hilly, or mountainous regions of Mexico, Central America, and parts of northern South America.

plasma Yellowish fluid in which red blood cells are suspended.

pleiotropy Common effects of alleles on more than one trait.

point mutation Change of a single base pair of DNA. Provided that both the mutated site and the original site are maintained, a point mutation produces a single nucleotide polymorphism (SNP).

polygenesis (see **polygenism**)

polygenic Model which posits many genetic loci affecting a trait (*see* **quantitative genetics**).

polygenism (polygenesis) An eighteenth- and nineteenth-century doctrine that held that the human races were separate biological species that were descended from different "Adams." This doctrine was in contrast with monogenesis.

polygynous Mating system characterized by an adult male having a sexual bond with more than one adult female.

polymerase chain reaction Molecular genetic technique for making many copies of a targeted section of DNA. The reaction involves repeated denaturing of the double-stranded DNA, annealing of primers, and extension onto the primers.

polymorphism Occurrence of two or more genetic (or phenotypic) variants within a population. Often refers to genetic variation where the rarest allele is maintained at a frequency >1% in the population.

pooled variance Average (weighted by sample size) squared deviation (around each group's average).

population history Study of how the history of a set of populations has affected the pattern of genetic distances between them.

population-specific allele (PSA) DNA marker used in admixture analysis where either an allele is absent in one of the parental populations or there are large differences in allele frequency between the parental populations.

population structure Factors that affect mate choice and genetic relationships between individuals and subpopulations within a population.

postnatal Life after childbirth.

prenatal Period from conception until childbirth, including development of the embryo and growth of the fetus.

previtamin D Chemical precursor to vitamin D. Ultraviolet radiation converts 7-dehydrocholesterol into previtamin D, which then is converted into vitamin D over several days.

primer Short section of synthetically produced nucleotides (also known as an *oligonucleotide*) that complements a particular DNA sequence and thus provides a foundation on which DNA polymerase can add bases by complementation.

principal coordinates analysis Multivariate statistical method used to produce plots representing genetic distances.

probability Relative frequency of an event or the degree of belief that an event will occur.

promoter site DNA sequence that signals the beginning of a gene so that transcription can start at the proper location.

pseudogene Gene that is not expressed or the product of which is not synthesized. Mutational alterations inactivate its transcription.

pyogenic Pus-producing.

quantitative genetics Study of the quantitative effects of one or more loci on one or more traits.

quantitative trait linkage Association between a putative (and not located) quantitative trait locus and a marker locus with a known position in the genome.

quantitative trait locus Largely synonymous with *major gene*. The primary distinction is that a quantitative trait locus has been localized to a particular segment of a chromosome (by quantitative trait linkage analysis), while a major gene's effects are observed without being able to ascertain where the genetic locus is located.

reading frame-shift Mutation that involves a deletion or insertion of DNA bases not in a multiple of three. Consequently, the codons are misread, often leading to radical changes in the protein and ultimately in the phenotype.

recessive allele An allele at a locus that is "masked" by a dominant allele in the production of the phenotype.

recombination Because of crossing-over between homologous chromosomes the loci can recombine to form new haplotypes.

recurrence relationship Mathematical equation that gives the value of something at time t + 1 as a function of the value at time t.

reflectance spectrophotometry Method of objectively measuring color by measuring the amount of light reflected back from a substance at selected wavelengths.

relative fitness Contribution by genotype of individuals to the next generation, scaled so that the maximum relative fitness is 1.0.

repeatability Extent to which a trait varies on being remeasured. Quantified on a scale from 0 to 1, with 0 meaning that the trait varies independently of the previous measurement and 1 meaning that it is identical to the previous measurement.

replacement model Model proposing that modern humans arose first in Africa as a new species and then dispersed across the Old World, replacing preexisting archaic populations outside of Africa. This model contrasts with the continuity model.

replication Process by which a cell makes two copies of double-stranded DNA from one original double-stranded molecule.

restriction endonuclease Bacterially produced enzyme that cleaves DNA molecules at particular short sequences. The term *restriction* refers to the fact that these enzymes restrict the replication of viral DNA within the bacterial cell, thus restricting viral growth.

restriction fragment length polymorphism (RFLP) Variation at a specific DNA site that arises because a "recognition sequence" for a specific restriction endonuclease is or is not present. If the recognition site is not present, then the endonuclease will not digest ("cut") the DNA, while if the site is present, the DNA will be "cut." This leads to variation in DNA fragment length.

ribonucleic acid (RNA) Molecule similar to DNA, but consisting of a single-strand with the nitrogenous bases adenine, guanine, cytosine, and uracil (instead of DNA's thymine) and a slightly different sugar in the sugar-phosphate backbone.

ribosomal RNA (rRNA) RNA found in a ribosome. Ribosomes also contain proteins, and are extra-nuclear organelles that function in the translation of mRNA into proteins.

rickets Disease characterized by defective bone growth due to vitamin D deficiency.

Robertsonian translocation Mutational event in which the long arms of two different acrocentric chromosomes fuse together to form a new metacentric or submetacentric chromosome.

secular change Change in a growth measure, such as weight or stature, over several generations.

serum Clear liquid that separates from the blood after blood is allowed to clot. *See also* **plasma**.

sex chromosomes X and Y chromosomes that determine the sex of an individual (with XX being female and XY being male).

sexual dimorphism Difference in size between males and females.

shifting balance theory Sewall Wright's model of evolution that incorporates all four evolutionary forces.

short interspersed nuclear elements (SINES) DNA sequences of about 80 to 400 base pairs that are dispersed throughout the genome. The most common SINE is the Alu insert which is present in up to a million copies.

sickle cell anemia Genetic disease causing the red blood cells to lyse or form into sickle-shaped cells. Oxygen transport is hampered, and other complications from anemia arise.

silent mutation Type of mutation which has no effect on the phenotype either because it does not change the amino acid, because the mutation occurs in a non-expressed portion of a gene, or because the mutation occurs in an extra-genic area of a chromosome.

single nucleotide polymorphism (SNP) Allelic variation that arose by a point mutation such that both the mutation and the original base pair are maintained. Single base pair in DNA that is variable (for example, some individuals may have A–T at the site while others may have C–G).

sister chromatids Paired arms of a chromosome formed by DNA replication and held together at the centromere.

sitting height Measure of upper body height made when the subject sits on a table, from the table to the top of the subject's head.

skinfold thickness Measure of fat content made by pinching fat and skin and measuring the thickness with a caliper. Some common skinfold measures are subscapular skinfold, triceps skinfold, and calf skinfold.

slash and burn horticulture Method used widely in the tropics where the underbrush is cut and large trees are either killed or felled, then the whole area is burned and crops are planted.

spermatid Haploid gamete formed by males in their testes.

spliceosomes Collections of RNAs and proteins that splice the exons of a gene together while the introns are removed.

stable equilibrium Point (usually a specific allele frequency) to which the population will return if perturbed away from the point.

standard error Single number that gives the level of certainty on an estimate. Typically, ±2 standard errors around an estimate is expected to contain the true value about 95% of the time.

stature Height.

stop codon Sequences TAA, TAG, and TGA of the non-template (coding) strand of DNA will be represented in a mRNA as UAA, UAG, and UGA. These codons do not match any tRNA anti-codons, so these codons act as "stop" points for the translation of mRNA sequence into amino acids of a protein.

subcutaneous fat Fat lying immediately below the skin.

subischial length Estimate of lower body length made by subtracting sitting height from stature.

submetacentric Condition of having a centromere displaced in a chromosome toward the end of each sister chromatid.

swidden *See* **slash and burn horticulture**.

TAQ polymerase Heat-stable DNA polymerase from the bacterium *Thermus aquaticus*, used for the extension phase of polymerase chain reaction.

telomeres Specialized ends of chromosomes that protect the chromosomes from losing DNA sequence information with each replication event.

template strand Single strand of DNA that serves as a template from which RNA is produced. Because the RNA bases are complementary to the template strand the RNA strand has the same sequence (save for substitution of uracil for thymine) as the non-template strand. *See also* non-template strand.

thermocycler Machine used to perform polymerase chain reaction. It holds DNA samples in individual tubes and cycles repetitively through three temperatures (usually 94°C for denaturation, 55°C for annealing, and 72°C for extension) at programmed intervals.

threshold trait Discrete (present/absent) trait that is influenced by an underlying continuous distribution, which can be modeled using quantitative genetic theory.

thymine One of the four nitrogenous bases found in DNA. Pairs with adenine.

total genetic variance Average squared difference around the average trait value, due to genetic effects.

transcription Process of encoding RNA from DNA.

transfer RNA (tRNA) RNA molecules that bind specific amino acids and bring them to the ribosomes, where they are matched to mRNA sequences in the production of proteins.

transition Single-base mutation where a purine is replaced by another purine or a pyrimidine by another pyrimidine.

transmission genetics Study of how alleles are passed from parents to offspring.

transposition Occurrence whereby a DNA sequence inserts itself in a new location in the genome. In the case of SINES and LINES the sequence is first copied to RNA, then the RNA is "back copied" to DNA, and then the DNA inserts at a new location. Consequently, the copy number for SINES and LINES increases with evolutionary time.

transversion Single-base mutation where a purine is replaced by a pyrimidine or a pyrimidine is replaced by a purine.

treponema (treponemal diseases) Term used for a suite of diseases caused by bacteria of the genus *Treponema*. Diseases include pinta, yaws, endemic syphilis (bejel), and venereal syphilis.

unique event polymorphism (UEP) Any mutation found in some, but not all individuals, and thought by researchers to have arisen only once in evolutionary history. For example, the presence/absence of a particular Alu insert would be classified as a UEP because it is highly unlikely that the Alu element would insert more than once in exactly the same location.

untranslated region (UTR) Portions of a messenger RNA prior to the initiation site for translation and after the stop sequence. The UTR is not translated into protein, but differs from introns in that introns are untranslated sequences *within* genes.

uracil Nitrogenous base found in RNA that is analogous to thymine found in DNA. *See also* **ribonucleic acid.**

variance Average squared deviation around the average.

velocity curve Graph used in analyzing human growth that shows the rate of growth as a function of age.

virulence Severity of a disease.

vitamin D Essential nutrient, primarily obtained by ultraviolet radiation–induced synthesis during human evolution.

yaws One of the treponemal diseases, caused by *Treponema pertenue*. Widespread among children living in humid tropical regions of the world.

zoonose (zoonotic) Disease transmitted directly to humans from other animals.

zygote Single-celled diploid stage formed by the union of a spermatid and an ovum.

References

Abbey D M (1999) Scientific correspondence: The Thomas Jefferson paternity case. *Nature* 397:32.

Adachi B (1937) Das Ohrenschmalz als Rassenmerkmal und der Rassengeruch ("Achselgeruch") nebst dem Rassenunterschied der Schweissdrüsen. *Zeitschrift für Rassenkunde* 6:273–307.

Adams N R and Cronjé P B (2003) A review of the biology linking fibre diameter with fleece weight, liveweight, and reproduction in Merino sheep. *Australian Journal of Agricultural Research* 54:1–10.

Adcock G J, Dennis E S, Easteal S, Huttley G A, Jermiin L S, Peacock W J, and Thorne A (2001) Mitochondrial DNA sequences in ancient Australians: Implications for modern human origins. *Proceedings of the National Academy of Sciences USA* 98:537–542.

Agarwal A, Guindo A, Cissoko U, Taylor J G, Coulibaly D, Koné A, Kayentao K, Djimde A, Plowe C V, Doumbo O, Wellems T E, and Diallo D (2000) Hemoglobin C associated with protection from severe malaria in the Dogon of Mali, a West African population with a low prevalence of hemoglobin S. *Blood* 96:2358–2362.

Agassiz L (1962) *Essay on Classification*, edited by Lurie E. Cambridge, MA: Belknap Press of Harvard University Press.

Ahnini R T, Henry J, and Pontarotti P (1997) Geography and history of the genes in the human MHC: Can we predict MHC organization in nonhuman primates? In: *Molecular Biology and Evolution of Blood Group and MHC Antigens in Primates*, edited by Blancher A, Klein J, and Socha W W. Berlin: Springer-Verlag, pp 325–338.

Aird I, Bentall H H, and Roberts J A F (1953) A relationship between cancer of the stomach and the ABO groups. *British Medical Journal* 1:799–801.

Akide-Ndunge O B, Ayi K, and Arese P (2003) The Haldane malaria hypothesis: Facts, artifacts, and a prophecy. *Redox Report* 8(5):311–316.

Ali S C, Bunker H C E, Aston F A, Ukoli R, and Kamboh M I (1998) Apolipoprotein A Kringle polymorphism and serum lipoprotein (a) concentrations in African blacks. *Human Biology* 70:477–490.

Alkout A M, Backwell C C, and Weir D M (2000) Increased inflammatory responses of persons of blood group O to *Helicobacter pylori*. *Journal of Infectious Diseases* 181:1364–1369.

Allen S J, O'Donnell A, Alexander N D E, Alpers M P, Peto T E A, Clegg J B, and Weatherall D J (1997) α$^+$-Thalassemia protects children against disease caused by other infections as well as malaria. *Proceedings of the Natlional Academy of Sciences USA* 94:14736–14741.

Allison A C (1956) The sickle-cell and hemoglobin-C genes in some African populations. *Annals of Human Genetics* 21:67–89.

Alper C A, Colton H R, Rosen S F, Rabson A R, MacNab G M, and Gear J S S (1972) Homozygous deficiency of C3 in a patient with repeated infections. *Lancet* II:1179–1181.

Amiel J L (1967) Study of the leucocyte phenotypes in Hodgkin's disease. In: *Histocompatibility Testing 1967*, edited by Curtoni E S, Mattuiz P C, and Tosi R M. Copenhagen: Munksgaard, pp 79–81.

Ammerman A J and Cavalli-Sforza L L (1971) Measuring the rate of spread of early farming in Europe. *Man* 6(4):674–688.

Ammerman A J and Cavalli-Sforza L L (1973) A population model for the diffusion of early farmers in Europe. In: *The Explanation of Culture Change: Models in Prehistory*, edited by Renfrew C. London: Duckworth, pp 343–358.

Ammerman A J and Cavalli-Sforza L L (1984) *The Neolithic Transition and the Genetics of Populations in Europe.* Princeton, NJ: Princeton University Press.

Amouyel P, Vidal O, Launay J M, and Laplanche J L (1994) The apolipoprotein E alleles as major susceptibility factors for Creutzfeldt-Jakob disease. *Lancet* 344:1315–1318.

Anderson S, Bankier A T, Barrell B G, de Bruijn M H, Coulson A R, Drouin J, Eperon I C, Nierlich D P, Roe B A, Sanger F, Schreier P H, Smith A J, and Staden Rand Young I G (1981) Sequence and organization of the human mitochondrial genome. *Nature* 290:457–465.

Antonarakis S E, Boehm C D, Sergeant G R, Theisen C D, Dover G J, and Kazazian H H Jr (1984) Origin of the β^S-globin gene in blacks: The contribution of recurrent mutation or gene conversion or both. *Proceedings of the National Academy of Sciences USA* 81:853–856.

Antonarakis S E, Orkin S H, Kazazian H H Jr, Goff S C, Boehm C D, Waber P G, Sexton J P, Ostrer H, Fairbanks V F, and Chakravarti A (1982) Evidence for multiple origins of the β^E-globin gene in Southeast Asia. *Proceedings of the National Academy of Sciences USA* 79:6608–6611.

Aoki K (2002) Sexual selection as a cause of human skin colour variation: Darwin's hypothesis revisited. *Annals of Human Biology* 29:589–608.

Armelagos G J, Carlson D S, and Van Gerven D (1982) The theoretical foundations and development of skeletal biology. In: *A History of American Physical Anthropology 1930–1980*, edited by Spencer F. New York: Academic Press, pp 305–328.

Arnaiz-Villena A, Elaiwa N, Silvera C, Rostom A, Moscoso J, Gomez-Casado E, Allende L, Varela P, and Martinez-Laso J (2001) The origin of Palestinians and their genetic relatedness with other Mediterranean populations. *Human Immunology* 62:889–900. [Retracted in: Suciu-Foca N and Lewis R (2001) Editorial. Anthropology and genetic markers (Retraction of Arnaiz-Villena et al.). *Human Immunology* 62:1063.]

Augstein H F, ed. (1996) *Race: The Origins of an Idea, 1760–1850.* Bristol, UK: Thoemmes Press.

Awadall P, Eyre-Walker A, and Maynard Smith J (1999) Linkage disequilibrium and recombination in hominid mitochondrial DNA. *Science* 286:2524–2525.

Baker L E (2001) *Mitochondrial DNA haplotype and sequence analysis of historic Choctaw and Menominee hair shaft samples.* PhD diss. University of Tennessee, Knoxville.

Ball S P, Tongue N, Gibaud A, Le Pendu J, Mollicone R, Gerard G, and Oriol R (1991) The human chromosome 19 linkage group *FUT1 (H)*, *FUT2 (SE)*, *LE*, *LU*, *PEPD*, *C3*, *APOC2*, *D19ST*, and *D19S9. Annals of Human Genetics* 55:225–233.

Bamshad M J, Wooding S, Watkins W S, Ostler C T, Batzer M A, and Jorde L B (2003) Human population genetic structure and inference of group membership. *American Journal of Human Genetics* 72:578–589.

Bandelt H-J and Forster P (1997) The myth of bumpy hunter–gatherer mismatch distributions. *American Journal of Human Genetics* 61:980–983.

Barbujani G and Bertorelle G (2001) Genetics and the population history of Europe. *Proceedings of the National Academy of Sciences USA* 98:22–25.

Barbujani G and Sokal R R (1990) Zones of sharp genetic change in Europe are also linguistic boundaries. *Proceedings of the National Academy of Sciences, USA* 87:1816–1819.

Barton N H and Keightley P D (2002) Understanding quantitative genetic variation. *Nature Genetics* 3:11–21.

Bartoshuk L M (1979) Bitter taste of saccharin related to the genetic ability to taste the bitter substance 6-*n*-propylthiouracil. *Science* 205:934–935.

Baru D and Paguio A S (1977) ABO blood groups and cholera. *Annals of Human Biology* 4:489–492.

Bass W M (1995) *Human Osteology: A Laboratory and Field Manual*, fourth edition. Columbia: Missouri Archaeological Society.

Bauer K H (1927) Homoisotransplantation van Epidermis bei einiigen Zwillingen. *Brun's Beitrage zur klinischen Chirurgie* 141:442–447.

Beall C M and Steegmann A T Jr (2000) Human adaptation to climate: Temperature, ultraviolet radiation, and altitude. In: *Human Biology: An Evolutionary and Biocultural Perspective*, edited by Stinson S, Bogin B, Huss-Ashmore R, and O'Rourke D. New York: John Wiley & Sons, pp 163–224.

Beals K L (1972) Head form and climatic stress. *American Journal of Physical Anthropology* 37:85–92.

Beals K L, Smith C L, and Dodd S M (1984) Brain size, cranial morphology, climate and time machines. *Current Anthropology* 25:301–330.

Beals K L, Smith C L, and Kelso A J (1992) ABO phenotype and morphology. *Current Anthropology* 33:221–224.

Beauchamp G K and Yamazaki K (1997) HLA and mate selection in humans: Commentary. *American Journal of Human Genetics* 61:494–496.

Beeson J G (1999) *Plasmodium falciparum* isolates from infected pregnant women and children are associated with distinct adhesive and antigenic properties. *Journal of Infectious Diseases* 180:464–472.

Beja-Pereira A, Luikart G, England P R, Bradley D G, Jann O C, Bertorelle G, Chamberlain A T, Nunes T P, Metodiev S, Ferrand N, and Erhardt G (2003) Gene–culture coevolution between cattle milk protein genes and human lactase genes. *Nature Genetics* 35:311–313.

Bell J I, Todd J A, and McDevitt, H O (1989) The molecular basis of HLA–disease association. *Advances in Human Genetics* 18:1–41. Series edited by H Harris and K Hirschhorn, NY: Plenum Press.

Bellwood P (2001) Early agriculturalist population diasporas? Farming, languages and genes. *Annual Reviews in Anthropology* 30:181–207.

Bertorelle G and Excoffier L (1998) Inferring admixture proportions from molecular data. *Molecular Biology and Evolution* 15:1298–1311.

Beutler E (1983) Glucose-6-phosphate dehydrogenase deficiency. In: *The Metabolic Basis of Inherited Disease*, fifth edition, edited by Stanbury J B, Wyngaarden J, Fredrickson D S, and Goldstein J. New York: McGraw-Hill, pp 1629–1653.

Beutler E (1990) The genetics of glucose-6-phosphate dehydrogenase deficiency. *Seminars in Hematology* 27:137–164.

Bindon J R and Zansky S M (1986) Growth patterns of height and weight among three groups of Samoan preadolescents. *Annals of Human Biology* 13:171–178.

Bittles A H and Smith M T (1994) Religious differentials in postfamine marriage patterns, Northern Ireland 1840–1915. *Human Biology* 66:59–76.

Black F L and Hedrick, P W (1997) Strong balancing selection at HLA loci: Evidence from segregation in South Amerindian families. *Proceedings of the National Academy of Sciences USA* 94:12452–12456.

Black J A and Dixon G H (1968) Amino-acid sequence of alpha chains of human haptoglobins. *Nature* 218:736–741.

Blakeslee A F (1932) Genetics of sensory thresholds: Taste for phenylthiocarbamide. *Proceedings of the National Academy of Sciences USA* 18:120–130.

Blakeslee A F and Salmon M R (1931) Odor and taste blindness. *Eugenical News* 16:105–109.

Blakeslee A F and Salmon T N (1935) Genetics of sensory thresholds: Individual taste reactions for different substances. *Proceedings of the National Academy of Sciences USA* 21:84–90.

Blancher A and Socha W W (1997a) The ABO, Hh and Lewis blood group in humans and non-human primates. In: *Molecular Biology and Evolution of Blood Group and MHC Antigens in Primates*, edited by Blancher A, Klein J, and Socha W W. Berlin: Springer-Verlag, pp 30–92.

Blancher A and Socha W W (1997b) The rhesus system. In: *Molecular Biology and Evolution of Blood Group and MHC Antigens in Primates*, edited by Blancher A, Klein J, and Socha W W, Berlin: Springer-Verlag, pp 147–218.

Blangero J (1995) Statistical genetic approaches to human adaptability. *Human Biology* 65:941–966.

Blangero J, Almasy L, Duggirala R, Williams-Blangero S, O'Connell P, and Stern M P (1999) Mapping quantitative trait loci influencing normal variation: The genetics of skin reflectance. *American Journal of Physical Anthropology* (Suppl) 28:93–94.

Block N (1996) How heritability misleads about race. *Boston Review* 20:30–35.

Blumenbach J F (1865) *The Anthropological Treatises of Johann Friedrich Blumenbach*, translated and edited by Bendyshe T. London: Longman, Green, Longman, Roberts, & Green.

Boas F (1911) *Changes in the Bodily Form of Descendents of Immigrants*, 61st Cong, 2d sess, S Doc 208. Reprinted from the Reports of the US Immigration Commission. New York: Columbia University Press, 1912.

Boas F (1912) *Changes in the Bodily Form of Descendants of Immigrants*. New York: Columbia University Press.

Boas F (1928) *Materials for the Study of Inheritance in Man*. New York: Columbia University Press.

Boas F (1940) *Race, Language and Culture*. New York: Macmillan.

Bodmer J (1996) World distribution of HLA alleles and implications for disease. In: *Variation in the Human Genome. Ciba Foundation Symposium 197*, edited by Chadwick D and Cardew G. Chichester, UK: John Wiley & Sons, pp 233–258.

Bodmer W F (1972) Population genetics of the HL-A system: Retrospect and prospect. In: *Histocompatibility Testing*, edited by Dausset J, and Colombani J. Copenhagen: Munksgaard, pp 611–617.

Bodmer W F (1991) *HLA 1991*. Oxford: Oxford University Press.

Bodmer W F and Cavalli-Sforza L L (1976) *Genetics, Evolution, and Man*. San Francisco: W H Freeman.

Boehm C D, Dowling C D, Antonarakis S E, Honig G R, and Kazazian H H Jr (1985) Evidence supporting a single origin of the β^C-globin gene in blacks. *American Journal of Human Genetics* 37:771–777.

Bogin B (1988) *Patterns of Human Growth*. Cambridge: Cambridge University Press.

Bogin B A and Smith H (1996) The evolution of the human life cycle. *American Journal of Human Biology* 8:703–716.

Borén T, Falk P, Roth K A, Göran L, and Normark S (1993) Attachment of *Helicobacter pylori* to human gastric epithelium mediated by blood group antigens. *Science* 262:1892–1895.

Bouchard T J Jr (1998) Genetic and environmental influences on adult intelligence and special mental abilities. *Human Biology* 70:257–279.

Bouchard T J Jr, Lykken D T, McGue M, Segal N, and Tellegen A (1990) Sources of human psychological differences: The Minnesota study of twins raised apart. *Science* 250:223–228.

Boyd W C (1950) *Genetics and the Races of Man: An Introduction to Modern Physical Anthropology*. Boston: D C Heath and Company.

Boyd W C (1958) *Genetics and the Races of Man*. Boston: Boston University Press.

Brace C L (1964) On the race concept. *Current Anthropology* 5:313–320.

Brace C L (1982) The roots of the race concept in American physical anthropology. In: *A History of American Physical Anthropology 1930–1980*, edited by Frank Spencer. New York: Academic Press, pp 11–29.

Brace C L (2000) *Evolution in an Anthropological View*. Walnut Creek, CA: AltaMira Press.

Brace C L and Hunt K D (1990) A nonracial craniofacial perspective on human variation: A(ustralia) to Z(uni). *American Journal of Physical Anthropology* 82:341–360.

Brace C L and Montagu A (1977) *Human Evolution: An Introduction to Biological Anthropology*, second edition. New York: Macmillan.

Branda R F and Eaton J W (1978) Skin color and nutrient photolysis: An evolutionary hypothesis. *Science* 201:625–626.

Braüer G and Chopra V P (1980) Estimating the heritability of hair colour and eye colour. *Journal of Human Evolution* 9:625–630.

Briggs R and King T J (1952) Transplantation of living nuclei from blastula cells into enucleated frog eggs. *Proceedings of the National Academy of Sciences USA* 38:455–463.

Broberg G (1983) *Homo sapiens*—Linnaeus's classification of man. In: *Linnaeus: The Man and His Work*, edited by Frängsmyr T. Berkeley: University of California Press, pp 156–194.

Brown K A and Pluciennik M (2001) Archaeology and human genetics: Lessons for both. *Antiquity* 75:101–106.

Browning M and McMichael A, eds. (1996) *HLA and MHC: Genes, Molecules and Function*. Oxford: BIOS Scientific Publishers.

Brues A M (1954) Selection and polymorphism in the A-B-O blood groups. *American Journal of Physical Anthropology* 12:559–597.

Brues A M (1977) *Peoples and Races*. New York: Macmillan.

Burchard E G, Ziv E, Coyle N, Gomez S L, Tang H, Karter A J, Mountain J L, Pérez-Stable E J, Sheppard D, and Risch N (2003) The importance of race and ethnic background in biomedical research and clinical practice. *New England Journal of Medicine* 348:1170–1175.

Butler J M (2001) *Forensic DNA Typing*. New York: Academic Press.

Byard P J (1981) Quantitative genetics of human skin color. *Yearbook of Physical Anthropology* 24:123–137.

Byard P J and Lees F C (1981) Estimating the number of loci determining skin colour in a hybrid population. *Annals of Human Biology* 8:49–58.

Cabana T, Jolicoeur P, and Michaud J (1993) Prenatal and postnatal growth and allometry of stature, head circumference, and brain weight in Québec children. *American Journal of Human Biology* 5:93–99.

Calderón R, Perez-Miranda A, Peña J A, Vidales C, Aresti U, and Dogoujon J M (2000) The genetic position of the autochthonous subpopulation of northern Navarre (Spain) in relation to other Basque subpopulations: A study based on Gm and Km immunoglobulin allotypes. *Human Biology* 72:619–640.

Calderón R, Vidales C, Peña J A, Perez-Miranda A, and Dogoujo J M (1998) Immunoglobulin allotypes (*Gm* and *Km*) in Basques from Spain: Approach to the origin of the Basque population. *Human Biology* 70:667–698.

Cann R L, Stoneking M, and Wilson A (1987) Mitochondrial DNA and human evolution. *Nature* 325:31–36.

Capelli C, Redhead N, Abernathy J K, Gratrix F, Wilson J F, Moen T, Hervig T, Richards M, Stumpf M P H, Underhill P A, Bradshaw P, Shaha A, Thomas M G, Bradman N, and Goldstein D B (2003) A Y chromosome census of the British Isles. *Current Biology* 13:979–984.

Capelli C, Wilson J F, Richards M, Stumpf M P H, Gatrix F, Oppenheimer S, Underhill P, Pascali V L, Ko T-M, and Goldstein D B (2001) A predominantly indigenous paternal heritage for Austronesian-speaking peoples of insular Southeast Asia and Oceania. *American Journal of Human Genetics* 68:432–443.

Cappadoro M, Giribaldi G, O'Brien E, Turrini F, Mannu F, Ulliers D, Simula G, Luzzatto L, and Arese P (1998) Early phagocytosis of glucose-6-phosphate dehydrogenase (G6PD)-deficient erythrocytes parasitized by *Plasmodium falciparum* may explain malaria protection in G6PD deficiency. *Blood* 92:2527–2534.

Caramelli D, Lalueza-Fox C, Vernes C, Lari M, Casoli A, Mallegni F, Chiarelli B, Dupanloup I, Bertanpetit J, Barbujani G, and Bertorelle G (2003) Evidence for a genetic discontinuity between Neandertals and 24,000-year-old anatomically modern Europeans. *Proceedings of the National Academy of Sciences USA* 100:6593–6597.

Carey J W and Steegmann A T Jr (1981) Human nasal protrusion, latitude, and climate. *American Journal of Physical Anthropology* 56:313–319.

Cartmill M (1998) The status of the race concept in physical anthropology. *American Anthropologist* 100:651–660.

Cartron J-P and Agre P (1995) RH blood groups and Rh-deficiency syndrome. In: *Blood Cell Biochemistry, Molecular Basis of Human Blood Group Antigens*, volume 6, edited by Cartron J-P, and Rouger P. New York: Plenum Press, pp 189–225.

Caspari R (2003) From types to populations: A century of race, physical anthropology, and the American Anthropological Association. *American Anthropologist* 105:65–76.

Catsimpoolas N (1980) Basic principles of different types of electrophoresis. In: *Handbook of Electrophoresis*, volume I, edited by Lewis L A and Opplt J J. Boca Raton FL: CRC Press, pp 11–23.

Cavalli-Sforza L L (1997) Genes, peoples, and languages. *Proceedings of the National Academy of Sciences of the USA* 94:7719–7724.

Cavalli-Sforza L L and Bodmer W F (1971) *The Genetics of Human Populations*. San Francisco: W H Freeman.

Cavalli-Sforza L L, Menozzi P, and Piazza A (1994) *The History and Geography of Human Genes*. Princeton: Princeton University Press.

Cavalli-Sforza L L, Piazza A, Menozzi P, and Mountain J (1988) Reconstruction of human evolution: Bringing together genetic, archaeological and linguistic data. *Proceedings of the National Academy of Sciences of the USA* 85:6002–6006.

Chakraborty R (1986) Gene admixture in human populations: Models and predictions. *Yearbook of Physical Anthropology* 29:1–43.

Chakraborty R, Kamboh M I, Nwankwo M, and Ferrell R E (1992) Caucasian genes in American blacks: New data. *American Journal of Human Genetics* 50:145–155.

Chakravartti M R, Verma B K, Hanurav T V, and Vogel F (1966) Relation between smallpox and the ABO blood groups in a rural population of West Bengal. *Humangenetik* 2:78–80.

Chapelle J-P, Albert A, Smeets J-P, Heusghem C, and Kulbertus H E (1982) Effect of the haptoglobin phenotype on the size of a myocardial infarct. *New England Journal of Medicine* 307:457–463.

Chaudhuri A (1977) Cholera and blood-group. *Lancet* 2:404–405.

Chaudhuri A and Pogo A O (1995) The Duffy blood group system and malaria. In: *Blood Cell Biochemistry. Molecular Basis of Human Blood Group Antigens*, volume 6, edited by Cartron J-P and Rouger P. New York: Plenum Press, pp 243–265.

Chautard-Freire-Maia E A (1974) Linkage relationships between 22 autosomal markers. *Annals of Human Genetics* 38:191–198.

Chebloune Y, Pagnier J, Tabuchet G, Faure C, Verdier G, Labie D, and Nigon V (1988) Structural analysis of the 5′ flanking region of the β-globin gene in African sickle cell anemia patients: Further evidence for three origins of the sickle cell mutation in Africa. *Proceedings of the National Academy of Sciences USA* 85:4431–4435.

Chikhi L, Destro-Bisol G, Bertorelle G, Pascali V, and Barbujani G (1998) Clines of nuclear DNA markers suggest a largely Neolithic ancestry of the European gene pool. *Proceedings of the National Academy of Sciences USA* 95:9053–9058.

Chikhi L, Nichols R A, Barbujani G, and Beaumont M A (2002) Y genetic data support the Neolithic demic diffusion model. *Proceedings of the National Academy of Sciences USA* 99:11008–11013.

Chotivanich K, Udomsangpetch R, Pattanapanyasat K, Chierakul W, Simpson J, Looareesuwan S, and White N (2002) Hemoglobin E: A balanced polymorphism protective against high parasitemias and thus severe *P. falciparum* malaria. *Blood* 100:1172–1176.

Clark P, Stark A E, Walsh R J, Jardine R, and Martin N G (1981) A twin study of skin reflectance. *Human Biology* 8:529–541.

Clausen H and Hakomori S I (1989) ABH and related histo-blood group antigens: Immunochemical differences in carrier isotypes and their distribution. *Vox Sanguinis* 56:1–20.

Cleve H (1973) The variants of the group-specific component: A review of their distribution in human populations. *Israel Journal of Medical Sciences* 9:1133–1146.

Cole T M III (1996) Early anthropological contributions to "geometric morphometrics." *American Journal of Physical Anthropology* 101:291–296.

Comas D, Plaza S, Calafell F, Sajantila A, and Bertranpetit J (2001) Recent insertion of an *Alu* element within a polymorphic human-specific *Alu* insertion. *Molecular Biology and Evolution* 18:85–88.

Comuzzie A G, Hixson J E, Almasy L, Mitchell B D, Mahaney M C, Dyer T D, Stern M P, MacCluer J W, and Blangero J (1997) A major quantitative trait locus determining serum leptin levels and fat mass is located on human chromosome 2. *Nature Genetics* 15:273–276.

Conneally P M, Dumont-Driscoll M, Huntzinger R S, Nance W E, and Jackson C E (1976) Linkage relations of the loci for Kell and phenylthiocarbamide (PTC) taste sensitivity. *Human Heredity* 26:267–271.

Cook G C and Al-Torki M T (1975) High intestinal lactase concentration in adult Arabs in Saudi Arabia. *British Medical Journal* 3:135–136.

Coombs R R A, Mourant A E, and Race R R (1946) *In vivo* isosensitization of red cells in babies with haemolytic disease. *Lancet* i:264–266.

Coon C S (1962) *The Origin of Races.* New York: Alfred A Knopf.

Coon C S, Garn S M, and Birdsell J B (1950) *Races: A Study of the Problems of Race Formation in Man.* Springfield, IL: Charles C Thomas.

Cooper R S, Kaufman J S, and Ward R (2003) Race and genomics. *New England Journal of Medicine* 348:1166–1170.

Corbo R M and Scacchi R (1999) Apolipoprotein E (*APOE*) allele distribution in the world. Is *APOE**4 a "thrifty" allele? *Annals of Human Genetics* 63:301–310.

Corbo R M, Scacchi R, Mureddu L, Mulas G, Castrechini S, and Rivasi P (1999) Apolipoprotein B, apolipoprotein E and angiotensin-converting enzyme polymorphisms in 2 Italian populations at different risk for coronary artery disease and comparison of allele frequencies among European populations. *Human Biology* 71:933–945.

Corder E H, Saunders A M, Strittmatter W J, Schmechel D E, Gaskell P C, Small G W, Roses A D, Haines J L, and Pericak-Vance M A (1993) Gene dose of apolipoprotein E type 4 allele and the risk of Alzheimer's disease in late onset families. *Science* 261:921–923.

Crawford M H (1973) The use of genetic markers of the blood in the study of the evolution of human populations. In: *Methods and Theories of Anthropological Genetics*, edited by Crawford M H and Workman P L. Albuquerque: University of New Mexico Press, pp 19–38.

Crawford M H (1975) Genetic affinities and origin of the Irish travellers. In: *Biosocial Interrelations in Population Adaptations*, edited by Watts E S, Johnston F E, Lasker G W. The Hague: Mouton, pp 93–103.

Crawford M H (1998) *The Origins of Native Americans: Evidence from Anthropological Genetics.* Cambridge: Cambridge University Press.

Crawford M H, Dykes D D, and Polesky H F (1989) Genetic structure of Mennonite populations of Kansas and Nebraska. *Human Biology* 61:493–514.

Crawford M H, Koertevlyessy T, Huntsman R G, Collins M, Duggirala R, Martin L, and Keeping D (1995) Effects of religion, economics, and geography on genetic structure of Fogo Island, Newfoundland. *American Journal of Human Biology* 7:437–451.

Crichton J M (1966) A multiple discriminant analysis of Egyptian and African Negro crania. *Papers of the Peabody Museum of Archaeology and Ethnology* 57:47–67.

Cruciani F, Santolamazza P, Shen P, Macaulay V, Moral P, Olckers A, Modiano D, Holmes S, Destro-Bisol G, Coia V, Wallace D C, Oefner P J, Torroni A, Cavalli-Sforza L L, Scozzari R, and Underhill P A (2002) A back migration from Asia to sub-Saharan Africa is supported by high-resolution analysis of human Y-chromosome haplotypes. *American Journal of Human Genetics* 70:1197–1214.

Cucca F and Todd J A (1996) HLA susceptibility to type 1 diabetes: Methods and mechanisms. In: *HLA and MHC: Genes, Molecules and Function*, edited by Browning M and McMichael A. Oxford: BIOS Scientific Publishers, pp 383–406.

Currat M, Trabuchet G, Rees D, Perrin P, Harding R M, Clegg J B, Langaney A, and Excoffier L (2002) Molecular analysis of the β-globin gene cluster in the Niokholo Mandenka population reveals a recent origin of the β^s Senegal mutation. *American Journal of Human Genetics* 70:207–223.

Custodio R and R G Huntsman (1984) Abnormal hemoglobins among the black Caribs. In: *Current Developments in Anthropological Genetics. Black Caribs: A Case Study in Biocultural Adaptation*, volume 3, edited by Crawford M H. New York: Plenum Press, pp 335–343.

Daiger S P, Schanfield M S, and Cavalli-Sforza L L (1975) Group-specific component (Gc) proteins bind vitamin D and 25-hydroxyvitamin D. *Proceedings of the National Academy of Sciences USA* 72:2076–2080.

Daniels J, McGuffin P, Owen M J, and Plomin R (1998) Molecular genetic studies of cognitive ability. *Human Biology* 70:281–296.

Dausset J (1958) Iso-leuco-anticorps. *Acta Haemotologica* 20:156–166.

Davis G (1999) Scientific correspondence: The Thomas Jefferson paternity case. *Nature* 397:32.

Dean M, Carrington M, and O'Brien S J (2002) Balanced polymorphism selected by genetic versus infectious human disease. *Annual Review of Genomics and Human Genetics* 3:263–292.

de Bruijn M H L and Fey G H (1985) Human complement component C3: cDNA coding sequence and derived primary structure. *Proceeding of the National Academy of Sciences USA* 82:708–712.

DeFries C, Fulker D W, and LaBuda M C (1987) Evidence for a genetic aetiology in reading disability of twins. *Nature* 329:537–539.

DeFries C, Olsen R K, Pennington B F, and Smith S D (1991) Colorado Reading Project: An update. In: *The Reading Brain: The Biological Basis of Dyslexia*, edited by Duane DD and Gray DB. Parkton, MD: York Press, pp 53–87.

De Knijff P, van den Maagdenberg A M J M, Frants R R, and Havekes L M (1994) Genetic heterogeneity of apolipoprotein E and its influence on plasma lipid and lipoprotein levels. *Human Mutation* 4:178–194.

De la Chapelle A and Wright F A (1998) Linkage disequilibrium mapping in isolated populations: The example of Finland revisited. *Proceedings of the National Academy of Sciences USA* 95:12416–12423.

Delusion of race (1936) *The Nature* 137:635–637.

Dempster A P, Laird N M, and Rubin D B (1977) Maximum likelihood from incomplete data via the EM algorithm. *Journal of the Royal Statistical Society B* 39:1–38.

Deniker J (1900) The races of man. Reprinted in Count E W (1950) *This Is Race: An Anthology Selected from the International Literature on the Races of Man*. New York: Henry Schuman, pp 207–221.

Derenko M V, Grzybowski T, Malyarchuk B A, Czarny J, Miscicka-Slwika D, and Zakharov I A (2001) The presence of mitochondrial haplogroup X in Altaians from south Siberia. *American Journal of Human Genetics* 69:237–241.

Devlin B, Daniels M, and Roeder K (1997) The heritability of IQ. *Nature* 388:468–471.

Devor E J (1987) Transmission of human craniofacial dimensions. *Journal of Craniofacial Genetics and Developmental Biology* 7:95–106.

Devor E J, McGue M, Crawford M H, and Lin P M (1986) Transmissible and nontransmissible components of anthropometric variation in the Alexanderwohl Mennonites. II. Resolution by path analysis. *American Journal of Physical Anthropology* 69:83–92.

Diamond J (1988) Express train to Polynesia. *Nature* 336:307–308.

Disotell T R (2000) Molecular anthropology and race. *Annals of the New York Academy of Sciences* 925:9–24.

Dobzhansky T (1944) On species and races of living and fossil man. *American Journal of Physical Anthropology* n.s. 2:251–265.

Dobzhansky T (1963) Possibility that *Homo sapiens* evolved independently 5 times is vanishingly small. *Current Anthropology* 4:360, 364–367.

Dow M M, Cheverud J M, and Friedlaender J S (1987) Partial correlation of distance matrices in studies of population structure. *American Journal of Physical Anthropology* 72:343–352.

Drayna D, Coon H, Kim U-K, Elsner T, Cromer K, Otterud B, Baird L, Peiffer A P, and Leppert M (2003) Genetic analysis of a complex trait in the Utah Reference Project: A major locus for PTC taste ability on chromosome 7q and a secondary locus on chromosome 16p. *Human Genetics* 112:567–572.

Duke B D (1998) *Jesse James Lived and Died in Texas*. Austin, TX: Eaken Press.

Dunn L C and Dobzhansky T (1952) *Heredity, Race and Society*. New York: New American Library.

Dupanloup I and Bertorelle G (2001) Inferring admixture proportions from molecular data: Extension to any number of parental groups. *Molecular Biology and Evolution* 18:672–675.

Durham W H (1991) *Coevolution: Genes, Culture, and Human Diversity*. Palo Alto CA: Stanford University Press.

Eaton G G (1976) The social order of Japanese macaques. *Scientific American* 235:96–106.

Eaton J W and Gavan J A (1965) Sensitivity to P-T-C among primates. *American Journal of Physical Anthropology* 23:381–388.

Eaton J W, Brandt P, Mahoney J R, and Lee J T (1982) Haptoglobin: A natural bacteriostat. *Science* 215:691–693.

Edington G M (1959) *Annual Report of the Department of Pathology*. Ibadan, Nigeria: University College.

Edwards E A and Duntley Q (1939) The pigments and color of living human skin. *American Journal of Anatomy* 65:1–33.

Edwards M C and Gibbs R A (1992) A human dimorphism resulting from loss of an *Alu*. *Genomics* 14:590–597.

Eichner E R, Finn R, and Krevans J R (1963) Relationship between serum antibody-levels and the ABO blood group polymorphism. *Nature* 198:164–165.

Embury S H, Miller J A, Dozy A M, Kan Y W, Chan V, and Todd D (1980) Two different molecular organizations account for the single α-globin gene of the α-thalassemia-2 genotype. *Journal of Clinical Investigation* 66:1319–1325.

European Bioinformatics Institute (EMBL-EBL) (2003) www.ebi.ac.uk/imgt/hla/stats.html.

Evans A G and Wellems T E (2002) Coevolutionary genetics of *Plasmodium* malaria parasites and their human hosts. *Integrative and Comparative Biology* 42:401–407.

Eveleth P B and Tanner J M (1990) *Worldwide Variation in Human Growth*, second edition. Cambridge: Cambridge University Press.

Evett I W and Weir B S (1998) *Interpreting DNA Evidence: Statistical Genetics for Forensic Scientists*. Sunderland, MA: Sinauer Associates.

Excoffier L, Smouse P E, and Quattro J M (1992) Analysis of molecular variance inferred from metric distances among DNA haplotypes: Application to human mitochondrial DNA restriction data. *Genetics* 131:479–491.

Fairchild H P (1944) Truth about race. *Harper's Magazine* 189:418–425.

Falconer D S and Mackay T F C (1996) *Introduction to Quantitative Genetics*, fourth edition. Essex, UK: Longman.

Falk C T and Li C C (1969) Negative assortative mating: Exact solution to a simple model. *Genetics* 62:215–223.

Fedigan L M (1983) Dominance and reproductive success in primates. *Yearbook of Physical Anthropology* 26:91–129.

Feldenzer J, Mears J G, Burns A L, Natta C, and Bank A (1979) Heterogeneity of DNA fragments associated with the sickle-globin gene. *Journal of Clinical Investigation* 64:751–755.

Ferrell R E (1993) Obesity: Choosing genetic approaches from a mixed menu. *Human Biology* 65:967–975.

Firschein I L (1961) Population dynamics of the sickle-cell trait in the Black Caribs of British Honduras, Central America. *American Journal of Human Genetics* 13:233–254.

Firschein I L (1984) Demographic patterns of the Garifuna (Black Caribs) of Belize. In: *Current Developments in Anthropological Genetics. Black Caribs: A Case Study in Biocultural Adaptation*, volume 3, edited by Crawford MH. New York: Plenum Press, pp 67–94.

Fisher N L, ed. (1996) *Cultural and Ethnic Diversity: A Source for Genetics Professionals*. Baltimore, MD: Johns Hopkins University Press.

Fisher R A (1918) The correlations between relatives on the supposition of Mendelian inheritance. *Transactions of the Royal Society of Edinburgh* 52:399–433.

Fisher R A and Race R R (1946) *Rh* frequencies in Britain. *Nature* 157:48–49.

Fisher R A, Ford E B, and Huxley J (1939) Taste-testing the anthropoid apes. *Nature* 144:750.

Fix A (1996) Gene frequency clines in Europe: Demic diffusion or natural selection? *Journal of the Royal Anthropological Institute* (n.s.) 2:625–643.

Flatz G (1967) Haemoglobin *E*: Distribution and population dynamics. *Humangenetik* 3:189–234.

Flatz G (1987) Genetics of lactose digestion in humans. In: *Advances in Human Genetics*, volume 16, edited by Harris H and Hirschhorn K. New York: Plenum Press, pp 1–77.

Flatz G and Rotthauwe H W (1973) Lactose nutrition and natural selection. *Lancet* 2:76–77.

Flint J, Harding R M, Boyce A J, and Clegg J B (1998) The population genetics of the haemoglobinopathies. *Baillière's Clinical Haematology* 11:1–51.

Flint J, Harding R M, Clegg J B, and J Boyce A J (1993) Why are some genetic disorders common? Distinguishing selection from other processes by molecular analysis of globin gene variants. *Human Genetics* 91:91–117.

Flint J, Hill A V S, Bowden D K, Oppenheimer S J, Sill P R, Serjeantson S W, Bana-Koiri J, Bhatia K, Alpers M P, Boyce A J, Weatherall D J, and Clegg J B (1986) High frequencies of α-thalassemia are the result of natural selection by malaria. *Nature* 321:744–749.

Flynn J R (1980) *Race, IQ and Jensen*. London: Routledge.

Flynn J R (1984) The mean IQ of Americans: Massive gains 1932 to 1978. *Psychological Bulletin* 95:29–51.

Flynn J R (1987) Massive IQ gains in 14 nations: What IQ tests really measure. *Psychological Bulletin* 101:171–191.

Forster P (2003) To err is human. *Annals of Human Genetics* 67:2–4.

Forster P, Harding R, Torroni A, and Bandelt H-J (1996) Origin and evolution of Native American mtDNA variation: A reappraisal. *American Journal of Human Genetics* 59:935–945.

Forster P, Röhl A, Lünnemann P, Brinkmann C, Zerjal T, Tyler-Smith C, and Brinkmann B (2000) A short tandem repeat-based phylogeny for the human Y chromosome. *American Journal of Human Genetics* 67:182–196.

Foster E A, Jobling M A, Taylor P, Donnelly P, de Knijff P, Mieremet R, Zerjal T, and Tyler-Smith C (1998) Jefferson fathered slave's last child. *Nature* 396:27–28.

Foster E A, Jobling M A, Taylor P G, Donnelly P, de Knijff P, Mieremet R, Zerjal T, and Tyler-Smith C (1999) Scientific correspondence: The Thomas Jefferson paternity case. *Nature* 397:32.

Foster M W and Sharp R R (2002) Race, ethnicity, and genomics: Social classifications as proxies of biological heterogeneity. *Genome Research* 12:844–850.

Fox A L (1931a) Six in ten "tasteblind" to bitter chemical. *Science News Letter* 9:249.

Fox A L (1931b) Tasteblindness. *Science* 73:14.

Fox A L (1932) The relationship between chemical constitution and taste. *Proceedings of the National Academy of Sciences USA* 18:115–120.

Foy H, Kondi A, Timms G L, Brass W, and Bushra F (1954) The variability of sickle cell rates in the tribes of Kenya and the southern Sudan. *British Medical Journal* 1:294–297.

Franciscus R G and Long J C (1991) Variation in human nasal height and breadth. *American Journal of Physical Anthropology* 85:419–427.

Friedlaender J S (1975) *Patterns of Human Variation: The Demography, Genetics, and Phenetics of Bougainville Islanders*. Cambridge, MA: Harvard University Press.

Friedlaender J S, Costa P T Jr, Bosse R, Ellis E, Rhoads J G, and Stoudt H W (1977) Longitudinal physique changes among healthy white veterans at Boston. *Human Biology* 49:541–558.

Friedman M J (1978) Erythrocytic mechanism of sickle cell resistance to malaria. *Proceedings of the National Academy of Sciences of the USA* 75:1994–1997.

Friedman M J and Trager W (1981) The biochemistry of resistance to malaria. *Scientific American* 244:154–164.

Frisancho A R (1993) *Human Adaptation and Accommodation*. Ann Arbor: University of Michigan Press.

Frisancho A R and Baker P T (1970) Altitude and growth: A study of the patterns of physical growth of a high altitude Peruvian Quechua population. *American Journal of Physical Anthropology* 32:279–292.

Frisancho A R, Wainwright R, and Way A (1981) Heritability and components of phenotypic expression in skin reflectance of Mestizos from the Peruvian lowlands. *American Journal of Physical Anthropology* 55:203–208.

Gabunia L, Vekua A, Lordkipanidze D, Swisher C C, Ferring R, Justus A, Nioradze M, Tvalchrelidze M, Antón S C, Bosinski G, Jöris O, de Lumley M A, Majsuradze G, and Mouskhelishvili A (2000) Earliest Pleistocene hominid cranial remains from Dmanisi, Republic of Georgia: Taxonomy, geological setting, and age. *Science* 288:1019–1025.

Gahmberg C G (1992) Red cell membrane proteins carrying the MN, ABH, and Rh antigen activities. In: *Protein Blood Group Antigens of the Human Red Cell*, edited by Agre PC and Cartron J-P. Baltimore, MD: Johns Hopkins University Press, pp 3–19.

Galton F (1889) *Natural Inheritance*. London: Macmillan.

Gardner R J (1977) A new estimate of the achondroplasia mutation rate. *Clinical Genetics* 11:31–38.

Garn S M (1961) *Human Races*. Springfield, IL: Charles C Thomas.

Garn S M and Coon C S (1955) On the number of races of mankind. *American Anthropologist* 57:996–1001.

Garrard G, Harrison G A, and Owen J J T (1967) Comparative spectrophotometry with E.E.L. and Photovolt instruments. *American Journal of Physical Anthropology* 27:389–396.

Gelpi, A (1973) Migrant populations and the diffusion of the sickle-cell gene. *Annals of Internal Medicine* 79:258–264.

Gerard G, Vitrac D, Le Pendu J, Muller A, and Oriol R (1982) H-deficient blood groups (Bombay) of Reunion Island. *American Journal of Human Genetics* 34:937–947.

Gerdes L U, Gerdes C, Hansen P S, Klausen I C, Færgeman O, and Dyerberg J (1996) The apolipoprotein E polymorphism in Greenland Inuit in its global perspective. *Human Genetics* 98:546–550.

Gilger J W (2000) Contributions and promise of human behavioral genetics. *Human Biology* 72:229–255.

Gimbutas M (1973) The beginning of the Bronze Age in Europe and the Indo-Europeans 3500–2500 BC. *Journal of Indo-European Studies* 1:163–214.

Gimbutas, M (1977) First wave of European steppe pastoralists into Copper Age Europe. *Journal of Indo-European Studies* 5(4): 277–338.

Gimbutas M (1980) The Kurgan wave #2 (c. 3400–3200 BC) into Europe and following transformation of culture. *Journal of Indo-European Studies* 8:273–315.

Gimbutas M (1985) Primary and secondary homeland of the Indo-Europeans: Comments on Gamkrelidze-Ivanov articles. *Journal of Indo-European Studies* 13:185–202.

Gimbutas M (1997) *The Kurgan Culture and the Indo-Europeanization of Europe*. Edited by Miriam Robbins Dexter and Karlene Jones-Bley, Monograph 18, Washington, DC, Institute for the Study of Man.

Gimbutas M (1999) *The Living Goddesss*. Edited and supplemented by Miriam Robbins Dexter, University of California Press, Berkeley.

Glass R I, Holmgren J, Haley C E, Khan M R, Svennerholm A-M, Stoll B J, Belayet K M, Hossain K, Black R E, Yunus M, and Baru D (1985) Predisposition for cholera of individuals with O blood group: Possible evolutionary significance. *American Journal of Epidemiology* 121:791–796.

Glendinning J I (1994) Is the bitter rejection response always adaptive? *Physiology and Behavior* 56:1217–1227.

Goebel T (1999) Pleistocene human colonization of Siberia and peopling of the Americas: An ecological approach. *Evolutionary Anthropology* 8:208–227.

Goldstein D B and Chikhi L (2002) Human migrations and population structure: What we know and why it matters. *Annual Review of Genomics and Human Genetics* 3:129–152.

Goldstein D B, Linares A R, Cavalli-Sforza L L, and Feldman M W (1995) An evaluation of genetic distances for use with microsatellite loci. *Genetics* 139:463–471.

Gómez-Casado E, Martínez-Laso J, Moscoso J, Zamora J, Martin-Villa M, Perez-Blas M, Lopez-Santalla M, Lucas Gramajo P, Silvera C, Lowry E, and Arnaiz-Villena A (2003) Origins of

Mayans according to HLA genes and the uniqueness of Amerindians. *Tissue Antigens* 61:425–436.

Goodman A H and Armelagos G J (1996) The resurrection of race: The concept of race in physical anthropology in the 1990s. In: *Race and Other Misadventures: Essays in Honor of Ashley Montagu in His Ninetieth Year*, edited by Lieberman L and Reynolds L T. Dix Hills, NY: General Hall.

Gould S J (1981) *The Mismeasure of Man*. New York: W W Norton.

Gould S J (1994) Special issue: The science of race. Race: What is it good for? Science looks at flesh and bones, genes and behavior. *Discover*.

Gould S J (1996) *The Mismeasure of Man*, second edition. New York: W W Norton.

Gower J C (1966) Some distance properties of latent root and vector methods used in multivariate analysis. *Biometrika* 53:325–338.

Gravlee C C, Bernard H R, and Leonard W R (2003a) Boas's *Changes in Bodily Form*: The immigrant study, cranial plasticity, and Boas's physical anthropology. *American Anthropologist* 105:326–332.

Gravlee C C, Bernard H R, and Leonard W R (2003b) Heredity, environment, and cranial form: A reanalysis of Boas's immigrant data. *American Anthropologist* 105:125–138.

Greene J C (1959) *The Death of Adam: Evolution and Its Impact on Western Thought*. Ames Iowa State University Press.

Greene J C (1981) *Science, Ideology, and World View: Essays in the History of Evolutionary Ideas*. Berkeley: University of California Press.

Greene L S (1993) G6PD deficiency as protection against *falciparum* malaria: An epidemiologic critique of population and experimental studies. *Yearbook of Physical Anthropology* 36:153–178.

Greene L S, McMahon L, and diIorio J (1993) Co-evolution of glucose-6-phosphate dehydrogenase deficiency and quinine taste sensitivity. *Annals of Human Biology* 20:497–500.

Greksa L P (1988) Effect of altitude on the stature, chest depth, and forced vital capacity of low-to-high altitude migrant children of European ancestry. *Human Biology* 60:23–32.

Greksa L P (1996) Evidence for a genetic basis to the enhanced total lung capacity of Andean highlanders. *Human Biology* 68:119–129.

Guggenheim J A, Kirov G, and Hodson S A (2000) The heritability of high myopia: A reanalysis of Goldschmidt's data. *Journal of Medical Genetics* 37:227–231.

Gunn R B, Gargus J J, and Fröhlich O (1992) The Kidd antigens and urea transport. In: *Protein Blood Group Antigens of the Human Red Cell*, edited by Agre PC and Cartron J-P. Baltimore, MD: Johns Hopkins University Press, pp 88–100.

Guo S-W and Reed D R (2001) The genetics of phenylthiocarbamide perception. *Annals of Human Biology* 28:111–142.

Gurdon J B, Laskey R A, and Reeves O R (1975) The developmental capacity of nuclei transplanted from keratinised skin cells of adult frogs. *Journal of Embryology and Experimental Morphology* 34:93–112.

Gutierrez G, Sánchez D, and Marin A (2002) A reanalysis of the ancient mitochondrial DNA sequences recovered from Neandertal bones. *Molecular Biology and Evolution* 19:1359–1366.

Hackett W E R, Dawson G W P, and Dawson C J (1956) The pattern of the ABO blood group frequencies in Ireland. *Heredity* 10:69–84.

Hadley T J and Miller L H (1992) Red cell antigens as receptors for malaria parasites. In: *Protein Blood Group Antigens of the Human Red Cell*, edited by Agre PC and Cartron J-P. Baltimore, MD: Johns Hopkins University Press, pp 228–245.

Hadley T J, Klotz F W, and Miller L H (1986) Invasion of erythrocytes by malaria parasites: A cellular and molecular overview. *Annual Review of Microbiology* 40:451–477.

Haldane J B S (1949) The rate of mutation of human genes. *Hereditas* 8(Suppl):267–273.

Hall F C and Bowness P (1996) HLA and disease: From molecular function to disease association. In: *HLA and MHC: Genes, Molecules and Function*, edited by Browning M and McMichael A. Oxford: BIOS Scientific Publishers, pp 353–381.

Hamblin M T and Di Rienzo A (2000) Detection of the signature of natural selection in humans: Evidence from the Duffy blood group locus. *American Journal of Human Genetics* 66:1669–1679.

Hamer D H, Hu S, Magnuson V L, Hu N, and Pattatucci A M L (1993) A linkage between DNA markers on the X chromosome and male sexual orientation. *Science* 261:321–327.

Hammer M F, Redd A J, Wood E T, Bonner M R, Jarjanazi H, Karafet T, Santachiara-Benerecetti S, Oppenheim A, Jobling M A, Jenkins T, Sotrer H, and Bonné-Tamir B (2000) Jewish and Middle Eastern non-Jewish populations share a common pool of Y-chromosome biallellic haplotypes. *Proceedings of the National Academy of Sciences USA* 97:6769–6774.

Handoko H Y, Lum J K, Gustiana, Rismalia, Kartapradja H, Sofro A S M, and Marzuki S (2001) Length variations in the COII-tRNA^{Lys} intergenic region of mitochondrial DNA in Indonesian populations. *Human Biology* 73:205–223.

Hanna B (2001) Jesse James dug up again: Second disinterment of James alias debated. *Fort-Worth Star Telegram*, 21 January 2001, p 7E.

Harding R M, Healy E, Ray A J, Ellis N S, Flanagan N, Todd C, Dixon C, Sajantila A, Jackson I J, Birch-Machin M A, and Rees J L (2000) Evidence for variable selective pressures at MC1R. *American Journal of Human Genetics* 66:1351–1361.

Harpending H and Jenkins T (1973) Genetic distance among Southern African populations. In: *Methods and Theories of Anthropological Genetics*, edited by Crawford M H and Workman P L. Albuquerque: University of New Mexico Press, pp 177–199.

Harpending H C, Batzer M A, Gurven M, Jorde L B, Rogers A R, and Sherry S T (1998) Genetic traces of ancient demography. *Proceedings of the National Academy of Sciences USA* 95:1961–1967.

Harpending H C, Relethford J H, and Sherry S T (1996) Methods and models for understanding human diversity. In: *Molecular Biology and Human Diversity*, edited by Boyce A J and Mascie-Taylor C G N. Cambridge: Cambridge University Press, pp 283–299.

Harris H (1966) Enzyme polymorphisms in man. *Proceedings of the Royal Society*, Ser. B 164: 298–310.

Harris H (1971) *The Principles of Human Biochemical Genetics*. Amsterdam: North-Holland.

Harris H and Hopkinson D A (1972) Average heterozygosity in man. *Annals of Human Genetics* 36:9–20.

Harris H, Hopkinson D A, and Edwards Y H (1977) Polymorphism and the subunit structure of enyzymes: A contribution to the neutralist-selectionist controversy. *Proceedings of the National Academy of Sciences, USA* 74:698–701.

Harris H and Kalmus H (1949) The measurement of taste sensitivity to phenylthiourea (PTC). *Annals of Eugenics (London)* 15:24–31.

Harris M (1968) *The Rise of Anthropological Theory*. New York: Harper & Row.

Harris R, Harrison G A, and Rondle C J M (1963) Vaccina virus and human blood-group-A substance. *Acta Genetica (Basel)* 13:44–57.

Harrison G A and Owen J J T (1964) Studies on the inheritance of human skin colour. *Annals of Human Genetics* 28:27–37.

Harrison G A, Owen J J T, Da Rocha F J, and Salzano F M (1967) Skin colour in southern Brazilian populations. *Human Biology* 39:21–31.

Harrison G A, Tanner J M, Pilbeam D R, and Baker P T (1988) *Human Biology: An Introduction to Human Evolution, Variation, Growth, and Adaptability*, third edition. Oxford: Oxford University Press.

Hartl D L (2000) *A Primer of Population Genetics*, third edition. Sunderland, MA: Sinauer Associates.

Harvey C B, Hollox E J, Poulter M, Wang Y, Rossi M, Auricchio S, Iqbal T H, Cooper B T, Barton R, Sarner M, Korpela R, and Swallow D M (1993) Regional localization of the lactase-phlorizin hydrolase gene, *LCT*, to chromosome 2q21. *Annals of Human Genetics* 57:179–185.

Hassan F A (1981) *Demographic Archaeology*. New York: Academic Press.

Hawks J, Hunley K, Lee S-H, and Wolpoff M (2000) Population bottlenecks and Pleistocene human evolution. *Molecular Biology and Evolution* 17:2–22.

Heard E, Clerc P, and Avner P (1997) X-chromosome inactivation in mammals. *Annual Review of Genetics* 31:571–610.

Heard R (1994) HLA and autoimmune disease. In: *HLA and Disease*, edited by Lechler R. San Diego: Academic Press, pp 123–151.

Hedrick P W and Thomson G (1983) Evidence for balancing selection at HLA. *Genetics* 104:449–456.

Hedrick P W, Whittam T S, and Parham P (1991) Heterozygosity at individual amino acid sites: Extremely high levels for *HLA-A* and *–B* genes. *Proceedings of the National Academy of Sciences USA* 88:5897–5901.

Helgason A, Hickey E, Goodacre S, Bosnes V, Stefánsson K, Ward R, and Sykes B (2001) mtDNA and the islands of the North Atlantic: Estimating the proportions of Norse and Gaelic ancestry. *American Journal of Human Genetics* 68:723–737.

Herrick J B (1910) Peculiar elongated and sickle-shaped red blood corpuscles in a case of severe anemia. *Archives of Internal Medicine* 6:517–521.

Herrnstein R J and Murray C (1994) *The Bell Curve: Intelligence and Class Structure in American Life.* New York: Free Press.

Heyerdahl T (1950) *Kontiki: Across the Pacific by Raft.* Chicago: Rand McNally.

Hill A V S, Allsopp C E M, Kwiatkowski D K, Anstey N M, Twumasi P, Rowe P A, Bennett S, Brewster D, McMichael A J, and Greenwood B M (1991) Common West African HLA antigens are associated with protection from severe malaria. *Nature* 352:595–600.

Hill A V S, Bowen D K, Trent R J, Higgs D R, Oppenheimer S J, Thein S L, Mickleson K N P, Weatherall D J, and Clegg J B (1985) Melanesians and Polynesians share a unique α-thalassemia mutation. *American Journal of Human Genetics* 37:571–580.

Himes J H (1980) Subcutaneous fat thickness as an indicator of nutritional stress. In: *Social and Biological Predictors of Nutritional Status, Physical Growth, and Neurological Development*, edited by Greene L S and Johnston F E. New York: Academic Press, pp 9–32.

Hirschfeld J (1959) Immune-electrophoretic demonstration of qualitative differences in human sera and their relation to the haptoglobins. *Acta Pathologica et Microbiologica Scandinavica* 47:160–168.

Hirschfeld L and Hirschfeld H (1919) Serological differences between the blood of different races: The result of researches on the Macedonian front. *Lancet* 197:675–679.

Hladik C-M and Simmen B (1996) Taste perception and feeding behavior in nonhuman primates and human populations. *Evolutionary Anthropology* 5:58–71.

Hladik C-M, Pasquet P, and Simmen B (2002) New perspectives on taste and primate evolution: The dichotomy in gustatory coding for perception of beneficent versus noxious substances as supported by correlations among human thresholds. *American Journal of Physical Anthropology* 117:342–348.

Hoff C, Thorneycroft I, Wilson F, and Williams-Murphy M (2001) Protection afforded by sickle-cell trait (Hb AS): What happens when malarial selection pressures are alleviated? *Human Biology* 73:583–586.

Hoffmann A A (2000) Laboratory and field heritabilities: Some lessons from *Drosophila*. In: *Adaptive Genetic Variation in the Wild*, edited by Mousseau T A, Sinervo B, and Endler J A. New York: Oxford University Press, pp 200–218.

Holden C and Mace R (1997) Phylogenetic analysis of the evolution of lactase digestion in adults. *Human Biology* 69:605–628.

Holick M F, MacLaughlin J A, and Doppelt S H (1981) Regulation of cutaneous previtamin D_3 photosynthesis in man: Skin pigment is not an essential regulator. *Science* 211:590–593.

Honig G R and Adams J G III (1986) *Human Hemoglobin Genetics.* Vienna: Springer-Verlag.

Hooton E A (1926) Methods of racial analysis [From his AAAS vice presidential address to section H in December 1925]. *Science* 63:75–81.

Hooton E A (1931) *Up from the Ape.* New York: Macmillan.

Hooton E A (1936) Plain statements about race. *Science* 83:511–513.

Hooton E A, Dupertuis C W, and Dawson H (1955) *The Physical Anthropology of Ireland. Papers of the Peabody Museum*, volume 30, Numbers 1–2. Cambridge: Peabody Museum.

Houghton P (1996) *People of the Great Ocean: Aspects of Human Biology of the Early Pacific.* New York: Cambridge University Press.

Houle D (1992) Comparing evolvability and variability of quantitative traits. *Genetics* 130:195–204.

Howells W W (1996) Howells' craniometric data on the internet. *American Journal of Physical Anthropology* 101:441–442.

Hu S, Pattatucci A M L, Patterson C, Li L, Fulker D W, Cherny S S, Kruglyak L, and Hamer D H (1995) Linkage between sexual orientation and chromosome Xq28 in males but not in females. *Nature Genetics* 11:248–256.

Huang C-H and Blumenfeld O O (1995) MNSs blood groups and major glycophorins: Molecular basis for allelic variation. In: *Blood Cell Biochemistry, Molecular Basis of Human Blood Group Antigens*, volume 6, edited by Cartron J-P and Rouger P. New York: Plenum Press, pp 153–188.

Hubert W, Matsumoto H, and de Stefano G F (1991) Gm and Km allotypes in four Sardinian population samples. *American Journal of Physical Anthropology* 86:45–50.

Hudson R R (1990) Gene genealogies and the coalescent process. *Oxford Surveys in Evolutionary Biology* 7:1–42.

Hughes A L and Nei M (1989) Nucleotide substitution at major histocompatibility complex class II loci: Evidence for overdominant selection. *Proceedings of the National Academy of Sciences USA* 86:958–962.

Hughes-Jones N C (1992) Hemolytic disease of the newborn. In: *Protein Blood Group Antigens of the Red Cell: Structure, Function, and Clinical Significance*, edited by Agre P C and Cartron J-P. Baltimore, MD: Johns Hopkins University Press, pp 185–191.

Hulse F (1962) Race as an evolutionary episode. *American Anthropologist* 64: 929–945.

Hunt E (1995) The role of intelligence in modern society. *American Scientist* 83:356–368.

Hurles M E, Maund E, Nicholson J, Bosch E, Renfrew C, Sykes B C, and Jobling M A (2003) Native American Y chromosome in Polynesia: The genetic impact of the Polynesian slave trade. *American Journal of Human Genetics* 72:1282–1287.

Hurles M E, Nicholson J, Bosch E, Renfrew C, Sykes B C, and Jobling M A (2002) Y chromosomal evidence for the origins of Oceanic-speaking peoples. *Genetics* 160:289–303.

Hutagalung R, Wilairatana P, Looareesuwan S, Brittenham G M, Aikawa M, and Gordeuk V R (1999) Influence of hemoglobin E trait on the severity of *falciparum* malaria. *Journal of Infectious Diseases* 179:283–286.

Huxley J S and Haddon A C (1936) *We Europeans*. New York: Harper.

Huxley T H (1865) On the methods and results of ethnology. In: *Man's Place in Nature and Other Anthropological Essays*. London: Macmillan.

Hyslop N E (1971) Ear wax and host defense [editorial]. *New England Journal of Medicine* 284:1099–1100.

Ibraimov A I (1991) Cerumen phenotypes in certain populations of Eurasia and Africa. *American Journal of Physical Anthropology* 84:209–211.

Ibraimov A I and Mirrakhimov M M (1979) PTC-tasting ability in populations living in Kirghizia with special reference to hypersensitivity: Its relation to age and sex. *Human Genetics* 46:97–105.

Ing R, Petrakis N L, and Ho H C (1973) Evidence against association between wet cerumen and breast cancer. *Lancet* I:41.

Ingram V M (1956) A specific chemical difference between the globins of normal human and sickle cell anaemia haemoglobin. *Nature* 178:792–794.

International Human Genome Sequencing Consortium (2001) Initial sequencing and analysis of the human genome. *Nature* 409:860–921.

Jablonski N G and Chaplin G (2000) The evolution of human skin coloration. *Journal of Human Evolution* 39:57–106.

Jablonski N G and Chaplin G (2002) Skin deep. *Scientific American* 287:74–81.

Jackson F L C (1993) The influence of dietary cyanogenic glycosides from cassava on human metabolic biology and microevolution. In: *Tropical Forests, People and Foods*, edited by Hladik C M. Paris: UNESCO and Parthenon Publishing Group, pp 321–338.

Jantz R L (1995) Franz Boas and Native American biological variability. *Human Biology* 67:345–353.

Jensen A R (1969) How much can we boost IQ and scholastic achievement? *Harvard Educational Review* 39:1–123.

Jensen A R (1998) *The g factor: The Science of Mental Ability.* Wesport, CT: Praeger.

Jobling M A (2001) In the name of the father: Surnames and genetics. *Trends in Genetics* 17:353–357.

Jobling M A and Tyler-Smith C (1995) Fathers and sons: The Y chromosome and human evolution. *Trends in Genetics* 11:449–456.

Jobling M A and Tyler-Smith C (2003) The human Y chromosome: An evolutionary marker comes of age. *Nature Reviews Genetics* 4:598–612.

Jobling M A, Bouzekri N, and Taylor P G (1998) Hypervariable digital DNA codes for human paternal lineages: MVR-PCR at the Y-specific minisatellite MSY1 (DYF155S1). *Human Molecular Genetics* 7:643–653.

Jobling M A, Hurles M E, and Tyler-Smith C (2004) *Human Evolutionary Genetics: Origins, Peoples and Disease.* New York: Garland.

Jokenin M, Andersson L C, and Gahnberg C G (1985) Identification of the major human sialoglycoprotein from red cell, glycophorin A^M, as the receptor for *Escherichia coli* IH 11165 and characterization of the receptor site. *European Journal of Biochemistry* 147:47–52.

Jones D (1996) An evolutionary perspective on physical attractiveness. *Evolutionary Anthropology* 5:97–109.

Jones E A, Goodfellow P N, Bodmer J G, and Bodmer W F (1975) Serological identification of *HL-A* linked human "Ia-type" antigens. *Nature* 256:650–652.

Jordan D P (1999) "To follow truth wherever it may lead": Dealing with the DNA controversy at Monticello. *Cultural Resource Management* 22:13–15.

Jorde L B (1980) The genetic structure of subdivided human populations: A review. In: *Current Developments in Anthropological Genetics. Theory and Methods,* volume 1, edited by Mielke J H and Crawford M H. New York: Plenum Press, pp 135–208.

Jorde L B (1985) Human genetic distance studies: Present status and future prospects. *Annual Review of Anthropology* 14:343–373.

Jorde L B (1995) Linkage disequilibrium as a gene mapping tool [editorial]. *American Journal of Human Genetics* 56:11–14.

Jorde L B, Bamshad M, and Rogers A R (1998) Using mitochondrial and nuclear DNA markers to reconstruct human evolution. *BioEssays* 20:126–136.

Jorde L B, Watkins W S, Carlson M, Groden J, Albertson H, Thliveris A, and Leppert M (1994) Linkage disequilibrium predicts physical distance in the adenomatous polyposis coli region. *American Journal of Human Genetics* 54:884–898.

Jorde L B, Workman P L, and Eriksson A W (1982) Genetic microevolution in the Åland Islands, Finland. In: *Current Developments in Anthropological Genetics. Ecology and Population Structure,* volume 2, edited by Crawford M H and Mielke J H. New York: Plenum Press, pp 333–365.

Kalla AK (1974) Human skin pigmentation: Its genetics and variation. *Humangenetik* 21:289–300.

Kalmus H (1957) Defective colour vision, P.T.C. tasting and drepanocytosis in samples from fifteen Brazilian populations. *Annals of Human Genetics* 21:313–317.

Kalmus H, Garay A L, Rodarte U, and Lourdes C (1964) The frequency of PTC tasting, hard ear wax, colour blindness and other genetical characteristics in urban and rural Mexican populations. *Human Biology* 36:134–145.

Kamboh M I, Crawford M H, Aston C E, and Leonard W R (1996) Population distributions of *APOE, APOH,* and *APOA4* polymorphisms and their relationship with quantitative plasma lipid levels among the Evenki herders of Siberia. *Human Biology* 68:231–243.

Kamel K and Awny A Y (1965) Origin of the sickling gene. *Nature* 205:919.

Kan Y W and Dozy A M (1978) Polymorphism of DNA sequence adjacent to human β-globin structural gene: Relationship to sickle mutation. *Proceedings of the National Academy of Sciences USA* 75:5631–5635.

Karafet T M, Zegura S L, Posukh O, Osipova L, Bergen A, Long J, Goldman D, Klitz W, Harihara S, de Knijff P, Wiebe V, Griffiths R C, Templeton A R, and Hammer M F (1999) Ancestral Asian source(s) of New World Y-chromosome founder haplotypes. *American Journal of Human Genetics* 64:817–831.

Kataura A and Kataura K (1967a) The comparison of free and bound amino acids between dry and wet types of cerumen. *Tohoku Journal of Experimental Medicine* 91:215–225.

Kataura A and Kataura K (1967b) The comparison of lipids between dry and wet types of cerumen. *Tohoku Journal of Experimental Medicine* 91:227–237.

Katzmarzyk P T and Leonard W R (1998) Climatic influences on human body size and proportions: Ecological adaptations and secular trends. *American Journal of Physical Anthropology* 106:483–503.

Kayser M, Brauer S, Weiss G, Underhill P A, Roewer L, Schiefenhövel W, and Stoneking M (2000) Melanesian origin of Polynesian Y chromosomes. *Current Biology* 10:1237–1246.

Kayser M, Brauer S, Weiss G, Underhill P A, Roewer L, Schiefenhövel W, and Stoneking M (2001) Independent histories of human Y chromosomes from Melanesia and Australia. *American Journal of Human Genetics* 68:173–190.

Kearsey M J and Pooni H S (1996) *The Genetical Analysis of Quantitative Traits*. New York: Chapman & Hall.

Kelso A J (1962) Dietary differences: A possible selective mechanism in ABO blood group frequencies. *Southwestern Lore* 28:48–56.

Kelso A J and Armelagos G J (1963) Nutritional factors as selective agencies in the determination of ABO blood groups. *Southwestern Lore* 29:44–48.

Kennedy K A R (1976) *Human Variation in Time and Space*. Dubuque, IA: Wm C Brown Company.

Kennedy R E Jr (1973) *The Irish: Emigration, Marriage, and Fertility*. Berkeley: University of California Press.

Kidd, K (1999) Appendix B, Opinions of Scientists Consulted, pp. 15–16, In: *Report of the Research Committee on Thomas Jefferson and Sally Hemings* (2000) Thomas Jefferson Foundation, .PDF® Version of report at: http://www.monticello.org/plantation/hemingscontro/hemings_report.html

Kihm A J, Kong Y, Hong W, Russell J E, Rouda S, Adachi K, Simon M C, Blobel G A, and Weiss M J (2002) An abundant erythroid protein that stabilizes free α-haemoglobin. *Nature* 417:758–763.

Kim U, Jorgenson E, Coon H, Leppert M, Risch N, and Drayna D (2003) Positional cloning of the human quantitative locus underlying taste sensitivity of phenylthiocarbamide. *Science* 299:1221–1225.

King J P, Kimmel M, and Chakraborty R (2000) A power analysis of microsatellite-based statistics for inferring past population growth. *Molecular Biology and Evolution* 17:1859–1868.

Kirk R L (1968) *The Haptoglobin Groups in Man. Monographs in Human Genetics*, volume 4, edited by Beckman L and Hauge M. Basel: S Karger.

Klass M and Hellman H (1971) *The Kinds of Mankind: An Introduction to Race and Racism*. Philadelphia: JB Lippincott.

Klitz W, Maiers M, Spellman S, Baxter-Lowe L A, Schmeckpeper B, Williams T M, and Fernandez-Viña (2003) New HLA haplotype frequency reference standards: High-resolution and large sample typing of *HLA DR-DQ* haplotypes in a sample of European Americans. *Tissue Antigens* 62:296–307.

Koertvelyessy T (2000) Phenylthiocarbamide tasting ability, aging, and thyroid function. In: *Different Seasons: Biological Aging Among the Mennonites of the Midwestern United States. Publications in Anthropology 21*, edited by Crawford MH. Lawrence: University of Kansas, pp 41–54.

Kofler A, Braun A, Jenkins T, Sergeanston S W, and Cleve H (1995) Characterization of mutants of the vitamin-D-binding protein/group specific component: GC Aborigine (1A1) from Australian Aborigines and South African blacks, and 2A9 from South Germany. *Vox Sanguinis*. 68:50–54.

Konigsberg L W (2000) Quantitative variation and genetics. In: *Human Biology: An Evolutionary and Biocultural Perspective*, edited by Stinson S, Bogin B, Huss-Ashmore R, and O'Rourke D. New York: John Wiley & Sons, pp 135–162.

Konigsberg L W and Ousley S D (1995) Multivariate quantitative genetics of anthropometric traits from the Boas data. *Human Biology* 67:481–498.

Korey K A (1980) Skin colorimetry and admixture measurements: Some further considerations. *American Journal of Physical Anthropology* 53:123–128.

Kraytsberg Y, Schwartz M, Brown T A, Ebralidse K, Kunz W S, Clayton D A, Vissing J, and Khrapko K (2004) Recombination of human mitochondrial DNA. *Science* 304:981.

Krieger H J and Vicente A T (1969) Smallpox and the ABO system in southern Brazil. *Human Heredity* 19:654–657.

Krimsky S (2002) For the record. *Nature Genetics* 30:139.

Krings M, Capelli C, Tschentscher F, Geisert H, Meyer S, von Haeseler A, Grossschmidt K, Possnert G, Paunovic M, and Pääbo S (2000) A view of Neandertal genetic diversity. *Nature Genetics* 26:144–146.

Krings M, Geisert H, Schmitz R W, Krainitzki H, and Pääbo S (1999) DNA sequence of the mitochondrial hypervariable region II from the Neandertal type specimen. *Proceedings of the National Academy of Sciences USA* 96:5581–5585.

Krings M, Stone A, Schmitz R W, Krainitzki H, Stoneking M, and Pääbo S (1997) Neandertal DNA sequences and the origin of modern humans. *Cell* 90:19–30.

Kudo S, Onda M, and Fukuda M (1994) Characterization of glycophorin A transcripts: Control by the common erythroid-specific promoter and alternative usage of different polyadenylation signals. *Journal of Biochemistry* 116:183–192.

Kuby J (1992) *Immunology*. New York: W.H. Freeman and Company.

Kurnit D M (1979) Evolution of sickle variant gene. [letter] *Lancet* 1:104.

Kwok C, Tyler-Smith C, Mendonca B B, Berkovitz G D, Goodfellow P N, and Hawkins J R (1996) Mutation analysis of the 2kb 5′ to SRY in XY females and XY intersex subjects. *Journal of Medical Genetics* 33:465–468.

Labie D, Pagnier J, Wajcman H, Fabry M E, and Nagel R L (1986) The genetic origin of the variability of the phenotypic expression of the *Hb S* gene. In: *Genetic Variation and Its Maintenance*, edited by Roberts D F and de Stefano G F. Cambridge: Cambridge University Press, pp 149–155.

Labuda D, Ziekiewicz E, and Yotova V (2000) Archaic lineages in the history of modern humans. *Genetics* 156:799–808.

Lander E S and Ellis J J (1998) Founding father. *Nature* 396:13–14.

Lander E S and Green P (1987) Construction of multilocus genetic maps in humans. *Proceedings of the National Academy of Sciences USA* 84:2363–2367.

Landsteiner K (1900) Zur Kenntnis der antifermentativen, lytischen und agglutinierenden Wirkungen des Blutserums und der Lymphe. *Zentralblatt Für Bakteriologie* 27:357–362.

Landsteiner K (1901) Uber agglutinationsercheinungen normalen menschlichen. *Blut Wiener Klinische Wochenschrift* 14:1132–1134.

Landsteiner K and Levine P A (1927a) A new agglutinable factor differentiating individual human bloods. *Proceedings of the Society for Experimental Biology and Medicine* 24:600–602.

Landsteiner K and Levine P A (1927b) Further observations on individual differences of human blood. *Proceedings of the Society for Experimental Biology and Medicine* 24:941–942.

Landsteiner K and Wiener A S (1940) An agglutinable factor in human blood recognized by immune sera for rhesus blood. *Proceedings of the Society for Experimental Biology and Medicine* 43:223.

Lanier S and Feldman J (2000) *Jefferson's Children: The Story of One American Family*. New York: Random House.

Lapouméroulie C, Dunda O, Ducrocq R, Trabuchet G, Mony-Lobé M, Bodo J M, Carnevale P, Labie D, Elion J, and Krishnamoorthy R (1992) A novel sickle cell mutation of yet another origin in Africa: The Cameroon type. *Human Genetics* 89:333–337.

Lasker G W (1954) Photoelectric measurement of skin color in a Mexican Mestizo population. *American Journal of Physical Anthropology* 12:115–121.

Lavinha J, Gonçalves J, Faustino P, Romão L, Osório-Almeida L, João Peres M, Picanço I, Carmo Martins M, Ducrocq R, Labie D, and Krishnamoorthy R (1992) Importation route of the sickle cell trait into Portugal: Contribution of molecular epidemiology. *Human Biology* 64:891–901.

Lechler R, ed. (1994) *HLA and Disease*. San Diego: Academic Press.

Lees F C and Byard P J (1978) Skin colorimetry in Belize. I. Conversion formulae. *American Journal of Physical Anthropology* 48:515–521.

Lees F C and Crawford M H (1976) Anthropometric variation in Tlaxcaltecan populations. In: *The Tlaxcaltecans: Prehistory, Demography, Morphology, and Genetics, University of Kansas Publications in Anthropology 7*, edited by Crawford M H. Lawrence: University of Kansas, pp 61–80.

Lees F C, Byard P J, and Relethford J H (1978) Interobserver error in human skin colorimetry. *American Journal of Physical Anthropology* 49:35–38.

Lees F C, Byard P J, and Relethford J H (1979) New conversion formulae for light-skinned populations using Photovolt and E.E.L. reflectometers. *American Journal of Physical Anthropology* 51:403–408.

Leffell M S, Donnenberg A D, and Rose N R, eds (1997) *Handbook of Human Immunology*. Boca Raton: CRC Press.

Lehmann H (1954) Origin of the sickle cell. *South African Journal of Science* 50:140–141.

Lehmann H and Huntsman R G (1974) *Man's Haemoglobins*, second edition. Amsterdam: North-Holland.

Leigh S R (1996) Evolution of human growth spurts. *American Journal of Physical Anthropology* 101:455–474.

Lell J T and Wallace D C (2000) The peopling of Europe from the maternal and paternal perspectives. *American Journal of Human Genetics* 67:1376–1381.

Leonard W R, Leatherman T L, Carey J W, and Thomas R B (1990) Contributions of nutrition versus hypoxia to growth in rural Andean populations. *American Journal of Human Biology* 2:613–626.

Le Pennec P-Y and Rouger P (1995) Nomenclature of blood group antigens. In: *Blood Cell Biochemistry. Molecular Basis of Human Blood Group Antigens*, volume 6, edited by Cartron J-P and Rouger P. New York: Plenum Press, pp 477–486.

Leutenegger W (1982) Sexual dimorphism in nonhuman primates. In: *Sexual Dimorphism in* Homo sapiens: *A Question of Size*, edited by Hall R L. New York: Praeger, pp 11–36.

LeVay S (1991) A difference in hypothalamic structure between heterosexual and homosexual men. *Science* 253:1034–1037.

Levine P (1943) Serological factors as possible causes in spontaneous abortions. *Journal of Heredity* 34:71–80.

Levine P and Stetson R E (1939) An unusual case of intragroup agglutination. *Journal of the American Medical Association* 113:126–127.

Levitan M and Montagu A (1971) *Textbook of Human Genetics*. New York: Oxford University Press.

Li C C (1976) *First Course in Population Genetics*. Pacific Grove, CA: Boxwood Press.

Li H, Zhao X, Qin F, Li H, Li L, He X, Chang X, Li Z, Liang K, Xing F, Chang W, Wong R, Yang I, Li F, Zhang T, Tian R, Webber B B, Wilson J B, and Huisman T H J (1990) Abnormal hemoglobins in the Silk Road region of China. *Human Genetics* 86:231–235.

Lieberman L (2001) How "Caucasoids" got such big crania and why they shrank. *Current Anthropology* 42:69–95.

Lieberman L and Jackson F L C (1995) Race and three models of human diversity. *American Anthropologist* 97:231–242.

Lieberman L and Kirk R C (2000) "Race" in anthropology in the 20th century: The decline and fall of a core concept, www.chsbs.cmich.edu/rod_kirk/norace/tables.htm.

Linné C von (1806) *A General System of Nature*, translated by Turton W. London: Lackington, Allen, and Co.

Linstedt R, Larson G, Falk P, Jodal U, Leffler H, and Svanborg C (1991) The receptor repertoire defines the host range for attaching *Escherichia coli* strains that recognize globo-A. *Infection and Immunity* 59:1086–1092.

Little C C and Tyzzer E E (1916) Further experimental studies on the inheritance of susceptibility to a transplantable tumor, carcinoma (JWA) of the Japanese waltzing mouse. *Journal of Medical Research* 33:393–453.

Little M A and Wolff M E (1981) Skin and hair reflectances in women with red hair. *Annals of Human Biology* 8:231–241.

Livingstone F B (1957) Sickling and malaria. *British Medical Journal* 1:762–763.

Livingstone F B (1958) Anthropological implications of sickle cell gene distribution in West Africa. *American Anthropologist* 60:533–562.

Livingstone F B (1962) On the non-existence of human races. *Current Anthropology* 3:279–281.

Livingstone F B (1967) *Abnormal Hemoglobins in Human Populations*. Chicago: Aldine.

Livingstone F B (1984) The Duffy blood groups, vivax malaria, and malaria selection in human populations: A review. *Human Biology* 56:413–425.

Livingstone F B (1989) Simulation of the diffusion of the β-globin variants in the Old World. *Human Biology* 61: 297–309.

Loehlin J C, Lindzey G, and Spuhler J N (1975) *Race Differences in Intelligence*. San Francisco: W H Freeman.

Lomberg H, Cedergren B, Leffler H, Nilsson B, Carlström A-S, and Svanborg-Edén C (1986) Influence of blood group on the availability of receptors for attachment of uropathogenic *Escherichia coli*. *Infection and Immunity* 51:919–926.

Lonjou C, Collins A, and Morton N E (1999) Allelic association between marker loci. *Proceedings of the National Academy of Sciences USA* 96:1621–1626.

Loomis W F (1967) Skin-pigment regulation of vitamin-D biosynthesis in man. *Science* 157:501–506.

Lovejoy AO (1933) *The Great Chain of Being: A Study of the History of an Idea*. The William James lectures. Cambridge, MA: Harvard University Press.

Luzzatto L and Mehta A (2001) Glucose-6-phosphate dehydrogenase deficiency. In: *The Metabolic and Molecular Bases of Inherited Disease*, edited by Scriver C R, Baudet A L, Sly W S, and Valle D. New York: McGraw-Hill, pp 4517–4541.

Lynch M and Crease T J (1990) The analysis of population survey data on DNA sequence variation. *Molecular Biology and Evolution* 3:377–394.

Lynch M and Walsh B (1998) *Genetics and Analysis of Quantitative Traits*. Sunderland, MA: Sinauer Associates.

MacEachern S (2000) Genes, tribes, and African history. *Current Anthropology* 41:357–384.

MacKintosh N J (1998) *IQ and Human Intelligence*. Oxford: Oxford University Press.

Madrigal L (1989) Hemoglobin genotypes, fertility, and the malaria hypothesis. *Human Biology* 61:311–325.

Maeda N (1991) DNA polymorphisms in the controlling region of the human haptoglobin genes: A molecular explanation for the haptoglobin 2-1 modified phenotype. *American Journal of Human Genetics* 49:158–166.

Malina R M (1979) Secular changes in size and maturity: Causes and effects. *Monographs of the Society for Research in Child Development* 44:59–102.

Mange E J and Mange A P (1999) *Basic Human Genetics*, second edition. Sunderland, MA: Sinauer Associates.

Marks J (1995) *Human Biodiversity: Genes, Race, and History*. New York: Aldine de Gruyter.

Marsh S G E, Albert E D, Bodmer W F, Bontrop R E, Dupont B, Erlich H A, Geraghty D E, Hansen J A, Mach B, Mayr W R, Parham P, Petersdorf E W, Sasazuki T, Schreuder G M Th, Strominger J L, Svejgaard A, and Terasaki P I (2002) Nomenclature for factors of the HLA system. *Tissue Antigens* 60:407–464.

Marsh S G E, Parham P, and Barber L D (2000) *The HLA FactsBook*. San Diego: Academic Press.

Marsh W L and Redman C M (1990) The Kell blood group system: A review. *Transfusion* 30:158–167.

Martin K , Stevenson J C, Crawford M H, Everson P M, and Schanfield M S (1996) Immunoglobulin haplotype frequencies in anabaptist population samples: Kansas and Nebraska Mennonites and Indiana Amish. *Human Biology* 68:45–62.

Martin L M and Jackson J F (1969) Cerumen types in Choctaw Indians. *Science* 163:677–678.

Martorell R (1980) Interrelationships between diet, infectious disease, and nutritional status. In: *Social and Biological Predictors of Nutritional Status, Physical Growth, and Neurological Development*, edited by Greene L S and Johnston F E. New York: Academic Press, pp 81–106.

Matisoo-Smith E, Roberts R M, Irwin G J, Allen J S, Penny D, and Lambert D M (1998) Patterns of prehistoric mobility in Polynesia indicated by mtDNA from the Pacific rat. *Proceedings of the National Academy of Sciences USA* 95:15145–15150.

Matsunaga E (1962) The dimorphism in human normal cerumen. *Annals of Human Genetics* 25:273–286.

McCracken R D (1971) Lactase deficiency: An example of dietary evolution. *Current Anthropology* 12:479–500.

McCullough J M and Giles E (1970) Human cerumen types in Mexico and New Guinea: A humidity-related polymorphism in "Mongoloid" peoples. *Nature* 226:460–462.

McGue M (1997) The democracy of the genes. *Nature* 388:417–418.

McKenzie R L and Elwood J M (1990) Intensity of solar ultraviolet radiation and its implications for skin cancer. *New Zealand Medical Journal* 103:152–154.

Mears J G, Lachman H M, Cabannes R, Amegnizin K P, Labie D, and Nagel R L (1981) Sickle gene: Its origin and diffusion from West Africa. *Journal of Clinical Investigation* 68:606–610.

Melton T, Clifford S, Kayser M, Nasidze I, Batzer M, and Stoneking M (2001) Diversity and hetero-geneity in mitochondrial DNA of North American populations. *Journal of Forensic Sciences* 46:46–52.

Menozzi P, Piazza A, and Cavalli-Sforza L L (1978) Synthetic maps of human gene frequencies in Europe. *Science* 201:786–792.

Michael J S (1988) A new look at Morton's craniological research. *Current Anthropology* 29:349–354.

Mielke J H, Devor E J, Kramer P L, Workman P L, and Eriksson A W (1982) Historical population structure of the Åland Islands, Finland. In: *Current Developments in Anthropological Genetics. Ecology and Population Structure*, volume 2, edited by Crawford M H and Mielke J H. New York: Plenum Press, pp 255–332.

Mielke J H, Relethford J H, and Eriksson A W (1994) Temporal trends in migration in the Åland Islands: Effects of population size and geographic distance. *Human Biology* 66:399–410.

Mielke J H, Workman P L, Fellman J, and Eriksson A W (1976) Population structure of the Åland Islands, Finland. *Advances in Human Genetics* 6:241–321.

Miller L H (1994) Impact of malaria on genetic polymorphism and genetic disease in Africans and African Americans. *Proceedings of the National Academy of Sciences USA* 91:2415–2519.

Miller O J and Therman E (2001) *Human Chromosomes*, fourth edition. New York: Springer-Verlag.

Mitchell R J, Howlett S, White N G, Federle L, Papiha S S, Briceno I, McComb J, Schanfield M S, Tyler-Smith C, Osipova L, Livshits G, and Crawford M H (1999) Deletion polymorphism in the human *COL1A2* gene: Genetic evidence of a non-African population whose descendants spread to all continents. *Human Biology* 71:901–914.

Modiano D, Gaia L, Sirima B S, Simporé J, Verra F, Konaté A, Rastrelli E, Olivieri A, Calissano C, Paganotti G M, D'Urbano L, Sanou I, Sawadogo A, Modiano G, and Coluzzi M (2001) Haemoglobin C protects against clinical *Plasmodium falciparum* malaria. *Nature* 414:305–308.

Modiano G, Morpurgo G, Terrenato L, Novelletto T, Di Reinzo A, Colombo B, Purpura M, Jariani M, Santachiara-Benerecetti S, Brega A, Dixit D A, Shrestha S L, Lania A, Wanachiwanawin W, and Luzzatto L (1991) Protection against malaria morbidity: Near fixation of the α-thalassemia gene in a Nepalese population. *American Journal of Human Genetics* 48:390–397.

Mohr J (1951a) Estimation of linkage between the Lutheran and the Lewis blood groups. *Acta Pathologica et Microbiologica Scandinavica* 29:339–344.

Mohr J (1951b) Search for linkage between Lutheran blood group and other hereditary characters. *Acta Pathologica et Microbiologica Scandinavica* 28:207–210.

Montagu A (1942a) The genetical theory of race, and anthropological method. *American Anthropologist* 44:369–375.

Montagu A (1942b) *Man's Most Dangerous Myth: The Fallacy of Race.* New York: Columbia University Press.

Montagu A (1945) On the phrase "ethnic group" in anthropology. *Psychiatry* 8:27–33.

Montagu A (1950) *On Being Human.* New York: Henry Schuman.

Montagu A (1962) The concept of race. *American Anthropologist* 64:919–928.

Montagu A (1963) What is remarkable about varieties of man is likeness, not differences. *Current Anthropology* 4:361–364.

Morton N E (1982) *Outline of Genetic Epidemiology*. New York: S. Karger.

Morton S G (1849) Observations on the size of the brain in various races and families of man. *Proceedings of the Academy of Natural Sciences Philadelphia* 4:221–224.

Mourant A E, Kopec A C, and Domaniewska-Sobczak K (1976a) *The Distribution of the Human Blood Groups and Other Polymorphisms*. London: Oxford University Press.

Mourant A E, Kopec A C, and Domaniewska-Sobczak K (1978) *Blood Groups and Diseases: A Study of Associations of Diseases with Blood Groups and Other Polymorphisms*. London: Oxford University Press.

Mourant A E, Tills D, and Domaniewska-Sobczak K (1976b) Sunshine and the geographical distribution of the alleles of the Gc system of plasma proteins. *Human Genetics* 33:307–314.

Mukhopadhyay C K, Attieh Z K, and Fox P L (1998) Role of ceruloplasmin in cellular iron uptake. *Science* 279:714–717.

Murray-McIntosh R P, Scrimshaw B J, Hatfield P J, and Penny D (1998) Testing migration patterns and estimating founding population size in Polynesia by using human mtDNA sequences. *Proceedings of the National Academy of Sciences USA* 95:9047–9052.

Myers J S, Vincent B J, Udall H, Watkins W S, Morrish T A, Kilroy G E, Swergold G D, Henke J, Henke L, Moran J V, Jorde L B, and Batzer M A (2002) A comprehensive analysis of recently integrated human Ta L1 elements. *American Journal of Human Genetics* 71:312–326.

Myers R H, Schaefer E J, Wilson P W F, D'Agostino R, Ordovas J M, Espino A, Au R, White R F, Knoefel J E, Cobb J L, McNulty K A, Beiser A, and Wolf P A (1996) Apolipoprotein E epsilon-4 association with dementia in a population-based study: The Framingham Study. *Neurology* 46:673–677.

Nachman M W and Crowell S L (2000) Estimate of the mutation rate per nucleotide in humans. *Genetics* 156:297–304.

Nagel R L and Labie D (1985) The consequences and implications of the multicentric origin of the *Hb S* gene. In: *Experimental Approaches for the Study of Hemoglobin Switching*, edited by Stamatoyannopoulos G and Nienhuis A W. New York: Alan R. Liss, pp 93–103.

Naik S N, Ishwad C S, Nadkarni J S, Nadarni J J, and Advani S H (1979) Study of haptoglobin polymorphism and its significance in human leukemias. *European Journal of Cancer* 15:1463–1469.

Nasidze I and Stoneking M (2001) Mitochondrial DNA variation and language replacements in the Caucasus. *Proceedings of the Royal Society of London B* 268:1197–1206.

Nasidze I, Risch G M, Robichaux M, Sherry S T, Batzer M A, and Stoneking M (2001) *Alu* insertion polymorphisms and the genetic structure of human populations from the Caucasus. *European Journal of Human Genetics* 9:267–272.

Nebel A, Filon D, Brinkmann B, Majumder P P, Faerman M, and Oppenheim A (2001) The Y chromosome pool of Jews as part of the genetic landscape of the Middle East. *American Journal of Human Genetics* 69:1095–1112.

Nebel A, Filon D, Weiss D A, Weale M, Faerman M, Oppenheim A, and Thomas M G (2000) High-resolution Y chromosome haplotypes of Israeli and Palestinian Arabs reveal geographic substructure and substantial overlap with haplotypes of Jews. *Human Genetics* 107:630–641.

Neel J V (1962) Diabetes mellitus: A "thrifty" genotype rendered detrimental by "progress"? *American Journal of Human Genetics* 14:353–362.

Neiman F D (2000) Coincidence or causal connection: The relationship between Thomas Jefferson's visits to Monticello and Sally Hemings's conceptions. *William and Mary Quarterly* 57:198–210.

Neisser U (1997) Rising scores on intelligence tests. *American Scientist* 85:440–447.

Nelson K and Holmes L B (1989) Malformations due to presumed spontaneous mutations in newborn infants. *New England Journal of Medicine* 320:19–23.

North K E, Martin J L, and Crawford M H (2000) The origins of the Irish travellers and the genetic structure of Ireland. *Annals of Human Biology* 27:453–465.

Ober C, Weitkamp L R, Cox N, Dytch H, Kostyu D and Elias S (1997) HLA and mate choice in humans. *American Journal of Human Genetics* 61:497–504.

Oberklaid F, Danks D M, Jensen F, Stace L, and Rosshandler S (1979) Achondroplasia and hypochondroplasia: Comments on frequency, mutation rate, and radiological features in skull and spine. *Journal of Medical Genetics* 16:140–146.

Oiso T (1975) A historical review of nutritional improvement in Japan after World War II. In: *Physiological Adaptability and Nutritional Status of the Japanese. B. Growth, Work Capacity and Nutrition of Japanese*, edited by Asahina K and Shigiya R. Tokyo: University of Tokyo Press.

Olson J M, Boehnke M, Neiswanger K, Roche A F, and Siervogel R M (1989) Alternative genetic models for the inheritance of the phenylthiocarbamide taste deficiency. *Genetic Epidemiology* 6:423–434.

OMIM (1998a) Online Mendelian Inheritance in Man OMIM™ Johns Hopkins University Baltimore MD. MIM entry 177500 1998, http://www.ncbi.nlm.nih.gov/omim/.

OMIM (1998b) Online Mendelian Inheritance in Man OMIM™ Johns Hopkins University Baltimore MD. MIM entry 147020 1998, http://www.ncbi.nlm.nih.gov/omim/.

OMIM (2001) Online Mendelian Inheritance in Man OMIM™ Johns Hopkins University Baltimore MD. MIM entries 177400 107741 and 190000 2001, http://www.ncbi.nlm.nih.gov/omim/.

On J F Blumenbach's *On the Native Varieties of the Human Species*. (1796) *Monthly Review* 20:515–523. Reprinted in *Race: The Origins of an Idea, 1760–1850*, edited by Augstein H F. Bristol, UK: Thoemmes Press, 1996, pp 58–67.

Oriol R (1995) ABO, Hh, Lewis, and Secretion: Seriology, genetics, and tissue distribution. In: *Blood Cell Biochemistry. Molecular Basis of Human Blood Group Antigens*, volume 6, edited by Cartron J-P and Rouger P. New York: Plenum Press, pp 37–73.

Oriol R, Danilovs J, and Hawkins B R (1981) A new genetic model proposing that the *Se* gene is a structural gene closely linked to the *H* gene. *American Journal of Human Genetics* 33:421–431.

Orioli I M, Castilla E E, and Barbosa-Neto J G (1986) The birth prevalence rates for the skeletal dysplasias. *Journal of Medical Genetics* 23:328–332.

Ott J (1991) *Analysis of Human Genetic Linkage*. Baltimore, MD: Johns Hopkins University Press.

Otten C M (1967) On pestilence, diet, natural selection, and the distribution of microbial and human blood group antigens and antibodies. *Current Anthropology* 8:209–226.

Ottenberg R (1925) A classification of human races based on geographic distribution of the blood groups. *Journal of the American Medical Association* 84:1393–95.

Ousley S D (1997) The quantitative genetics of epidermal ridges using multivariate maximum likelihood estimation. PhD diss, University of Tennessee, Knoxville.

Ovchinnikov I, Rubin A, and Swergold G D (2002) Tracing the LINEs of human evolution. *Proceedings of the National Academy of Sciences USA* 99:10522–10527.

Ovchinnikov I V, Götherström A, Romanova G P, Kharitonov V M, Lidén K, and Goodwin W (2000) Molecular analysis of Neanderthal DNA from the northern Caucasus. *Nature* 404:490–493.

Pagnier J, Mears J G, Dunda-Belkhodja O, Schaefer-Rego K E, Beldjord C, Nagel R L, and Labie D (1984) Evidence for the multicentric origin of the sickle cell hemoglobin gene in Africa. *Proceedings of the National Academy Sciences USA* 81:1771–1773.

Paller A S (2004) Piecing together the puzzle of cutaneous mosaicism. *Journal of Clinical Investigation* 114:1407–1409.

Palmer L J, Burton P R, James A L, Musk A W, and Cookson W O (2000) Familial aggregation and heritability of asthma-associated quantitative traits in a population-based sample of nuclear families. *European Journal of Human Genetics* 8:853–860.

Pante-De-Sousa G, De Cassia Mousinho-Ribeiro R, Melo Dos Santos E J and Guerreiro J F (1999) β-Globin haplotype analysis in Afro-Brazilians from the Amazon region: Evidence for a significant gene flow from Atlantic West Africa. *Annals of Human Biology* 26:365–373.

Parham P and Ohta T (1996) Population biology of antigen presentation by MHC class I molecules. *Science* 272:67–73.

Park M S, Terasaki P I, Bernoco D, and Iwaki Y (1978) Evidence for a second B-cell locus separate from the DR locus. *Transplantation Proceedings* 10:823–828.

Parra E J, Kittles R A, Argyropoulos G, Pfaff C L, Hiester K, Bonilla C, Sylvester N, Parrish-Gause D, Garvey W T, Jin L, McKeigue P M, Kamboh M I, Ferrell R E, Pollitzer W S, and Shriver M D (2001) Ancestral proportions and admixture dynamics in geographically defined African Americans living in South Carolina. *American Journal of Physical Anthropology* 114:18–29.

Parra E J, Marcini A, Akey J, Martinson J, Batzer M A, Cooper R, Forrester T, Allison D B, Deka R, Ferrell R E, and Shriver M D (1998) Estimating African American admixture proportions by use of population-specific alleles. *American Journal of Human Genetics* 63:1839–1851.

Pasvol G, Waincoat J S, and Weatherall D J (1982) Erythrocytes deficient in glycophorin resist invasion by the malarial parasite *Plasmodium falciparum*. *Nature* 297:64–66.

Pattatucci A M (1998) Molecular investigations into complex behavior: Lessons from sexual orientation studies. *Human Biology* 70:367–386.

Paul W E, editor (1989) *Fundamental Immunology*, 2nd Edition, New York: Raven Press.

Paulesu E, Démonet J-F, Fazio F, McCrory E, Chanoine V, Brunswick N, Cappa SF, Cossu G, Habib M, Frith C D, and Frith U (2001) Dyslexia: Cultural diversity and biological unity. *Science* 291:2165–2167.

Pennington B F, Gilger J W, Pauls D, Smith S A, Smith S D, and DeFries J C (1991) Evidence for major gene transmission of developmental dyslexia. *Journal of the American Medical Association* 266:1527–1534.

Pereira L, Dupanloup I, Rosser Z H, Jobling M A, and Barbujani G (2001) Y-chromosome mismatch distributions in Europe. *Molecular Biology and Evolution* 18:1259–1271.

Peterson G M, Rotter J I, Cantor R M, Field L L, Greenwald S, Lim J S T, Roy C, Schoenfeld V, Lowden J A, and Kaback M M (1983) The Tay-Sachs disease gene in North American Jewish populations: Geographic variations and origin. *American Journal of Human Genetics* 35:1258–1269.

Petrakis N L (1969) Dry cerumen—a prevalent genetic trait among American Indians. *Nature* 222:1080–1081.

Petrakis N L (1971) Cerumen genetics and human breast cancer. *Science* 173:347–349.

Petrakis N L, Doherty M, Lee R E, Smith S C, and Page N L (1971) Demonstration and implications of lysozyme and immunoglobulins in human ear wax. *Nature* 229:119–120.

Petrakis N L, Molohon K T, and Tepper D J (1967) Cerumen in American Indians: Genetic implications of sticky and dry types. *Science* 158:1192–1193.

Pettenkofer H J and Bicherich R (1960) Über Antigen-Gemeinshcaften zwischen den menschlichen Blutgruppen ABO und den Erregern gemeingefährlicher Krankheiten. *Zentralblatt für Bakteriologie, Parastenkunde, Infektionskrankheiten und Hygiene Erste Abteilung Medizinisch-Hyginenische Bakteriologie Virusforshung und Parastologie Originale* 179:433–436.

Pettenkofer H J, Stöss B, Helmbold W, and Vogel F (1962) Severe smallpox scars in A+AB. *Nature* 193:445–446.

Piazza A (1993) Who are the Europeans? *Science* 260(5115):1767–1769.

Piazza A, Rendine S, Minch E, Menozzi P, Mountain J, and Cavalli-Sforza L L (1995) Genetics and the origin of European Languages. *Proceedings of the National Academy of Sciences of the USA* 92:5836–5840.

Pillard R C and Bailey M (1998) Human sexual orientation has a heritable component. *Human Biology* 70:347–365.

Pittiglio D H (1986) Genetics and biochemistry of A, B, H and Lewis antigens. In: *Blood Group Systems: ABH and Lewis*, edited by Wallace M E and Gibbs F L. Arlington, VA: American Association of Blood Banks, pp 1–56.

Plomin R, Owen M J, and McGuffin P (1994) The genetic basis of complex human behaviors. *Science* 264:1733–1739.

Pogo A O and Chaudhuri A (1997) The Duffy blood group system and its extension in nonhuman primates. In: *Molecular Biology and Evolution of Blood Group and MHC Antigens in Primates*, edited by Blancher A, Klein J, and Socha W W. Berlin: Springer-Verlag, pp 219–235.

Poirier J, Davignon J, Bouthillier D, Kogan S, Bertrand P, and Gauthier S (1993) Apolipoprotein E polymorphism and Alzheimer's disease. *Lancet* 342: 697–699.

Pollitzer W S (1964) Analysis of a triracial isolate. *Human Biology* 36:362–373.

Post P W and Rao D C (1977) Genetic and environmental determinants of skin color. *American Journal of Physical Anthropology* 47:399–402.

Post P W, Daniels F Jr, and Binford R T Jr (1975) Cold injury and the evolution of "white" skin. *Human Biology* 47:65–80.

Post P W, Krauss A N, Waldman S, and Auld P A M (1977) Skin reflectance of newborn infants from 25 to 44 weeks gestational age. *Human Biology* 48:541–557.

Powell J F and Neves W A (1999) Craniofacial morphology of the first Americans: Pattern and process in the peopling of the New World. *Yearbook of Physical Anthropology* 42:153–188.

Pritchard J K, Stephens M, and Donnelly P (2000) Inference of population structure using multilocus genotype data. *Genetics* 155:945–959.

Quintana-Murci L and Fellous M (2001) The human Y chromosome: The biological role of a "functional wasteland." *Journal of Biomedicine and Biotechnology* 1:18–24.

Race R R (1944) An "incomplete" antibody in human serum. *Nature* 153:771–772.

Race R R and Sanger R (1950) *Blood Groups in Man.* Springfield, IL: Charles C Thomas.

Ragusa A, Lombardo M, Sortino G, Lombardo T, Nagel R L, and Labie D (1988) β^S gene in Sicily is in linkage disequilibrium with the Benin haplotype: Implications for gene flow. *American Journal of Hematology* 27:139–141.

Rana B K, Hewett-Emmett D, Jin L, Chang B H-J, Sambuughin N, Lin M, Watkins S, Bamshad M, Jorde L B, Ramsay M, Jenkins T, and Li W-H (1999) High polymorphism at the human melanocortin 1 receptor locus. *Genetics* 151:1547–1557.

Rao N and Telen M J (1995) Lutheran antigens, Lutheran regulatory genes, and Lutheran regulatory gene targets. In: *Blood Cell Biochemistry. Molecular Basis of Human Blood Group Antigens,* volume 6, edited by Cartron J-P and Rouger P. New York: Plenum Press, pp 281–297.

Redd A J, Takezaki N, Sherry S T, McGarvey S T, Sofro A S M, and Stoneking M (1995) Evolutionary history of the COII/tRNA^{Lys} intergenic 9 base pair deletion in human mitochondrial DNAs from the Pacific. *Molecular Biology and Evolution* 12:604–615.

Reddy B M and Rao D C (1989) Phenylthiocarbamide taste sensitivity revisited: Complete sorting test supports residual family resemblance. *Genetic Epidemiology* 6:413–421.

Reed R R, Bartoshuk L M, Duffy V, Marino S, and Price R A (1995) Propylthiouracil tasting: Determination of underlying threshold distributions using maximum likelihood. *Chemical Senses* 20:529–533.

Reed R R, Nanthakumar E, North M, Bell C, Bartoshuk L M, and Price R A (1999) Localization of a gene for bitter-taste perception to human chromosome 5p15. *American Journal of Human Genetics* 64:1478–1480.

Reed T E (1969) Caucasian genes in American Negroes. *Science* 165:762–768.

Rees J L and Flanagan N (1999) Pigmentation, melanocortins and red hair. *Quarterly Journal of Medicine* 92:125–131.

Reich D E and Goldstein D B (1998) Genetic evidence for a Paleolithic human population expansion in Africa. *Proceedings of the National Academy of Sciences USA* 95:8119–8123.

Reid M E and Lomas-Francis C (1997) *The Blood Group Antigen FactsBook.* San Diego: Academic Press.

Relethford J H (1985) Isolation by distance, linguistic similarity, and the genetic structure on Bougainville Island. *American Journal of Physical Anthropology* 66:317–326.

Relethford J H (1988) Estimation of kinship and genetic distance from surnames. *Human Biology* 60:475–492.

Relethford J H (1992) Cross-cultural analysis of migration rates: Effects of geographic distance and population size. *American Journal of Physical Anthropology* 89:459–466.

Relethford J H (1995) Review of *Race, Evolution, and Behavior: A Life History Perspective. American Journal of Physical Anthropology* 98:91–94.

Relethford J H (1997) Hemispheric difference in human skin color. *American Journal of Physical Anthropology* 104:449–457.

Relethford J H (2001a) Ancient DNA and the origins of modern humans. *Proceedings of the National Academy of Sciences USA* 98:390–391.

Relethford J H (2001b) *Genetics and the Search for Modern Human Origins.* New York: John Wiley & Sons.

Relethford J H (2002) Apportionment of global human genetic diversity based on craniometrics and skin color. *American Journal of Physical Anthropology* 118:393–398.

Relethford J H (2003a) *Reflections of Our Past: How Human History Is Revealed in Our Genes.* Denver: Westview Press.

Relethford J H (2003b) *The Human Species: An Introduction to Biological Anthropology,* fifth edition. New York: McGraw-Hill.

Relethford J H and Blangero J (1990) Detection of differential gene flow from patterns of quantitative variation. *Human Biology* 62:5–25.

Relethford J H and Brennan E R (1982) Temporal trends in isolation by distance on Sanday, Orkney Islands. *Human Biology* 54:315–327.

Relethford J H and Crawford M H (1995) Anthropometric variation and the population history of Ireland. *American Journal of Physical Anthropology* 96:25–38.

Relethford J H and Crawford M H (1998) Influence of religion and birthplace on the genetic structure of Northern Ireland. *Annals of Human Biology* 25:117–125.

Relethford J H, Crawford M H and Blangero J (1997) Genetic drift and gene flow in post-famine Ireland. *Human Biology* 69:443–465.

Relethford J H and Harpending H C (1994) Craniometric variation, genetic theory, and modern human origins. *American Journal of Physical Anthropology* 95:249–270.

Relethford J H and Harpending H C (1995) Ancient differences in population size can mimic a recent African origin of modern humans. *Current Anthropology* 36:667–674.

Relethford J H and Jorde L B (1999) Genetic evidence for larger African population size during recent human evolution. *American Journal of Physical Anthropology* 108:251–260.

Relethford J H and McKenzie R L (1998) Reply to Chaplin and Jablonski's comment on "Hemispheric difference in human skin color". *American Journal of Physical Anthropology* 107:223–224.

Relethford J H, Lees F C, and Byard P J (1985) Sex and age variation in the skin color of Irish children. *Current Anthropology* 26:396–397.

Relethford J H and Mielke J H (1994) Martial exogamy in the Åland Islands, 1750–1949. *Annals of Human Biology* 21:13–21.

Relethford J H, Stern M P, Gaskill S P, and Hazuda H P (1983) Social class, admixture, and skin color variation in Mexican Americans and Anglo Americans living in San Antonio, Texas. *American Journal of Physical Anthropology* 61:97–102.

Remington R D and Schork M A (1970) *Statistics with Applications to the Biological and Health Sciences.* Englewood Cliffs, NJ: Prentice Hall.

Renfrew C (1987) *Archaeology and Language: The Puzzle of Indo-European Origins.* New York: Cambridge University Press.

Renfrew C (1999) Time depth, convergence theory and innovation in Proto-Indo-European: "Old Europe" as a PIE linguistic area. *Journal of Indo-European Studies* 27(3&4):257–293.

Renfrew C (2000) At the edge of knowability: Towards a prehistory of languages. *Cambridge Archaeological Journal* 10(1):7–34.

Renfrew C (2001) From molecular genetics to archaeogenetics. *Proceedings of the National Academy of Sciences USA* 98:4830–4832.

Rensch B (1929) *Das Prinzip geographischer Rassenkreise und das Problem der Artbildung.* Berlin: Boentraeger.

Rich S M and Ayala F J (2000) Population structure and recent evolution of *Plasmodium falciparum.* *Proceedings of the National Academy of Sciences USA* 97:6994–7001.

Richards M and Macaulay V (2001) The mitochondrial gene tree comes of age. *American Journal of Human Genetics* 68:1315–1320.

Richards M, Macaulay V, Hickey E, Vega E, Sykes B, Guida V, Rengo C, Sellitto D, Cruciani F, Kivisild T, Villems R, Thomas M, Rychkov S, Rychkov O, Rychkov Y, Gölge M, Dimitrov D, Hill E, Bradley D, Romano V, Calì F, Vona G, Demaine A, Papiha S, Triantaphyllidis C,

Stefanescu G, Hatina J, Belledi M, Di Rienzo A, Novelletto A, Oppenheim Aa, Nørby S, Al-Zaheri N, Santachiara-Benerecetti S, Scozzari R, Torroni A, and Bandelt H-J (2000) Tracing European founder lineages in the Near Eastern mtDNA pool. *American Journal of Human Genetics* 67:1251–1276.

Richards M, Macaulay V, Torroni A, and Bandelt H-J (2002) In search of geographical patterns in European mitochondrial DNA. *American Journal of Human Genetics* 71:1168–1174.

Richards M, Rengo C, Cruciani F, Gratrix F, Wilson J F, Scozzari R, Macaulay V, and Torroni A (2003) Extensive female-mediated gene flow from sub-Saharan Africa into Near East Arab populations. *American Journal of Human Genetics* 72:1058–1064.

Rigters-Aris C A E (1973a) A reflectometric study of the skin in Dutch families. *Journal of Human Evolution* 2:123–136.

Rigters-Aris C A E (1973b) Reflectometrie cutanee des Fali (Cameroun). *Proceedings of the Koninklijke Nederlandse Akademie van Wetenschappen, Series C* 76:500–511.

Ringelhann R, Hathron M K S, Jilly P, Grant F, and Parniczky G (1976) A new look at the protection of hemoglobin AS and AC genotypes against *Plasmodium falciparum* infection: A census tract approach. *American Journal of Human Genetics* 28:270–279.

Roberts D F (1978) *Climate and Human Variability*, second edition. Menlo Park, CA: Cummings.

Roberts D F and Kahlon D P S (1976) Environmental correlations of skin colour. *Annals of Human Biology* 3:11–22.

Roberts D F, Billewicz W Z, and McGregor I A (1978) Heritability of stature in a west African population. *Annals of Human Genetics* 42:15–24.

Robins A H (1991) *Biological Perspectives on Human Pigmentation*. Cambridge: Cambridge University Press.

Robinson S J and Manning J T (2000) The ratio of 2nd to 4th digit length and male homosexuality. *Evolution and Human Behavior* 21:333–345.

Rogers A R (1992) Error introduced by the infinite sites model. *Molecular Biology and Evolution* 9:1181–1184.

Rogers A R (1995) Genetic evidence for a Pleistocene population explosion. *Evolution* 49:608–615.

Rogers A R (2001) Order emerging from chaos in human evolutionary genetics. *Proceedings of the National Academy of Sciences USA* 98:779–780.

Rogers A R and Harpending H C (1986) Migration and genetic drift in human populations. *Evolution* 40:1312–1327.

Rogers A R and Jorde L B (1995) Genetic evidence on modern human origins. *Human Biology* 67:1–36.

Rogers A R, Fraley A E, Bamshad M J, Watkins W S, and Jorde L B (1996) Mitochondrial mismatch analysis is insensitive to the mutational process. *Molecular Biology and Evolution* 13:895–902.

Rogers J, Mahaney M C, Almasy L, Comuzzie A G, and Blangero J (1999) Quantitative trait linkage mapping in anthropology. *Yearbook of Physical Anthropology* 42:127–151.

Romesburg H C (1984) *Cluster Analysis for Researchers*. Belmont, CA: Wadsworth.

Rosenberg N A, Pritchard J K, Weber J L, Cann H M, Kidd K K, Zhivotovsky L A, and Feldman M W (2002) Genetic structure of human populations. *Science* 298:2381–2385.

Rosenberg N A, Woolf E, Pritchard J K, Schaap T, Gefel D, Shpirer I, Lavi U, Bonné-Tamir B, Hillel J, and Feldman M W (2001) Distinctive genetic signatures in the Libyan Jews. *Proceedings of the National Academy of Sciences USA* 98:858–863.

Rosser Z H and 62 coauthors (2000) Y-chromosomal diversity in Europe is clinal and influenced primarily by geography, rather than by language. *American Journal of Human Genetics* 67:1526–1543.

Rossman I (1977) Anatomic and body composition changes with aging. In: *Handbook of the Biology of Aging*, edited by Finch C E and Hayflick L. New York: Van Nostrand Reinhold, pp 189–221.

Roychoudhury A K and Nei M (1988) *Human Polymorphic Genes: Worldwide Distribution*. New York: Oxford University Press.

Ruff C B (1994) Morphological adaptation to climate in modern and fossil hominids. *Yearbook of Physical Anthropology* 37:65–107.

Rushton J P (1995) *Race, Evolution, and Behavior: A Life History Perspective*. New Brunswick, NJ: Transaction Publishers.

Ruwando C, Khea S C, Snow R W, Yates S N R, Kwiatkoweld D, Gupta S, Warn P, Alisopp G E M, Gilbert S C, Peschu N, Newbold C I, Greenwood S M, Marsh K, and Hill A V S (1995) Natural selection of hemi- and heterozygotes for G6PD deficiency in Africa by resistance to severe malaria. *Nature* 376:246–249.

Sahi T, Launiala K, and Laitinen H (1983) Hypolactasia in a fixed cohort of young Finnish adults, a follow-up study. *Scandinavian Journal of Gastroenterology* 18:865–870.

Salamon H, Klitz W, Easteal S, Gao X, Erlich H A, Fernandez-Viña M, Trachtenberg E A, McWeeney S K, Nelson M P, and Thomson G (1999) Evolution of HLA class II molecules: Allelic and amino acid site variability across populations. *Genetics* 152:393–400.

Salem A-H, Myers J S, Otieno A C, Watkins W S, Jorde L B, and Batzer M A (2003) Line-1 preTa elements in the human genome. *Journal of Molecular Biology* 326:1127–1146.

Salmon C, Salmon D, Liberge G, Andre R, Tippett P, and Sanger R (1961) Un nouvel antigene de groupe sanguin erythrocytaire present chez 80% des sujets de race blanche. *Nouvelle Revue Française d'Hematologie* 1:649–661.

Salmon T N and Blakeslee A F (1935) Genetics of sensory thresholds: Variations within single individuals in taste sensitivity for PTC. *Proceedings of the National Academy of Sciences USA* 21:78–83.

Sanjek R (1994) The enduring inequalities of race. In: *Race*, edited by Gregory S and Sanjek R. New Brunswick, NJ: Rutgers University Press, pp 1–17.

Santos F R, Pandya A, Tyler-Smith C, Pena S D J, Schanfield M, Leonard W R, Osipova L, Crawford M H, and Mitchell R J (1999) The Central Siberian origin for native American Y chromosomes. *American Journal of Human Genetics* 64:619–628.

Santos F R, Pena S D J, and Tyler-Smith C (1995) PCR haplotypes for the human Y chromosome based on alphoid satellite DNA variants and heteroduplex analysis. *Gene* 165:191–198.

Saunders A M, Strittmatter W J, Schmechel D, St George-Hyslop P H, Pericak-Vance M A, Joo S H, Rosi B L, Gussella J F, Crapper-MacLachlan D R, Alberts M J, Hulette C, Crain B, Goldgaber D, and Roses A D (1993) Association of apolipoprotein E allele *E4* with late-onset familial and sporadic Alzheimer's disease. *Neurology* 43: 1467–1472.

Savitt T L and Goldberg M F (1989) Herrick's 1910 case report of sickle cell anemia. The rest of the story. *Journal of the American Medical Association* 261:266–271.

Schneider S and Excoffier L (1999) Estimation of past demographic parameters from the distribution of pairwise differences when the mutation rates vary among sites: Application to human mitochondrial DNA. *Genetics* 152:1079–1089.

Schoenemann P T, Budinger T F, Sarich V M, and Wang S-Y (2000) Brain size does not predict general cognitive ability within families. *Proceedings of the National Academy of Sciences USA* 97:4932–4937.

Schurr T G (2000) Mitochondrial DNA and the peopling of the New World. *American Scientist* 88:246–253.

Schwartz M and Vissing J (2002) Paternal inheritance of mitochondrial DNA. *New England Journal of Medicine* 347:576–580.

Seckler D (1980) Malnutrition: An intellectual odyssey. *Western Journal of Agricultural Economics* 5:219–227.

Segal N L and MacDonald K B (1998) Behavioral genetics and evolutionary psychology: Unified perspectives in personality research. *Human Biology* 70:159–184.

Serjeant G R and Serjeant B E (2001) *Sickle Cell Disease*. Oxford: Oxford University Press.

Shashok K (2003) Pitfalls of editorial miscommunication. *British Medical Journal* 326:1262–1264.

Sheen F M, Sherry S T, Risch G M, Robichaux M, Nasidze I, Stoneking M, Batzer M A, and Swergold G D (2000) Reading between the LINEs: Human genomic variation induced by LINE-1 retrotransposition. *Genome Research* 10:1496–1508.

Shriver M D and Parra E J (2000) Comparison of narrow-band reflectance spectroscopy and tristimulus colorimetry for measurements of skin and hair color in persons of different biological ancestry. *American Journal of Physical Anthropology* 112:17–27.

Simm G (1998) *Genetic Improvement of Cattle and Sheep*. Ipswich, UK: Farming Press.

Simmen B (1994) Taste discrimination and diet differentiation among New World primates. In: *The Digestive System in Mammals: Food, Form and Function*, edited by Chivers D J and Langer P. Cambridge: Cambridge University Press, pp 150–165.

Simoons F J (1969) Primary adult lactose intolerance and the milking habit. I. Review of medical research. *American Journal of Digestive Diseases* 14:819–836.

Simoons F J (1970) Primary adult lactose intolerance and the milking habit: A problem in biological and cultural interrelations, II. A culture historical hypothesis. *American Journal of Digestive Diseases* 15:695–710.

Singer S and Hilgard H R (1978) *The Biology of People*. San Francisco: W.H. Freeman and Company.

Skorecki K, Selig S, Blazer S, Bradman R, Bradman N, Waburton P J, Ismajlowicz M, and Hammer M F (1997) Y chromosomes of Jewish priests. *Nature* 385:32.

Slatkin M (1995) A measure of population subdivision based on microsatellite allele frequencies. *Genetics* 139:457–462.

Slotkin J S (1965) *Readings in Early Anthropology*. Chicago: Aldine.

Smedley A (1999) *Race in North America: Origin and Evolution of a Worldview*. Boulder, CO: Westview Press.

Smit M, De Knijff P, Frants R R, Klasen E C, and Havekes L M (1987) Familial dysbetalipoproteine-mic subjects with the E3/E2 phenotype exhibit an E2 isoform with only one cysteine residue. *Clinical Genetics* 32:335–341.

Smith H V, Kusel J R, and Girdwood R W A (1983) The production of human A and B blood group like substances by *in vitro* maintained second stage *Tosocara canis* larvae: Their presence on the outer larval surfaces and in their excretions/secretions. *Clinical and Experimental Immunology* 54:625–633.

Smith R (2003) Editorial misconduct. *British Medical Journal* 326:1224–1225.

Smith S D, Kelley P M, and Brower A M (1998) Molecular approaches to the genetic analysis of specific reading disability. *Human Biology* 70:239–256.

Smith S E and Davies P D O (1973) Quinine taste thresholds: A family study and twin study. *Annals of Human Genetics* 37:227–232.

Smith S S (1810) *An Essay on the Causes of the Variety of Complexion and Figure in the Human Species*. New Brunswick, NJ: J. Simpson and Co. Reprinted and edited by Jordan W D (1965) Cambridge, MA: Belknap Press of Harvard University Press.

Smithies O (1955) Zone electrophoresis in starch gels: Group variations in the serum proteins of normal human adults. *Biochemistry Journal* 61:629–641.

Smithies O (1957) Variations in human serum β-globulins. *Nature* 180:1482–1483.

Snyder L H (1926) Human blood groups: Their inheritance and racial significance. *American Journal of Physical Anthropology* 9:233–263.

Snyder L H (1930) The "laws" of serologic race-classification studies in human inheritance IV. *Human Biology* 2:128–133.

Snyder L H (1931) Inherited taste deficiency. *Science* 74:151–152.

Snyder L H (1932) Studies in human inheritance. IX. The inheritance of taste deficiency in man. *Ohio Journal of Science* 32:436–440.

Socha W W and Blancher A (1997) The MNSs blood group system. In: *Molecular Biology and Evolution of Blood Group and MHC Antigens in Primates*, edited by Blancher A, Klein J, and Socha W W. Berlin: Springer-Verlag, pp 93–112.

Sokal R R, Oden N L and Thomson B A (1992) Origins of the Indo-Europeans: Genetic evidence. *Proceedings National Academy of Science, USA* 89:7669–7673.

Solomon E and Bodmer W F (1979) Evolution of sickle variant gene [letter]. *Lancet* 1:923.

Soodyall H L, Vigilant L, Hill AV, Stoneking M, and Jenkins T (1996) mtDNA control-region sequence variation suggests multiple independent origins of an "Asian-specific" 9-bp deletion in sub-Saharan Africa. *American Journal of Human Genetics* 58:595–608.

Sparks C S and Jantz R L (2002) A reassessment of human cranial plasticity: Boas revisited. *Proceedings of the National Academy of Sciences USA* 99:14636–14639.

Sparks C S and Jantz R L (2003) Changing times, changing faces: Franz Boas's immigrant study in modern perspective. *American Anthropologist* 105:333–337.

Springer G F and Horton R E (1969) Blood group isoantibody stimulation in man by feeding blood group-active bacteria. *Journal of Clinical Investigation* 48:1280–1291.

Springer G F and Wiener A S (1962) Alleged causes of the present day world distribution of the human ABO blood groups. *Nature* 193:444–446.

Spuhler J N (1968) Assortative mating with respect to physical characteristics. *Eugenics Quarterly* 15:128–140.

Stewart T D (1951a) Scientific responsibility [editorial]. *American Journal of Physical Anthropology* 9:1–3.

Stewart T D (1951b) Objectivity in race classifications. *American Journal of Physical Anthropology* 9:470–472.

Stoll C, Dott B, Roth M P, and Alembik Y (1989) Birth prevalence rates of skeletal dysplasias. *Clinical Genetics* 35:88–92.

Stone A C and Stoneking M (1998) mtDNA analysis of a prehistoric Oneota population: Implications for the peopling of the New World. *American Journal of Human Genetics* 62:1153–1170.

Stone A C, Starrs J E, and Stoneking M (2001) Mitochondrial DNA analysis of the presumptive remains of Jesse James. *Journal of Forensic Sciences* 46:173–176.

Strandskov H H and Washburn S L (1951) Genetics and physical anthropology [editorial]. *American Journal of Physical Anthropology* 9:261–263.

Strickberger M W (1995) *Evolution*, second edition. Boston: Jones and Bartlett.

Stringer C (2003) Out of Ethiopia. *Nature* 423:692–695.

Strobeck C (1987) Average number of nucleotide differences in a sample from a single subpopulation: A test for population subdivision. *Genetics* 117:149–153.

Sturm R A, Box N F, and Ramsay M (1998) Human pigmentation genetics: The difference is only skin deep. *BioEssays* 20:712–721.

Su B, Jin L, Underhill P, Martinson J, Saha N, McGarvey S T, Shriver M D, Chu J, Oefner P, Chakraborty R, and Deka R (2000) Polynesian origins: Insights from the Y chromosome. *Proceedings of the National Academy of Sciences USA* 97:8225–8228.

Suciu-Foca N and Lewis R (2001) Editorial. Anthropology and genetic markers (Retraction of Arnaiz-Villena et al.), *Human Immunology* 62:1063.

Sunderland E, Tills D, Bouloux C, and Doyl J (1973) Genetic studies in Ireland. In: *Genetic Variation in Britain*, edited by Roberts D F and Sunderland E. London: Taylor and Francis, pp 141–159.

Swisher C C, Curtis G H, Jacob T, Getty A G, Suprijo A, and Widiasmoro (1994) Age of the earliest known hominids in Java, Indonesia. *Science* 263:1118–1121.

Sykes B (1999) The molecular genetics of European ancestry. *Philosophical Transactions of the Royal Society of London* B 354:131–139.

Sykes B and Irven C (2000) Surnames and the Y chromosome. *American Journal of Human Genetics* 66:1417–1419.

Szabo G (1967) The regional anatomy of the human integument with special reference to the distribution of hair follicles, sweat glands and melanocytes. *Philosophical Transactions of the Royal Society of London Series B Biological Sciences* 252:447–485.

Szathmary E J E (1993) Genetics of aboriginal North Americans. *Evolutionary Anthropology* 1:202–220.

Takahata N, Lee S-H, and Satta Y (2001) Testing multiregionality of modern human origins. *Molecular Biology and Evolution* 18:172–183.

Talbot C, Lendon C, Craddock N, Shears S, Morris J C, and Goate A (1994) Protection against Alzheimer's disease with apoE epsilon-2. *Lancet* 343:1432–1433.

Taliaferro W H and Huck J G (1923) The inheritance of sickle-cell anaemia in man. *Genetics* 8:594–598.

Tang Y, Schon E A, Wilichowski E, Vazquez-Memije M E, Davidson E, and King M P (2000) Rearrangements of human mitochondrial DNA (mtDNA): New insights into the regulation of mtDNA copy number and gene expression. *Molecular Biology of the Cell* 11:1471–1485.

Tanner J M (1989) *Foetus into Man: Physical Growth from Conception to Maturity*, revised and enlarged edition. Cambridge, MA: Harvard University Press.

Tasa G L, Murray C J, and Boughton J M (1985) Reflectometer reports on human pigmentation. *Current Anthropology* 26:511–512.

Taylor G L and Prior A M (1938) Blood groups in Britain II: Distribution in the population. *Annals of Eugenics* 8:356–361.

Telen M J (1992) The Lutheran antigens and proteins affected by Lutheran regulatory genes. In: *Protein Blood Group Antigens of the Human Red Cell*, edited by Agre P C and Cartron J-P. Baltimore, MD: Johns Hopkins University Press, pp 70–87.

Templeton A R (1982) Adaptation and the integration of evolutionary forces. In: *Perspectives on Evolution*, edited by Milkman R. Sunderland, MA: Sinauer Associates, pp 15–31.

Templeton A R (1998) Human races: A genetic and evolutionary perspective. *American Anthropologist* 100:632–650.

Templeton A R (2002) Out of Africa again and again. *Nature* 416:45–51.

Terrell J (1986) *Prehistory in the Pacific Islands*. New York: Cambridge University Press.

Terrell J E (2000) Essay: Anthropological and scientific fact. *American Anthropologist* 102:808–817.

Terrell J E, Kelly K M, and Rainbird P (2001) Foregone conclusions: In search of "Papuans" and "Austronesians." *Current Anthropology* 42:97–124.

Thieme F P (1952) The population as a unit of study. *American Anthropologist* 54:504–509.

Thomas M G, Parfitt T, Weiss D A, Skorecki K, Wilson J F, le Roux M, Bradman N, and Goldstein D B (2000) Y chromosomes traveling south: The Cohen modal haplotype and the origins of the Lemba—the "black Jews of southern Africa." *American Journal of Human Genetics* 66:674–686.

Thomas M G, Skorecki K, Ben-Ami H, Parfitt T, Bradman N, and Goldstein D B (1998) Origins of Old Testament priests. *Nature* 394:138–140.

Thomas M G, Weale M E, Jones A L, Richards M, Smith A, Redhead N, Torroni A, Scozzari R, Gratrix F, Tarekegn A, Wilson J F, Capelli C, Bradman N, and Goldstein D B (2002) Founding mothers of Jewish communities: Geographically separated Jewish groups were independently founded by very few female ancestors. *American Journal of Human Genetics* 70:1411–1420.

Thompson G R (1962) Significance of haemoglobins S and C in Ghana. *British Medical Journal* 1:682–685.

Thomson R, Pritchard J K, Shen P, Oefner P J, and Feldman M W (2000) Recent common ancestry of human Y chromosomes: Evidence from DNA sequence data. *Proceedings of the National Academy of Sciences USA* 97:7360–7365.

Tills D (1977) Red cells and serum proteins and enzymes of the Irish. *Annals of Human Biology* 4:23–34.

Tills D, Teesdale P, and Mourant A (1977) Blood groups of the Irish. *Annals of Human Biology* 4:23–34.

Tishkoff S A, Dietzsch E, Speed W, Pakstis A J, Kidd J R, Cheung K, Bonné-Tamir B, Santachiara-Benerecetti A S, Moral P, Krings M, Pääbo S, Watson E, Risch N, Jenkins T, and Kidd K K (1996) Global patterns of linkage disequilibrium at the CD4 locus and modern human origins. *Science* 271:1380–1387.

Tishkoff S A, Varkonyi R, Cahinhinan N, Abbes S, Argyropoulos G, Destro-Bisol G, Drousiotou A, Dangerfield B, Lefranc G, Loiset J, Piro A, Stoneking M, Tagarelli A, Tagarelli G, Touma E H, Williams S M, and Clark A G (2001) Haplotype diversity and linkage disequilibrium at human G6PD: Recent origin of alleles that confer malarial resistance. *Science* 293:455–462.

Tiwari J L and Terasaki P I (1985) *HLA and Disease Associations*. New York: Springer-Verlag.

Tiwari S C (1963) Studies of crossing between Indians and Europeans. *Annals of Human Genetics* 26:219–227.

Tobias P V (1975) Anthropometry among disadvantaged peoples: Studies in southern Africa. In: *Biosocial Interrelations in Population Adaptation*, edited by Watts E S, Johnston F E, and Lasker G W. Chicago: Aldine, pp 287–305.

Tomita H, Yamada K, Ghadami M, Ogura T, Yanai Y, Nakatomi K, Sadamatsu M, Masui A, Kato N, and Niikawa N (2002) Mapping of the wet/dry earwax locus to the pericentromeric region of chromosome 16. *Lancet* 359:2000–2002.

Torroni A, Bandelt H-J, D'Urbano L, Lahermo P, Moral P, Sellitto D, Rengo C, Forster P, Savontaus M L, Bonne-Tamir B, and Schozzari R (1998) MtDNA analysis reveals a major Paleolithic population expansion from southwestern to northeastern Europe. *American Journal of Human Genetics* 62:1137–1152.

Towne B, Guo S, Roche A F, and Siervogel R M (1993) Genetic analysis of patterns of growth in infant recumbent length. *Human Biology* 65:977–989.

Trabuchet G, Elion J, Dunda O, Lapoumeroulie C, Docrocq R, Nadifi S, Zohoun I, Chaventre A, Carnevale P, Nagel R L, Krishnamoorthy R, and Labie D (1991) Nucleotide evidence of the unicentric origin of the β^c mutation in Africa. *Human Genetics* 87:597–601.

Tsamantanis C, Delanassios JG, Kottaridis S, and Christodoulou C (1980) Haptoglobin types in breast carcinoma. *Human Heredity* 30:44–45.

Tseng M, Williams R C, Maurer K R, Schanfield M S, Knowler W C, and Everhart J E (1998) Genetic admixture and gallbladder disease in Mexican Americans. *American Journal of Physical Anthropology* 106:361–371.

Tsuneto L T, Probst C M, Hutz M H, Salzano F M, Rodriguez-Delfin L A, Zago M A, Hill K, Hurtado A M, Reibeiro-dos-Santos A K C, and Petzl-Erler M L (2003) HLA class II diversity in seven Amerindian populations. Clues about the origins of the Aché. *Tissue Antigens* 62:512–526.

Uinuk-Ool T, Takezaki N, and Klein J (2003) Ancestry and kinships of native Siberian populations: The HLA evidence. *Evolutionary Anthropology* 12:231–245.

Underhill P A, Passarino G, Lin A A, Marzuki S, Oefner P J, Cavalli-Sforza L L, and Chambers G K (2001) Maori origins, Y-chromosome haplotypes and implications for human history in the Pacific. *Human Mutation* 17:271–280.

Vandenberg S G (1972) Assortative mating, or who marries whom? *Behavior Genetics* 2:127–157.

Van den Berghe P L and Frost P (1986) Skin color preference, sexual dimorphism and sexual selection: A case of gene culture co-evolution? *Ethnic and Racial Studies* 9:87–113.

van Loon F P L, Clemens J D, Sack D A, Rao M R, Ahmed F, Chowdhury S, Harris J R, Ali M, Chakraborty J, Khan M R, Neogy P K, Svennerholm A M, and Holmgren J (1991) ABO blood groups and the risk of diarrhea due to enterotoxigenic *Escherichia coli*. *Journal of Infectious Diseases* 163:1243–1246.

Vogel F and Helmbold W (1972) Blutgruppen-Populations genetik und Statistik. In: *Humangenetik, ein kurzes Handbuch*, volume 1/4, edited by Becher P E. Stuttgart: Thieme, pp 129–557.

Vogel F and Motulsky A G (1997) *Human Genetics: Problems and Approaches*. Berlin: Springer-Verlag.

Vogel F, Pettenkofer H J, and Helmbold W (1960) Über die Populationsgenetik der ABO-Blutgruppen. *Actaicae Genet* 10:267–294.

Volkman S K, Barry A E, Lyons E J, Nielson K M, Thomas S M, Choi M, Thakore S S, Day K P, Wirth D F, and Hartl D L (2001) Recent origin of *Plasmodium falciparum* from a single progenitor. *Science* 293:482–484.

Vonnegut K Jr (1973) *Breakfast of Champions*. New York: Dell.

von Rood J J, Eernisse J G, and van Leeuwen A (1958) Leukocyte antibodies in sera of pregnant women. *Nature* 181:1735–1736.

Wall J D (2000) Detecting ancient admixture in humans using sequence polymorphism data. *Genetics* 154:1271–1279.

Walsh B (2001) Estimating the time to the most recent common ancestor for the Y chromosome or mitochondrial DNA for a pair of populations. *Genetics* 158:897–912.

Walsh R J and Montgomery C A (1947) A new isoagglutinin subdividing the MN blood groups. *Nature* 160:504.

Ward R H, Redd A, Valencia D, Frazier B L, and Pääbo S (1993) Genetic and linguistic diversity in the Americas. *Proceedings of the National Academy of Sciences USA* 90:10663–10667.

Washburn S L (1951) The new physical anthropology. *Transactions New York Academy of Sciences,* series 2, 13:298–304.

Wassermann H P and Heyl T (1968) Quantitative data on skin pigmentation in South African races. *South African Medical Journal* 42:98–101.

Watkins W S, Ricker C E, Bamshad M J, Carroll M L, Nguyen S V, Batzer M A, Harpending H C, Rogers R A, and Jorde L B (2001) Patterns of ancestral human diversity: An analysis of *Alu*-insertion and restriction-site polymorphisms. *American Journal of Human Genetics* 68:738–752.

Weatherall D J, Miller L H, Baruch D I, Marsh K, Doumbo O K, Casals-Pasual C, and Roberts D J (2002) Malaria and the red cell. *Hematology* 2002:35–57.

Weber J L, David D, Heil J, Fan Y, Zhao C, and Marth G (2002) Human diallelic insertion/deletion polymorphisms. *American Journal of Human Genetics* 71:854–862.

Weinberg R A, Scarr S, and Waldman I D (1992) The Minnesota transracial adoption study: A follow-up of IQ test performance at adolescence. *Intelligence* 16:117–135.

Weiner J S (1951) A spectrophotometer for measurement of skin color. *Man* 51:152–153.

Weiss K M (1993) *Genetic Variation and Human Diseases: Principles and Evolutionary Approaches.* New York: Cambridge University Press.

Weitz C A, Garruto R M, Chin C-T, Liu J-C, Liu R-L, and He X (2000) Morphological growth of Han boys and girls born and raised near sea level and at high altitude in western China. *American Journal of Human Biology* 12:665–681.

Welch S and Langmead L (1990) A comparison of the structure and properties of normal human transferrin and a genetic variant of human transferrin. *International Journal of Biochemistry* 22:275–282.

White T D, Asfaw B, DeGusta D, Gilbert H, Richards G D, Suwa G, and Howell F C (2003) Pleistocene *Homo sapiens* from Middle Awash, Ethiopia. *Nature* 423:742–747.

Whitney G and Harder D B (1994) Genetics of bitter perception in mice. *Physiology and Behavior* 56:1141–1147.

Whittingham S and Propert D N (1986) Gm and Km allotypes immune response and disease susceptibility. *Monographs in Allergy* 19:52–70.

Wiener A S (1943) Genetic theory of the Rh blood type. *Proceedings of the Society for Experimental Biology and Medicine* 54:316–319.

Wiener A S (1944) The Rh series of allelic genes. *Science* 100:595–597.

Wiener A S (1946) Recent developments in the knowledge of the Rh-Hr blood types; test for Rh sensitization. *American Journal of Clinical Pathology* 16:477–497.

Willerman L, Schultz R, Rutledge J N, and Bigler E D (1991) *In vivo* brain size and intelligence. *Intelligence* 15:223–228.

Williams T J, Pepitone M E, Christensen S E, Cooke B M, Huberman A D, Breedlove N J, Breedlove T J, Jordan C L, and Breedlove S M (2000) Finger-length ratios and sexual orientation. *Nature* 404:455–456.

Williams-Blangero S and Blangero J (1989) Anthropometric variation and the genetic structure of the Jirels of Nepal. *Human Biology* 61:1–12.

Williams-Blangero S and Blangero J (1991) Skin color variation in eastern Nepal. *American Journal of Physical Anthropology* 85:281–291.

Williams-Blangero S and Blangero J (1992) Quantitative genetic analysis of skin reflectance: A multivariate approach. *Human Biology* 64:35–49.

Wilmut I, Schnieke A E, McWhir J, Kind A J, and Campbell K H S (1997) Viable offspring derived from fetal and adult mammalian cells. *Nature* 385:810–813.

Wolfe L D and Gray J P (1982) A cross-cultural investigation into the sexual dimorphism of stature. In: *Sexual Dimorphism in Homo sapiens: A Question of Size,* edited by Hall R L. New York: Praeger, pp 197–230.

Wood C S (1974) Preferential feeding of *Anopheles gambiae* mosquitoes on human subjects of blood group O: A relationship between the ABO polymorphism and malaria vectors. *Human Biology* 46:385–404.

Workman P L and Niswander J D (1970) Population studies on southwestern Indian tribes. II. Local genetic differentiation in the Papago. *American Journal of Human Genetics* 22:24–49.

Workman P L, Blumberg B S, and Cooper A J (1963) Selection, gene migration and polymorphic stability in a U.S. white and Negro population. *American Journal of Human Genetics* 15:71–84.

Workman P L, Harpending H, Lalouel J M, Lynch C, Niswander J D, and Singleton R (1973) Population studies on southwestern Indian tribes. VI. Papago population structure: A comparison of genetic and migration analyses. In: *Genetic Structure of Populations*, edited by Morton N E. Honolulu: University Press of Hawaii, pp 166–194.

World Health Organization (1997) *Weekly Epidemiological Record* 72:269–276.

Wright S (1977) *Evolution and Genetics of Human Populations: Volume 3. Experimental Results and Evolutionary Deductions*. Chicago: University of Chicago Press.

Xiong M and Guo S-W (1997) Fine-scale genetic mapping based on linkage disequilibrium: Theory and applications. *American Journal of Human Genetics* 60:1513–1531.

Yaounq J, Perichon M, Chorney M, Yaouanq J, Perichon M, Chorney M, Pontarotti P, Le Treut A, El Kahloun A, Mauvieux V, Blayau M, Jouanolle A M, Chauvel B, Moirand R, Nouel O, Le Gall J Y, Feingold J and David V (1994) Anonymous marker loci within 400 kb of *HLA-A* generate haplotypes in linkage disequilibrium with the hemochromatosis gene (*HFE*). *American Journal of Human Genetics* 54:252–263.

Young F A, Baldwin W R, Box R A, Harris E, and Johnson C (1969) The transmission of refractive error within Eskimo families. *American Journal of Optometry* 46:676–685.

Zeng Y, Rogers G P, Huang S, Schechter A N, Salamah M, Perrine S, and Berg P E (1994) Sequence of the −530 region of the beta-globin gene of sickle cell anemia patients with the Arabian haplotype. *Human Mutation* 3:163–165.

Zerjal T, Beckman L, Beckman G, Mikelsaar A-V, Krumina A, Kuèinskas V, Hurles M E, and Tyler-Smith C (2001) Geographical, linguistic, and cultural influences on genetic diversity: Y-chromosomal distribution in northern European populations. *Molecular Biology and Evolution* 18:1077–1087.

Zhao T and Lee T D (1989) Gm and Km allotypes in 74 Chinese populations: A hypothesis of the origin of the Chinese nation. *Human Genetics* 83:101–110.

Zhivotovsky L A (2001) Estimating divergence time with the use of microsatellite genetic distances: Impacts of population growth and gene flow. *Molecular Biology and Evolution* 18:700–709.

Zhivotovsky L A, Rosenberg N A, and Feldman M W (2003) Features of evolution and expansion of modern humans, inferred from genomewide microsatellite markers. *American Journal of Human Genetics* 72:1171–1186.

Zoossmann-Diskin A (2000) Are today's Jewish priests descended from the old ones? *Homo* 51:156–162.

Zvelebil M (1995) Indo-European origins and the agricultural transition in Europe. *Journal of European Archaeology* 3(1):33–70.

Zvelebil M (1998) Genetic and cultural diversity of Europe: A comment on Cavalli-Sforza. *Journal of Anthropological Research* 54:411–417.

Index